T0181985

Birkhäuser

Modern Birkhäuser Classics

Many of the original research and survey monographs in pure and applied mathematics published by Birkhäuser in recent decades have been groundbreaking and have come to be regarded as foundational to the subject. Through the MBC Series, a select number of these modern classics, entirely uncorrected, are being re-released in paperback (and as eBooks) to ensure that these treasures remain accessible to new generations of students, scholars, and researchers.

Christian Okonek
Michael Schneider
Heinz Spindler

Vector Bundles on Complex Projective Spaces

With an Appendix by S. I. Gelfand

Corrected reprint of the 1988 Edition

Christian Okonek
Institut für Mathematik
Universität Zürich
Winterthurerstrasse 190
8057 Zürich
Switzerland
okonek@math.uzh.ch

Michael Schneider†
Universität Bayreuth
Germany

Heinz Spindler
Institut für Mathematik
Universität Osnabrück
Albrechtstraße 28a
49076 Osnabrück
Germany
heinz.spindler@mathematik.uni-osnabrueck.de

2010 Mathematics Subject Classification: Primary Classification: 14-02, 32-02
Secondary Classification: 14D20, 14D21, 14D22, 14J60, 14J81, 14N05, 14N25; 32L10, 32J25, 32G13, 32G81

ISBN 978-3-0348-0150-8 e-ISBN 978-3-0348-0151-5
DOI 10.1007/978-3-0348-0151-5

Library of Congress Control Number: 2011930254

© 1980 Birkhäuser Verlag
Originally published under the same title as volume 3 in the Progress in Mathematics series by Birkhäuser Verlag, Switzerland, ISBN 978-0-8176-3000-3
Corrected second printing 1988 by Birkhäuser Verlag, ISBN 978-0-8176-3385-1
Corrected reprint 2011 by Springer Basel AG

This work is subject to copyright. All rights are reserved, whether the whole or part of the material is concerned, specifically the rights of translation, reprinting, re-use of illustrations, broadcasting, reproduction on microfilms or in other ways, and storage in data banks. For any kind of use whatsoever, permission from the copyright owner must be obtained.

Cover design: deblik, Berlin

Printed on acid-free paper

Springer Basel AG is part of Springer Science+Business Media

www.birkhauser-science.com

Introduction

These lecture notes are intended as an introduction to the methods of classification of holomorphic vector bundles over projective algebraic manifolds X. To be as concrete as possible we have mostly restricted ourselves to the case $X = \mathbb{P}_n$. According to Serre (GAGA) the classification of holomorphic vector bundles is equivalent to the classification of algebraic vector bundles. Here we have used almost exclusively the language of analytic geometry. The book is intended for students who have a basic knowledge of analytic and (or) algebraic geometry. Some fundamental results from these fields are summarized at the beginning.

One of the authors gave a survey in the Séminaire Bourbaki 1978 on the current state of the classification of holomorphic vector bundles over $\mathbb{P}_n$. This lecture then served as the basis for a course of lectures in Göttingen in the Winter Semester 78/79. The present work is an extended and up-dated exposition of that course. Because of the introductory nature of this book we have had to leave out some difficult topics such as the restriction theorem of Barth. As compensation we have appended to each section a paragraph in which historical remarks are made, further results indicated and unsolved problems presented.

The book is divided into two chapters. Each chapter is subdivided into several sections which in turn are made up of a number of paragraphs. Each section is preceded by a short description of its contents.

In assembling the list of literature we have done our best to include all the articles about vector bundles over $\mathbb{P}_n$ which are known to us. On the other hand we have not thought it necessary to include works about the classification of holomorphic vector bundles over curves. The reader interested in this highly developed theory is recommended to read an article by Tjurin (Russian Math. Surveys 1974) or the lecture notes of a course held at Tata Institute by Newstead.

Part of the present interest in holomorphic vector bundles comes from the connection to physics. The mathematician who is interested in this connection is recommended to see the ENS-Séminaire of Douady and Verdier. In the final paragraph of the present lecture notes he will also find remarks about that topic and some literature citations.

R. M. Switzer has not only translated the manuscript of these notes into English but has also aided us in answering many mathematical questions. For this assistance we wish to thank him heartily. Furthermore we wish to thank Mrs. M. Schneider for doing such a good job with the unpleasant task of typing these notes and H. Hoppe for assembling the index and doing the difficult job of inserting the mathematical symbols.

Preface to the second printing

Since the appearance of this book there has been considerable activity concerning vector bundles on projective spaces and other model manifolds.

At first there were mainly two reasons for this:

i) The connection between algebraic bundles on $\mathbb{P}_n$ and subvarieties $X \subset \mathbb{P}_n$ of projective space.

ii) The connection between algebraic vector bundles on $\mathbb{P}_3$ and solutions of the Yang–Mills equations (via the Penrose transform) on S^4.

Through the spectacular work of Donaldson it became apparent that for algebraic surfaces there is a very close connection between algebraic and differential geometry and topology.

Recently Donaldson and Uhlenbeck, Yau solved the conjecture of Hitchin and Kobayashi (i.e., stable vector bundles admit a Hermite–Einstein metric). This shows once more how differential geometry has become more and more important in the theory of vector bundles.

For this new edition we compiled a list of references for the period 1980–1987 which (even if far from complete) should be helpful for future research. We also include some older papers which we were not aware of when the book was first written.

Bayreuth and Göttingen, August 1987.

Preface to the MBC-Series Edition

This release is an essentially unchanged edition of the second printing of the book "Vector Bundles on Complex Projective Spaces" originally published as Volume 3 of Birkhäuser's series "Progress in Mathematics". An english translation of S. I. Gelfand's appendix to the russian edition has been added.

Dave Benson of the University of Aberdeen retyped the original manuscript in AMSLaTeX and polished R. Zeinstra's english translation of Gelfand's appendix. He also suggested to contact Springer to re-release this book. We thank him heartily for all his efforts.

Osnabrück and Zürich, August 2010.

Contents

CHAPTER 1

Holomorphic vector bundles and the geometry of $\mathbb{P}_n$

§1. Basic definitions and theorems

In this section we shall establish the notation and assemble the most important facts about the cohomology of projective spaces with coefficients in an analytic coherent sheaf. Then we shall recall the definition of the Chern classes of a vector bundle and for holomorphic bundles we shall interpret them in some cases as the dual classes of appropriate submanifolds.

1.1. Serre duality, the Bott formula, Theorem A and Theorem B. For an $(n+1)$-dimensional complex vector space V we denote by $\mathbb{P}(V)$ the associated projective space of lines in V; in particular we have

$$\mathbb{P}_n = \mathbb{P}(\mathbb{C}^{n+1}).$$

$\mathbb{P}_n$ has a natural structure as compact complex manifold.

Let X be a complex space with structure sheaf $\mathcal{O}_X$. If F is a coherent analytic sheaf over X and $x \in X$ a point, then we denote the stalk of F at x by F_x and set

$$F(x) = F_x/\mathfrak{m}_x F_x = F_x \otimes_{\mathcal{O}_{X,x}} \mathcal{O}_{X,x}/\mathfrak{m}_x$$

where $\mathfrak{m}_x \subset \mathcal{O}_{X,x}$ denotes the maximal ideal of the local ring $\mathcal{O}_{X,x}$.

Let E be a holomorphic vector bundle over X. Then we have the sheaf $\mathcal{O}_X(E)$ of germs of holomorphic sections in E. $\mathcal{O}_X(E)$ is a locally free sheaf of rank $r = \operatorname{rk} E$. In what follows we shall not distinguish between a vector bundle E and the associated locally free sheaf $\mathcal{O}_X(E)$. With the notation introduced above we then have

$$E(x) = E_x/\mathfrak{m}_x E_x$$

for the fibre over x of a holomorphic vector bundle E. A homomorphism of sheaves $\phi\colon E \to F$ between two holomorphic vector bundles defines for every point x a $\mathbb{C}$-linear map

$$\phi(x)\colon E(x) \to F(x)$$

and a homomorphism of $\mathcal{O}_{X,x}$-modules

$$\phi_x\colon E_x \to F_x.$$

WARNING. For a homomorphism of sheaves $\phi\colon E \to F$ it can happen that $\phi_x\colon E_x \to F_x$ is injective but $\phi(x)\colon E(x) \to F(x)$ is not. If for example $0 \neq (f_1, \dots, f_r) \in H^0(X, \mathcal{O}_X^{\oplus r})$ is an r-tuple of holomorphic functions on X, X a complex manifold, and if in a point x one has $f_i(x) = 0$, i.e., $f_{i,x} \in \mathfrak{m}_x$, for $i = 1, \dots, r$, then

$$\phi\colon \mathcal{O}_X \to \mathcal{O}_X^{\oplus r}, \quad h \to (hf_1, \dots, hf_r)$$

is a monomorphism of sheaves, but in the point x

$$\phi(x)\colon \mathcal{O}_{X,x}/\mathfrak{m}_x \to (\mathcal{O}_{X,x}/\mathfrak{m}_x)^{\oplus r}$$

is the zero homomorphism.

On the other hand we have (Nakayama-Lemma): $\phi_x\colon E_x \to F_x$ is surjective precisely when $\phi(x)\colon E(x) \to F(x)$ is surjective.

Let D be a (Cartier-)divisor on a complex manifold X which is given with respect to an open covering (U_i) by a family of meromorphic functions f_i on U_i. D determines a holomorphic line bundle $[D]$ on X with transition functions

$$g_{ij} = f_i/f_j.$$

If D is effective, i.e., a (possibly nonreduced) complex subspace of codimension 1 in X which is given locally by one equation, then $[-D]$ is the sheaf of ideals of D in X:

$$J_D = [-D].$$

If $D \subset X$ is a divisor without singularities, then the analytic restriction $J_D|D$ of the sheaf of ideals to D is the conormal bundle $N^*_{D/X}$ of D in X,

$$J_D|D = J_D \otimes \mathcal{O}_X/J_D = J_D/J_D^2 = N^*_{D/X}.$$

Dualizing we thus have

$$N_{D/X} = [D]|D.$$

As usual we shall denote by $\mathcal{O}_{\mathbb{P}_n}(1)$ the hyperplane bundle over $\mathbb{P}_n$. $\mathcal{O}_{\mathbb{P}_n}(1)$ is defined by

$$\mathcal{O}_{\mathbb{P}_n}(1) = [H],$$

where $H \subset \mathbb{P}_n$ is some hyperplane. With respect to the standard covering $(U_i)_{i=0,\dots,n}$ then $\mathcal{O}_{\mathbb{P}_n}(1)$ is represented by the cocycle

$$g_{ij} = z_j/z_i$$

($(z_0 : \cdots : z_n)$ are the homogeneous coordinates). The dual bundle $\mathcal{O}_{\mathbb{P}_n}(1)^*$ of $\mathcal{O}_{\mathbb{P}_n}(1)$ will be denoted by $\mathcal{O}_{\mathbb{P}_n}(-1)$. $\mathcal{O}_{\mathbb{P}_n}(-1)$ is the tautological line bundle over $\mathbb{P}_n$,

$$\mathcal{O}_{\mathbb{P}_n}(-1) = \{(\ell, v) \in \mathbb{P}_n \times \mathbb{C}^{n+1} \mid v \in \ell\}.$$

For $k \in \mathbb{Z}$ we take

$$\mathcal{O}_{\mathbb{P}_n}(k) = \begin{cases} \mathcal{O}_{\mathbb{P}_n}(1)^{\otimes k} & \text{for } k \geq 0 \\ \mathcal{O}_{\mathbb{P}_n}(-1)^{\otimes |k|} & \text{for } k \leq 0, \end{cases}$$

and for any coherent analytic sheaf F over $\mathbb{P}_n$ we define

$$F(k) = F \otimes_{\mathcal{O}_{\mathbb{P}_n}} \mathcal{O}_{\mathbb{P}_n}(k).$$

For an effective divisor D of degree k we have

$$[D] = \mathcal{O}_{\mathbb{P}_n}(k);$$

the sections of the line bundle $\mathcal{O}_{\mathbb{P}_n}(k)$, $k \geq 0$, can be identified with the homogeneous polynomials $P \in \mathbb{C}[z_0, \ldots, z_n]$ of degree k, i.e.,

$$H^0(\mathbb{P}_n, \mathcal{O}_{\mathbb{P}_n}(k)) \simeq \{P \in \mathbb{C}[z_0, \ldots, z_n] \mid P \text{ homogeneous of degree } k\}.$$

Let $H \subset \mathbb{P}_n$ be a hyperplane; then we have the short exact sequence

$$0 \to [-H] \to \mathcal{O}_{\mathbb{P}_n} \to \mathcal{O}_H \to 0,$$

and thus also

$$0 \to \mathcal{O}_{\mathbb{P}_n}(-1) \to \mathcal{O}_{\mathbb{P}_n} \to \mathcal{O}_H \to 0. \tag{1}$$

If one regards $\mathcal{O}_{\mathbb{P}_n}(-1)$ as subbundle of $\mathcal{O}_{\mathbb{P}_n}^{\oplus(n+1)}$, then one obtains an exact sequence of vector bundles

$$0 \to \mathcal{O}_{\mathbb{P}_n}(-1) \to \mathcal{O}_{\mathbb{P}_n}^{\oplus(n+1)} \to Q \to 0. \tag{2}$$

The quotient Q is the twisted holomorphic tangent bundle (Griffiths and Harris [49], p. 409)

$$Q = T_{\mathbb{P}_n}(-1).$$

The sequence

$$0 \to \mathcal{O}_{\mathbb{P}_n}(-1) \to \mathcal{O}_{\mathbb{P}_n}^{\oplus(n+1)} \to T_{\mathbb{P}_n}(-1) \to 0 \tag{2}$$

is called the *Euler sequence.*

Let $\Omega^p_{\mathbb{P}_n}$ be the sheaf of germs of holomorphic p-forms on $\mathbb{P}_n$; thus

$$\Omega^1_{\mathbb{P}_n} = T^*_{\mathbb{P}_n}, \quad \Omega^p_{\mathbb{P}_n} = \Lambda^p \Omega^1_{\mathbb{P}_n}.$$

If one dualizes the Euler sequence and takes the pth exterior power (Hirzebruch [62], p. 55) one gets the following exact sequence of vector bundles

$$0 \to \Omega^p_{\mathbb{P}_n}(p) \to \mathcal{O}_{\mathbb{P}_n}^{\oplus\binom{n+1}{p}} \to \Omega^{p-1}_{\mathbb{P}_n}(p) \to 0. \tag{3}$$

For the canonical bundle $\omega_{\mathbb{P}_n} = \Omega^n_{\mathbb{P}_n}$ we have

$$\omega_{\mathbb{P}_n} = \mathcal{O}_{\mathbb{P}_n}(-n-1)$$

by taking $p = n+1$. If E is an r-bundle we shall usually write $\det E$ instead of $\Lambda^r E$ — e.g., $\omega_{\mathbb{P}_n} = \det \Omega^1_{\mathbb{P}_n}$.

The cohomology groups of a compact complex manifold X with values in a coherent analytic sheaf are finite dimensional $\mathbb{C}$-vector spaces (Grauert–Remmert [47], p. 188). We set

$$h^q(X, F) = \dim_{\mathbb{C}} H^q(X, F).$$

If X is an n-dimensional projective-algebraic complex manifold with canonical line bundle ω_X, then we have for any holomorphic vector bundle E over X

Serre duality: $H^q(X, E)^* \cong H^{n-q}(X, E^* \otimes \omega_X)$.

The formula (Hirzebruch [62], p. 47)

$$\Lambda^q E^* \cong \Lambda^r E^* \otimes \Lambda^{r-q} E \quad (r = \operatorname{rank} E)$$

gives us for $E = \Omega^1_{\mathbb{P}_n}$

$$\Omega^{p\,*}_{\mathbb{P}_n} \cong \Omega^{n-p}_{\mathbb{P}_n} \otimes \Omega^{n\,*}_{\mathbb{P}_n}.$$

Serre duality then implies

$$h^q(\mathbb{P}_n, \Omega^p_{\mathbb{P}_n}(k)) = h^{n-q}(\mathbb{P}_n, \Omega^{n-p}_{\mathbb{P}_n}(-k)).$$

The values of $h^q(\mathbb{P}_n, \Omega^p_{\mathbb{P}_n}(k))$ are given by the

Bott formula (Bott [23]):

$$h^q(\mathbb{P}_n, \Omega^p_{\mathbb{P}_n}(k)) = \begin{cases} \binom{k+n-p}{k}\binom{k-1}{p} & \text{for } q = 0,\ 0 \le p \le n,\ k > p \\ 1 & \text{for } k = 0,\ 0 \le p = q \le n \\ \binom{-k+p}{-k}\binom{-k-1}{n-p} & \text{for } q = n,\ 0 \le p \le n,\ k < p-n \\ 0 & \text{otherwise} \end{cases}$$

In particular for $p = 0$ we have:

$$h^q(\mathbb{P}_n, \mathcal{O}_{\mathbb{P}_n}(k)) = \begin{cases} \binom{n+k}{k} & \text{for } q = 0,\ k \ge 0 \\ \binom{-k-1}{-k-1-n} & \text{for } q = n,\ k \le -n-1 \\ 0 & \text{otherwise} \end{cases}$$

If we set $H^q(\mathbb{P}_n, \Omega^p_{\mathbb{P}_n}(*)) = \bigoplus_{k \in \mathbb{Z}} H^q(\mathbb{P}_n, \Omega^p_{\mathbb{P}_n}(k))$, then we have for $0 < q < n$

$$h^q(\mathbb{P}_n, \Omega^p_{\mathbb{P}_n}(*)) = \begin{cases} 1 & \text{if } p = q \\ 0 & \text{if } p \ne q. \end{cases}$$

It is a useful exercise to deduce the Bott formula by induction using equations (1), (3) and

$$H^q(\mathbb{P}_n, \mathcal{O}_{\mathbb{P}_n}) = 0, \quad q \geq 1.$$

The vanishing of $H^q(\mathbb{P}_n, \mathcal{O}_{\mathbb{P}_n})$ for $q > 0$ results from a clever Laurent separation, cf. Bănică and Stănăşilă [8].

If F is an arbitrary coherent analytic sheaf over $\mathbb{P}_n$, then the behavior of the k-fold twisted sheaf $F(k) = F \otimes \mathcal{O}_{\mathbb{P}_n}(k)$ for $k \to \infty$ is described by the Theorems A and B of Serre [110].

Theorem A. *For every coherent analytic sheaf F over $\mathbb{P}_n$ there is a $k_0 \in \mathbb{Z}$ so that for $k \geq k_0$ the sheaf $F(k)$ is generated by global sections.*

A coherent analytic sheaf F is said to be generated by global sections if the canonical homomorphism of sheaves

$$\phi\colon H^0(\mathbb{P}_n, F) \otimes_{\mathbb{C}} \mathcal{O}_{\mathbb{P}_n} \to F, \quad \phi_x(s \otimes h) = hs_x,$$

is surjective.

Theorem B. *For every coherent analytic sheaf F over $\mathbb{P}_n$ there is a $k_0 \in \mathbb{Z}$ such that for $k \geq k_0$ and all $q > 0$*

$$H^q(\mathbb{P}_n, F(k)) = 0.$$

If E is locally free, i.e., a vector bundle, then $h^0(\mathbb{P}_n, E(k)) = h^n(\mathbb{P}_n, E^*(-k-n-1))$ and thus by Theorem B $h^0(\mathbb{P}_n, E(k)) = 0$ for $k \ll 0$. On the other hand for $k \gg 0$ $E(k)$ has many sections. Thus there is a uniquely determined minimal $k_0 = k_0(E)$ with

$$h^0(\mathbb{P}_n, E(k_0)) \neq 0, \quad h^0(\mathbb{P}_n, E(k)) = 0 \quad \text{for} \quad k < k_0.$$

We shall also need the following fundamental theorems from the theory of analytic sheaves, whose proof can be found in Bănică–Stănăşilă [8].

Let $f\colon X \to Y$ be a proper holomorphic mapping of complex spaces, F a coherent analytic sheaf over X and E a holomorphic vector bundle over X.

Coherence theorem. *The i th direct image sheaf $R^i f_* F$ is a coherent analytic sheaf over Y for all $i \geq 0$.*

Semicontinuity theorem. *If f is a flat mapping then for all i, $s \geq 0$*

$$\{y \in Y \mid h^i(f^{-1}(y), E|f^{-1}(y)) \geq s\}$$

is a closed analytic subset of Y. Here the fibre $f^{-1}(y)$ is to have the canonical complex structure $\mathcal{O}_{f^{-1}(y)} = \mathcal{O}_X / \mathfrak{m}_y \cdot \mathcal{O}_X$.

Base-change theorem. *Let f be flat, Y reduced and for some $i \geq 0$ let the function $s(y) = h^i(f^{-1}(y), E|f^{-1}(y))$ be independent of $y \in Y$ — say $s(y) = s$ for all $y \in Y$. If*

$$\begin{array}{ccc} X' & \xrightarrow{\psi} & X \\ {\scriptstyle g}\downarrow & & \downarrow{\scriptstyle f} \\ Y' & \xrightarrow[\phi]{} & Y \end{array}$$

is an arbitrary change of base, then the canonical $\mathcal{O}_Y$-module homomorphism

$$\phi^* R^i f_*(E) \to R^i g_*(\psi^* E)$$

is an isomorphism. In particular for every point $y \in Y$

$$(R^i f_*(E))(y) = H^i(f^{-1}(y), E|f^{-1}(y)).$$

The image sheaf $R^i f_*(E)$ is locally free of rank s and thus an s-bundle over Y.

Finally we shall need the

Projection formula. *Let E' be a vector bundle over Y. Then*

$$R^i f_*(f^* E' \otimes F) \simeq E' \otimes R^i f_*(F) \quad (i \geq 0).$$

1.2. Chern classes and dual classes. As a compact complex manifold $\mathbb{P}_n$ has a uniquely determined fundamental class

$$o_{\mathbb{P}_n} \in H_{2n}(\mathbb{P}_n; \mathbb{Z}).$$

$H^2(\mathbb{P}_n; \mathbb{Z})$ has a canonical generator h_n, which is specified by the following conditions:

a) $h_{n-1} = i^* h_n$, where $i\colon \mathbb{P}_{n-1} \to \mathbb{P}_n$ is a linear embedding.

b) $h_n^n[\mathbb{P}_n] = 1$, where $h_n^n[\mathbb{P}_n]$ is the evaluation of h_n^n on the fundamental class $o_{\mathbb{P}_n}$.

For the singular cohomology ring of $\mathbb{P}_n$ we have

$$H^*(\mathbb{P}_n; \mathbb{Z}) = \mathbb{Z}[h_n] = \mathbb{Z}[t]/(t^{n+1}).$$

Hereafter we shall simply write h for h_n.

To associate to an arbitrary (continuous) complex vector bundle E of rank r over an admissible (Hirzebruch [62], p. 57) topological space X characteristic classes $c_i(E) \in H^{2i}(X; \mathbb{Z})$, $i = 0, \ldots, r$, one can proceed as follows: one sets $c_0(E) = 1$ and defines the first Chern class of the hyperplane bundle $[H] = \mathcal{O}_{\mathbb{P}_n}(1)$ by

$$c_1(\mathcal{O}_{\mathbb{P}_n}(1)) = h \in H^2(\mathbb{P}_n; \mathbb{Z}).$$

For every line bundle L over an admissible space X there is a classifying map $f\colon X \to \mathbb{P}_n$ for an appropriate $\mathbb{P}_n$, which is uniquely determined up to homotopy and such that $L = f^*\mathcal{O}_{\mathbb{P}_n}(1)$. Thus $c_1(L) = f^*h$ specifies the first Chern class for line bundles.

If E is an arbitrary complex vector bundle of rank r over an admissible base space X,

$$p\colon \mathbb{P}(E) \to X$$

the associated projective bundle of lines in E, then p^*E contains a canonical line bundle $L \subset p^*E$,

$$L = \{(\ell, v) \in \mathbb{P}(E) \times E \mid v \in \ell\}.$$

Let $s \in H^2(\mathbb{P}(E); \mathbb{Z})$ be the first Chern class of L^*. For every fibre $\mathbb{P}_{r-1} = p^{-1}(x)$ we have $s|\mathbb{P}_{r-1} = h \in H^2(\mathbb{P}_{r-1}; \mathbb{Z})$. By the Theorem of Leray and Hirsch (Switzer [120], p. 365) the cohomology $H^*(\mathbb{P}(E); \mathbb{Z})$ is a free $H^*(X; \mathbb{Z})$-module with basis

$$\{1, s, \dots, s^{r-1}\},$$

that is each element $y \in H^*(\mathbb{P}(E); \mathbb{Z})$ has a unique representation

$$y = \sum_{i=1}^{r} p^*(x_i) s^{r-i} \qquad (x_i \in H^*(X; \mathbb{Z})).$$

In particular there are welldefined cohomology classes

$$c_i \in H^{2i}(X; \mathbb{Z})$$

with

$$s^r = -\sum_{i=1}^{r} p^*(c_i) s^{r-i}.$$

One then defines the Chern classes of E by

$$c_i(E) = c_i, \qquad i = 1, \dots, r.$$

Thus the Chern classes describe how far the ring structure of $H^*(\mathbb{P}(E); \mathbb{Z})$ deviates from the ring structure of $H^*(X \times \mathbb{P}_{r-1}; \mathbb{Z})$.

$c(E) = 1 + c_1(E) + \cdots + c_r(E)$ is the *total Chern class* of E; it defines a natural transformation

$$c\colon \mathrm{Vect_{top}}(-) \to H^{2*}(-; \mathbb{Z})$$

from the isomorphism classes of complex vector bundles to the "even" cohomology ring. c is uniquely determined by the Cartan formula

$$c(E \oplus F) = c(E)c(F)$$

and the normalization

$$c(\mathcal{O}_{\mathbb{P}_n}(1)) = 1 + h$$

(Switzer [120], p. 376).

The Cartan formula is the most important tool for calculating the Chern classes of complicated bundles. For a given vector bundle E over X one can by repeated construction of the projective bundle find a space Y and a map $p\colon Y \to X$ such that p^*E decomposes over Y into a direct sum of line bundles (topologically!)

$$p^*E = L_1 \oplus \cdots \oplus L_r$$

and such that moreover

$$p^*\colon H^*(X;\mathbb{Z}) \to H^*(Y;\mathbb{Z})$$

is injective. This so called splitting principle will be applied as follows: one factors the total Chern class $c(E)$ "formally" in an appropriate ring extension of $H^*(X;\mathbb{Z})$ as

$$c(E) = \prod (1 + x_i(E))$$

and computes with the "classes" $x_i(E)$ as if they were the first Chern classes of line bundles L_i.

Example. In order to determine the Chern classes of $\Lambda^p E$ for an r-bundle E, one factors $c(E)$ as

$$c(E) = \prod_{i=1}^{r} (1 + x_i).$$

Because $\Lambda^p(L_1 \oplus \cdots \oplus L_r) = \bigoplus_{1 \le i_1 < \cdots < i_p \le r} (L_{i_1} \otimes \cdots \otimes L_{i_p})$ we get the formula

$$\begin{aligned} c(\Lambda^p E) &= \prod_{1 \le i_1 < \cdots < i_p \le r} (1 + x_{i_1} + \cdots + x_{i_p}) \\ &= 1 + \sum_{k=1}^{r} P_k(s_1(x), \ldots, s_k(x)) \end{aligned}$$

with certain universal polynomials P_k in the elementary symmetric functions $s_j(x)$ of the x_i. The kth Chern class of $\Lambda^p E$ is then

$$c_k(\Lambda^p E) = P_k(c_1(E), \ldots, c_k(E)).$$

Similarly one obtains formulae for the Chern classes of E^*, $E \otimes F$, $S^p E$. If L is a line bundle for example, then

$$c_k(E \otimes L) = \sum_{i=0}^{k} \binom{r-i}{k-i} c_i(E) c_1(L)^{k-i}.$$

Let $0 \to F \to E \to Q \to 0$ be an exact sequence of vector bundles over X. Such sequences always split (topologically!), i.e.,

$$E = F \oplus Q.$$

The Cartan formula permits us to compute the Chern classes of E from those of F and Q.

EXAMPLE. If one tensors the Euler sequence

$$0 \to \mathcal{O}_{\mathbb{P}_n}(-1) \to \mathcal{O}_{\mathbb{P}_n}^{\oplus(n+1)} \to T_{\mathbb{P}_n}(-1) \to 0$$

with $\mathcal{O}_{\mathbb{P}_n}(1)$, one gets the exact sequence

$$0 \to \mathcal{O}_{\mathbb{P}_n} \to \mathcal{O}_{\mathbb{P}_n}^{\oplus(n+1)}(1) \to T_{\mathbb{P}_n} \to 0$$

and thus the Chern classes of $\mathbb{P}_n$

$$c(\mathbb{P}_n) = c(T_{\mathbb{P}_n}) = (1+h)^{n+1}.$$

The isomorphism $H^*(\mathbb{P}_n;\mathbb{Z}) = \mathbb{Z}[h]/(h^{n+1})$ permits us to regard the Chern classes of bundles over $\mathbb{P}_n$ as integers. Thus one can write

$$c_i(\mathbb{P}_n) = \binom{n+1}{i} \quad \text{for} \quad i = 0, 1, \ldots, n.$$

If one puts stronger conditions on the base space and total space of a vector bundle (differentiable, holomorphic, algebraic), then there are various possibilities to characterize the Chern classes of the vector bundle with the help of this extra structure.

i) Let X be a differentiable manifold, E a differentiable line bundle over X. Then the defining cocycle (g_{ij}) associated to a suitable open covering defines a cohomology class $\xi \in H^1(X, \mathcal{C}^{\infty *})$. Here $\mathcal{C}^{\infty *}$ denotes the sheaf of germs of non-vanishing differentiable functions. It is not difficult to see that under this mapping the isomorphism classes of differentiable line bundles correspond exactly to the classes $\xi \in H^1(X, \mathcal{C}^{\infty *})$, cf. Hirzebruch [62]. The tensor product of line bundles corresponds to the sum in the group $H^1(X; \mathcal{C}^{\infty *})$ (H^1 is to be written additively). If X is even a complex manifold, then the isomorphism classes of holomorphic line bundles over X correspond to the elements in

$$H^1(X, \mathcal{O}_X^*),$$

where $\mathcal{O}_X^*$ is the sheaf of germs of non-vanishing holomorphic functions on X.

Over X we have the exponential sequence

$$0 \to \underline{\mathbb{Z}} \to \mathcal{C}^\infty \xrightarrow{\exp} \mathcal{C}^{\infty *} \to 0, \quad (\exp(f) = e^{2\pi i f})$$

($0 \to \underline{\mathbb{Z}} \to \mathcal{O}_X \to \mathcal{O}_X^* \to 0$ in the holomorphic case). The associated long exact cohomology sequence contains the segment

$$\cdots \to H^1(X, \mathcal{C}^\infty) \to H^1(X, \mathcal{C}^{\infty *}) \xrightarrow{\delta} H^2(X, \underline{\mathbb{Z}}) \to H^2(X, \mathcal{C}^\infty) \to \cdots$$

We can identify $H^i(X, \underline{\mathbb{Z}})$ with the singular cohomology group $H^i(X, \mathbb{Z})$. Because $\mathcal{C}^\infty$ is a fine sheaf, the groups $H^i(X, \mathcal{C}^\infty)$ vanish for $i > 0$. Thus one has an isomorphism

$$\delta \colon H^1(X, \mathcal{C}^{\infty *}) \to H^2(X; \mathbb{Z}).$$

If the line bundle L determines a class $\xi \in H^1(X, \mathcal{C}^{\infty *})$, then we have (Hirzebruch, [62] p. 62)

$$\delta(\xi) = c_1(L).$$

Example. Let $X = \mathbb{P}_n$. Then for every integer

$$k \in H^2(\mathbb{P}_n; \mathbb{Z})$$

there is exactly one differentiable line bundle with k as first Chern class. Over $\mathbb{P}_n$ however we also have

$$H^1(\mathbb{P}_n, \mathcal{O}_{\mathbb{P}_n}) = H^2(\mathbb{P}_n, \mathcal{O}_{\mathbb{P}_n}) = 0.$$

Thus out of the long exact cohomology sequence of the holomorphic exponential sequence

$$0 \to \underline{\mathbb{Z}} \to \mathcal{O}_{\mathbb{P}_n} \to \mathcal{O}_{\mathbb{P}_n}^* \to 0$$

we get an isomorphism

$$H^1(\mathbb{P}_n, \mathcal{O}_{\mathbb{P}_n}^*) \xrightarrow{\delta} H^2(\mathbb{P}_n, \mathbb{Z}).$$

Because the diagram

$$\begin{array}{ccc} H^1(\mathbb{P}_n, \mathcal{O}_{\mathbb{P}_n}^*) & \xrightarrow{\delta} & H^2(\mathbb{P}_n, \mathbb{Z}) \\ \downarrow & & \downarrow \\ H^1(\mathbb{P}_n, \mathcal{C}^{\infty *}) & \xrightarrow{\delta} & H^2(\mathbb{P}_n, \mathbb{Z}) \end{array}$$

commutes, it follows: to every $k \in \mathbb{Z}$ there is exactly one differentiable line bundle over $\mathbb{P}_n$ with k as first Chern class. This bundle has exactly one holomorphic (algebraic) structure namely $\mathcal{O}_{\mathbb{P}_n}(k)$, i.e.,

$$\operatorname{Pic} \mathbb{P}_n \cong \mathbb{Z}.$$

ii) With differential geometric methods one can associate with differentiable bundles over manifolds certain classes in the de Rham cohomology $H^*_{\mathrm{DR}}(X)$. Under the de Rham isomorphism $H^*_{\mathrm{DR}}(X) \to H^*(X;\mathbb{R})$ these classes are mapped onto the "real" Chern classes $c_i(E)\otimes 1 \in H^*(X,\mathbb{Z})\otimes_{\mathbb{Z}}\mathbb{R} = H^*(X,\mathbb{R})$. We shall not need this remark in what follows.

iii) Let $Y \underset{i}{\hookrightarrow} X$ be an r-codimensional closed complex submanifold of the compact complex manifold X. The *dual class of* Y *in* X

$$d_X(Y) \in H^{2r}(X;\mathbb{Z})$$

is the Poincaré dual of the image of the fundamental class $i_*o_Y \in H_{2n-2r}(X;\mathbb{Z})$:

$$d_X(Y) \cap o_X = i_*o_Y$$

If Y is 1-codimensional, i.e., a divisor in X without singularities, then (Hirzebruch, [62] p. 69)

$$c_1([Y]) = d_X(Y).$$

More generally for a holomorphic r-bundle E over a projective algebraic manifold X the following holds: if $s \in H^0(X,E)$ is a holomorphic section transversal to the zero section of E, $Y = (s = 0)$ the set of zeros of s, then the dual class of Y in X is the rth Chern class of E:

$$c_r(E) = d_X(Y)$$

(Grothendieck, [51], Kleiman, [75]).

§2. The splitting of vector bundles

In this paragraph we investigate under what circumstances a holomorphic vector bundle can be decomposed into a direct sum of holomorphic line bundles. The first result in this direction is the theorem of Grothendieck, which says that the only indecomposable bundles on $\mathbb{P}_1$ are the line bundles. This theorem is an important tool for investigating bundles over other spaces. If $\mathbb{P}_1 \subset X$ then every vector bundle over X splits after restriction to $\mathbb{P}_1$. The splitting behaviour depends very much on the embedding $\mathbb{P}_1 \subset X$ as we shall see in examples of 2-bundles over $\mathbb{P}_2$. Finally we shall prove a cohomological splitting criterion for vector bundles on projective spaces, from which it in particular follows that a holomorphic vector bundle E over $\mathbb{P}_n$, $n \geq 2$, splits as a direct sum of line bundles precisely when this is the case for the restriction to some projective plane.

2.1. The theorem of Grothendieck. We shall say that a holomorphic r-bundle splits when it can be represented as a direct sum of r holomorphic line bundles. On the projective line all holomorphic vector bundles split; we have the

THEOREM 2.1.1 (Grothendieck). *Every holomorphic r-bundle E over $\mathbb{P}_1$ has the form*

$$E = \mathcal{O}_{\mathbb{P}_1}(a_1) \oplus \cdots \oplus \mathcal{O}_{\mathbb{P}_1}(a_r)$$

with uniquely determined numbers $a_1, \ldots, a_r \in \mathbb{Z}$ with

$$a_1 \geq a_2 \geq \cdots \geq a_r.$$

PROOF. The proof is by induction on the rank r. For $r = 1$ there is nothing to prove (1.2). Suppose the assertion has already been proved for all r-bundles. For an $(r+1)$-bundle E there is ((1.1.)!) a uniquely determined number $k_0 \in \mathbb{Z}$ with

$$h^0(\mathbb{P}_1, E(k_0)) \neq 0, \quad h^0(\mathbb{P}_1, E(k)) = 0 \quad \text{for} \quad k < k_0.$$

Let $0 \neq s \in H^0(\mathbb{P}_1; E(k_0))$. We claim that s has no zeros. For if there were an $x \in \mathbb{P}_1$ with $s(x) = 0$, then s would be a section in $E(k_0) \otimes_{\mathcal{O}_{\mathbb{P}_1}} J_x = E(k_0 - 1)$ in contradiction to the choice of k_0. Here J_x is the sheaf of ideals of the point-divisor x, i.e., $J_x = \mathcal{O}_{\mathbb{P}_1}(-1)$.

The section s defines a trivial subbundle and thus an exact sequence of vector bundles

$$(*) \qquad 0 \to \mathcal{O}_{\mathbb{P}_1} \xrightarrow{s} E(k_0) \to F \to 0.$$

By the induction hypothesis F splits, i.e., there are well defined numbers $b_1 \geq \cdots \geq b_r$, $b_i \in \mathbb{Z}$, with

$$F = \mathcal{O}_{\mathbb{P}_1}(b_1) \oplus \cdots \oplus \mathcal{O}_{\mathbb{P}_1}(b_r).$$

It now suffices to show that the sequence (*) splits. The obstruction to its splitting lies in the group

$$\operatorname{Ext}^1_{\mathbb{P}_1}(F, \mathcal{O}_{\mathbb{P}_1}) = H^1(\mathbb{P}_1, F^* \otimes \mathcal{O}_{\mathbb{P}_1}) = \bigoplus_{i=1}^{r} H^1(\mathbb{P}_1, \mathcal{O}_{\mathbb{P}_1}(-b_i)).$$

Thus if $b_i < 2$ the entire obstruction group vanishes. But we even have $b_i \leq 0$ for $i = 1, 2, \ldots, r$! To see this tensor the sequence (*) with $\mathcal{O}_{\mathbb{P}_1}(-1)$ to obtain

$$0 \to \mathcal{O}_{\mathbb{P}_1}(-1) \to E(k_0 - 1) \to \bigoplus_{i=1}^{r} \mathcal{O}_{\mathbb{P}_1}(b_i - 1) \to 0,$$

a sequence whose long exact cohomology sequence looks as follows

$$0 = H^0(\mathbb{P}_1, \mathcal{O}_{\mathbb{P}_1}(-1)) \to H^0(\mathbb{P}_1, E(k_0 - 1)) \to \bigoplus_{i=1}^{r} H^0(\mathbb{P}_1, \mathcal{O}_{\mathbb{P}_1}(b_i - 1)) \to H^1(\mathbb{P}_1, \mathcal{O}_{\mathbb{P}_1}(-1)) = 0.$$

Thus by definition of k_0 we have

$$0 = h^0(\mathbb{P}_1, E(k_0 - 1)) = \sum_{i=1}^{r} h^0(\mathbb{P}_1, \mathcal{O}_{\mathbb{P}_1}(b_i - 1)).$$

This implies $b_i \leq 0$ for $i = 1, 2, \ldots, r$. Hence the sequence (*) must split and we obtain after tensoring with $\mathcal{O}_{\mathbb{P}_1}(-k_0)$

$$E = \mathcal{O}_{\mathbb{P}_1}(-k_0) \oplus \bigoplus_{i=1}^{r} \mathcal{O}_{\mathbb{P}_1}(b_i - k_0) = \bigoplus_{i=1}^{r+1} \mathcal{O}_{\mathbb{P}_1}(a_i)$$

with $a_1 \geq \cdots \geq a_{r+1}$. Thus the existence of the decomposition is proved.

Let us now suppose there were two different decompositions

$$(**) \qquad \mathcal{O}_{\mathbb{P}_1}(a_1) \oplus \cdots \oplus \mathcal{O}_{\mathbb{P}_1}(a_r) = \mathcal{O}_{\mathbb{P}_1}(b_1) \oplus \cdots \oplus \mathcal{O}_{\mathbb{P}_1}(b_r)$$

with $a_1 \geq \cdots \geq a_r$, $b_1 \geq \cdots \geq b_r$. Let a_j be the first of the a's which is not equal to b_j so that $a_1 = b_1, \ldots, a_{j-1} = b_{j-1}$. Without restriction we may suppose $a_j > b_j$. Tensoring (**) with $\mathcal{O}_{\mathbb{P}_1}(-a_j)$ we get

$$\bigoplus_{i=1}^{j-1} \mathcal{O}_{\mathbb{P}_1}(a_i - a_j) \oplus \mathcal{O}_{\mathbb{P}_1} \oplus \bigoplus_{i=j+1}^{r} \mathcal{O}_{\mathbb{P}_1}(a_i - a_j) = \bigoplus_{i=1}^{j-1} \mathcal{O}_{\mathbb{P}_1}(b_i - a_j) \oplus \bigoplus_{i=j}^{r} \mathcal{O}_{\mathbb{P}_1}(b_i - a_j).$$

Since $a_i = b_i$ for $i < j$ and $a_j > b_j$ one has more holomorphic sections on the left side of this equation than on the right — a contradiction. □

Here it is appropriate to utter the following warning: if

$$0 \to \bigoplus_{i=1}^{r} \mathcal{O}_{\mathbb{P}_1}(a_i) \to E \to \bigoplus_{i=1}^{s} \mathcal{O}_{\mathbb{P}_1}(b_i) \to 0$$

is an exact sequence of vector bundles over $\mathbb{P}_1$, it does *not* in general follow that

$$E = \bigoplus_{i} \mathcal{O}_{\mathbb{P}_1}(a_i) \oplus \bigoplus_{j} \mathcal{O}_{\mathbb{P}_1}(b_j).$$

The only topological invariant of an r-bundle over $\mathbb{P}_1$ is the first Chern class c_1, i.e., every continuous r-bundle E over $\mathbb{P}_1$ is of the form

$$E = \mathcal{O}_{\mathbb{P}_1}(c_1) \oplus \mathcal{O}_{\mathbb{P}_1}^{\oplus(r-1)}.$$

From the Cartan formula one sees immediately that for a holomorphic bundle over $\mathbb{P}_1$

$$E = \mathcal{O}_{\mathbb{P}_1}(a_1) \oplus \cdots \oplus \mathcal{O}_{\mathbb{P}_1}(a_r)$$

the first Chern class is

$$c_1(E) = \sum_{i=1}^{r} a_i.$$

Thus we have a complete classification of the holomorphic r-bundles over $\mathbb{P}_1$; every such bundle E determines a unique r-tuple

$$(a_1, \ldots, a_r) \in \mathbb{Z}^r \quad \text{with} \quad a_1 \geq \cdots \geq a_r.$$

Two such r-tuples determine topologically isomorphic bundles precisely when their sums are equal.

2.2. Jump lines and the first examples. Let G_n be the Grassmann manifold of lines in $\mathbb{P}_n$. We shall denote by ℓ the point of G_n which corresponds to a projective line $L \subset \mathbb{P}_n$.

Let E be a holomorphic r-bundle over $\mathbb{P}_n$. According to the theorem of Grothendieck there is for every $\ell \in G_n$ an r-tuple

$$a_E(\ell) = (a_1(\ell), \ldots, a_r(\ell)) \in \mathbb{Z}^r; \quad a_1(\ell) \geq \cdots \geq a_r(\ell)$$

with $E|L \cong \bigoplus_{i=1}^r \mathcal{O}_L(a_i(\ell))$. In this way the mapping

$$a_E \colon G_n \to \mathbb{Z}^r$$

is defined, $a_E(\ell)$ is called the *splitting type of E on L.*

DEFINITION 2.2.1. E is *uniform* if a_E is constant.

Uniform bundles will be more thoroughly investigated in the next paragraph. For the moment we only consider an

EXAMPLE. $E = T_{\mathbb{P}_n}$. Let $H \subset \mathbb{P}_n$ be a hyperplane. Then H has a normal bundle $N_{H/\mathbb{P}_n} = \mathcal{O}_H(1)$ and one has the exact sequence

$$0 \to T_H \to T_{\mathbb{P}_n}|H \to \mathcal{O}_H(1) \to 0.$$

Since the obstruction to the splitting of this sequence lies in the vanishing group (1.1.)

$$H^1(H, T_H(-1)) \cong H^{n-2}(H, \Omega_H^1(-n+1)),$$

one has for the restriction of $T_{\mathbb{P}_n}$ to H

$$T_{\mathbb{P}_n}|H \cong T_H \oplus \mathcal{O}_H(1).$$

Now $T_L = \omega_L^* = \mathcal{O}_L(2)$ for every line $L \subset \mathbb{P}_n$, so it follows by induction over n that the tangent bundle $T_{\mathbb{P}_n}$ is uniform of splitting type

$$a_{T_{\mathbb{P}_n}} = (2, 1, \ldots, 1).$$

DEFINITION 2.2.2. A holomorphic r-bundle E over $\mathbb{P}_n$ is *homogeneous* if for every projective transformation $t \in \mathrm{PGL}(n+1,\mathbb{C})$ we have

$$t^*E \cong E.$$

Since any line can be transformed onto any other by a projective transformation, homogeneous bundles are certainly uniform.

The tangent bundle $T_{\mathbb{P}_n}$ is a homogeneous bundle, for the differential of a projective transformation t defines an isomorphism

$$T_{\mathbb{P}_n} \cong t^*T_{\mathbb{P}_n}.$$

We now make the acquaintance of some bundles which are not uniform. We again consider the mapping

$$\begin{aligned} a_E\colon G_n &\to \mathbb{Z}^r \\ \ell &\mapsto (a_1(\ell),\dots,a_r(\ell)), \quad a_1(\ell) \geq \cdots \geq a_r(\ell) \end{aligned}$$

which is defined by an r-bundle E. We give $\mathbb{Z}^r$ the lexicographical ordering — i.e., $(a_1,\dots,a_r) \leq (b_1,\dots,b_r)$ if the first non-zero difference $b_i - a_i$ is positive. Let

$$\underline{a}_E = \inf_{\ell \in G_n} a_E(\ell).$$

DEFINITION 2.2.3. $S_E = \{\ell \in G_n | a_E(\ell) > \underline{a}_E\}$ is the set of jump lines, $\underline{a}_E$ is the *generic splitting type* of E.

We shall show in §3 that $U_E = G_n \setminus S_E$ is a non-empty Zariski-open subset of G_n. On the lines $L \subset \mathbb{P}_n$ with $\ell \in U_E$ the bundle E has the constant splitting type $\underline{a}_E$. In general $S_E \neq \varnothing$.

We wish to illustrate this last assertion with the example of 2-bundles over $\mathbb{P}_2$. In order to construct 2-bundles on $\mathbb{P}_n$, $n \geq 2$, one could try to obtain them as extensions of line bundles. However if

$$0 \to \mathcal{O}_{\mathbb{P}_n}(a) \to E \to \mathcal{O}_{\mathbb{P}_n}(b) \to 0$$

is an exact sequence of vector bundles over $\mathbb{P}_n$, then E is necessarily the direct sum of $\mathcal{O}_{\mathbb{P}_n}(a)$ and $\mathcal{O}_{\mathbb{P}_n}(b)$, for the obstruction to the splitting of the sequence lies in $H^1(\mathbb{P}_n, \mathcal{O}_{\mathbb{P}_n}(-b+a))$, and this group vanishes (1.1.). Thus we have shown:

LEMMA 2.2.4. *A 2-bundle E over $\mathbb{P}_n$, $n \geq 2$, which does not split contains no proper subbundles.*

By Theorem A the bundle $E(k)$ has many sections for k sufficiently large. Let us suppose that $s \in H^0(\mathbb{P}_2, E)$ is a section with m simple zeros

$$x_1,\dots,x_m \in \mathbb{P}_2.$$

We blow up $\mathbb{P}_2$ in these points and obtain a modification

$$\sigma\colon X \to \mathbb{P}_2,$$

on which the lifted bundle $\sigma^* E$ has a section $\sigma^*(s)$, which vanishes to first order precisely on the exceptional divisor C.

$$\sigma^*(s) \in H^0(X, \sigma^* E \otimes J_C)$$

is thus nonvanishing and defines a subbundle

$$J_C^* = [C] \subset \sigma^* E.$$

Therefore we obtain over X an exact sequence

$$(*) \qquad 0 \to [C] \to \sigma^* E \to Q \to 0$$

with a line bundle Q.

The exceptional set C consists of the m components

$$C_i = \sigma^{-1}(x_i) \cong \mathbb{P}_1 \quad \text{with} \quad C_i^2 = -1 = c_1(N_{C_i/X}).$$

If one restricts (*) to C_i, then because

$$[C]|C_i = [C_i]|C_i = N_{C_i/X} = \mathcal{O}_{C_i}(-1)$$

one gets the exact sequence

$$0 \to \mathcal{O}_{C_i}(-1) \to \mathcal{O}_{C_i}^{\oplus 2} \to Q|C_i \to 0$$

and thus $Q|C_i = \mathcal{O}_{C_i}(1)$.

Hence it is reasonable to investigate extensions

$$0 \to [C] \to E' \to [-C] \to 0$$

over the modification X whose restrictions to the exceptional curves C_i are of the form

$$0 \to \mathcal{O}_{C_i}(-1) \to \mathcal{O}_{C_i}^{\oplus 2} \to \mathcal{O}_{C_i}(1) \to 0$$

and then demonstrate that E' is the lifting

$$E' = \sigma^* E$$

of a 2-bundle E over $\mathbb{P}_2$. We shall now carry out this idea and see that the splitting behaviour of the bundle E thus constructed is easy to describe.

THEOREM 2.2.5. *Let $x_1, \ldots, x_m$ be points of the projective plane. There is a holomorphic 2-bundle over $\mathbb{P}_2$, whose restriction to any line L, on which exactly a points of the set $\{x_1, \ldots, x_m\}$ lie, splits in the form*

$$E|L = \mathcal{O}_L(a) \oplus \mathcal{O}_L(-a).$$

The generic splitting type of this bundle is $(0,0)$.

PROOF. Let $\sigma\colon X \to \mathbb{P}_2$ be the σ-process in the points x_i, $C_i = \sigma^{-1}(x_i)$, $C = C_1 + \cdots + C_m$, hence

$$[C] = \bigotimes_{i=1}^{m} [C_i].$$

The sheaf of ideals $[-C] = J_C \subset \mathcal{O}_X$ defines the sequence

$$0 \to [-C] \to \mathcal{O}_X \to \mathcal{O}_C \to 0$$

or after tensoring with $[C]$

$$(*) \qquad 0 \to \mathcal{O}_X \to [C] \to [C]|C \to 0.$$

The extensions of $[-C]$ by $[C]$ are classified by

$$\mathrm{Ext}^1_X([-C],[C]) = H^1(X,[2C]).$$

By tensoring (*) with $[C]$ we get the cohomology sequence

$$\cdots \to H^1(X,[C]) \to H^1(X,[2C]) \to H^1(C,[2C]|C) \\ \to H^2(X,[C]) \to \cdots$$

The group $H^1(C,[2C]|C)$ classifies the extensions

$$0 \to [C]|C \to \,?\, \to [-C]|C \to 0.$$

If we therefore show that $H^2(X,[C])$ vanishes, then each of these extensions can be extended from C to all of X. From the cohomology sequence of (*) we get

$$\cdots \to H^2(X,\mathcal{O}_X) \to H^2(X,[C]) \to H^2(C,[C]|C) \to \cdots$$

and $H^2(C,[C]|C)$ vanishes because C is 1-dimensional. However, because $\sigma\colon X \to \mathbb{P}_2$ is a σ-process, we have

$$H^2(X,\mathcal{O}_X) \cong H^2(\mathbb{P}_2,\mathcal{O}_{\mathbb{P}_2});$$

hence $H^2(X,\mathcal{O}_X)$ vanishes and thus also $H^2(X,[C])$. The restriction homomorphism

$$H^1(X,[2C]) \to H^1(C,[2C]|C)$$

is therefore surjective. With Serre duality we have

$$H^1(C,[2C]|C) = \bigoplus_{i=1}^{m} H^1(C_i,[2C_i]|C_i) = \bigoplus_{i=1}^{m} H^1(C_i,\mathcal{O}_{C_i}(-2))$$
$$= \bigoplus_{i=1}^{m} H^0(C_i,\mathcal{O}_{C_i}) = H^0(C,\mathcal{O}_C).$$

The 1 in $H^0(C, \mathcal{O}_C)$ corresponds to the extension $\eta \in H^1(C, [2C]|C)$ which over C_i is given by the Euler sequence

$$(**) \qquad 0 \to \mathcal{O}_{C_i}(-1) \to \mathcal{O}_{C_i}^{\oplus 2} \to \mathcal{O}_{C_i}(1) \to 0.$$

Let $\xi \in H^1(X, [2C])$ be an element with $\xi|C = \eta$,

$$(***) \qquad 0 \to [C] \to E' \to [-C] \to 0$$

the associated extension. By construction restriction of this sequence to C_i gives the sequence (**).

We must now show that $E' = \sigma^* E$ for some 2-bundle E over $\mathbb{P}_2$. This is certainly the case if each of the points x_i has a neighborhood U_i so that E' is trivial over $\sigma^{-1}(U_i)$. Thus it suffices to prove the following

LEMMA 2.2.6. *Let $U \subset \mathbb{C}^2$ be an open Stein neighborhood of* 0 *and*

$$V = \{(u : v, x, y) \in \mathbb{P}_1 \times U \mid xv = yu\} \xrightarrow{\sigma} U$$

the σ-process for U in the point 0, *$C = \sigma^{-1}(0)$ the exceptional divisor in V. Let*

$$\text{(i)} \qquad 0 \to [C] \to E' \to [-C] \to 0$$

be an extension over V, whose restriction to C is the Euler sequence

$$0 \to \mathcal{O}_C(-1) \to \mathcal{O}_C^{\oplus 2} \to \mathcal{O}_C(1) \to 0.$$

Then $E' = \mathcal{O}_C^{\oplus 2}$ is trivial over V.

PROOF. Besides the extension (i) we have a further canonical extension

$$\text{(ii)} \qquad 0 \to [C] \to \mathcal{O}_V^{\oplus 2} \to [-C] \to 0,$$

whose restriction to C gives the Euler sequence. (ii) is defined by the section

$$s \in H^0(V, \mathcal{O}_V^{\oplus 2}), \quad s(u : v, x, y) = (u : v, x, y, (x, y)),$$

which vanishes exactly on C to first order. If we now show that

$$\operatorname{Ext}^1([-C], [C]) \to \operatorname{Ext}^1([-C]|C, [C]|C),$$

i.e., $H^1(V, [2C]) \to H^1(C, [2C]|C)$, is injective, then it follows that (i) and (ii) are equal and in particular that $E' = \mathcal{O}_V^{\oplus 2}$, which is what we want.

Because of the exact sequence

$$0 \to [C] \to [2C] \to [2C]|C \to 0$$

$H^1(V, [2C]) \to H^1(C, [2C]|C)$ is injective if $H^1(V, [C]) = 0$. From the sequence

$$0 \to \mathcal{O}_V \to [C] \to [C]|C \to 0$$

one sees that it suffices to show $H^1(V, \mathcal{O}_V) = 0$. Because $\sigma\colon V \to U$ is a point-modification, we have $H^1(V, \mathcal{O}_V) \cong H^1(U, \mathcal{O}_U)$. Since U is Stein $H^1(V, \mathcal{O}_V)$ vanishes. □

We now continue with the proof of the theorem: according to the lemma there is a 2-bundle E over $\mathbb{P}_2$ with $\sigma^*E = E'$. We have the extension

$$0 \to [C] \to \sigma^*E \to [-C] \to 0. \tag{***}$$

In order to investigate the splitting behavior of E we consider for a line $L \subset \mathbb{P}_2$ its strict transform

$$\tilde{L} = \overline{\sigma^{-1}(L \setminus \{x_1, \ldots, x_m\})} \subset X.$$

Let $x_1, \ldots, x_a$ (possibly after renumbering) be the points which lie on L.

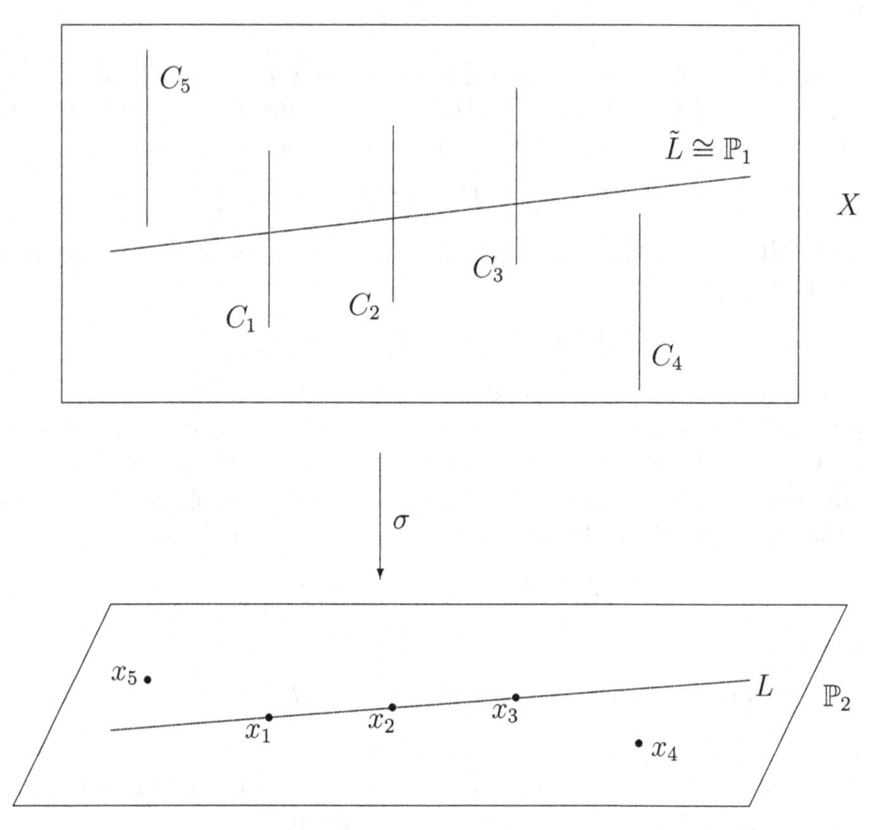

Then we have

$$\tilde{L} \cdot C = \tilde{L} \cdot (C_1 + \cdots + C_m) = \tilde{L} \cdot C_1 + \cdots + \tilde{L} \cdot C_a = a.$$

If we restrict the sequence (***) to $\tilde{L}$, then because

$$c_1([C]|\tilde{L}) = \tilde{L} \cdot C = a$$

we get

$$0 \to \mathcal{O}_{\tilde{L}}(a) \to \sigma^* E|\tilde{L} \to \mathcal{O}_{\tilde{L}}(-a) \to 0.$$

This sequence splits because $H^1(\tilde{L}, \mathcal{O}_{\tilde{L}}(2a)) = 0$ and thus

$$E|L \cong \sigma^* E|\tilde{L} \cong \mathcal{O}_{\tilde{L}}(a) \oplus \mathcal{O}_{\tilde{L}}(-a),$$

where we identify $\tilde{L}$ and L by means of σ.

The set of jump lines

$$S_E \subset G_2 = \mathbb{P}_2^*$$

consists of those points $\ell \in \mathbb{P}_2^*$ whose associated line contains at least one point x_i and is thus a union of m lines in $\mathbb{P}_2^*$ ($\{\ell \in \mathbb{P}_2^* \mid x \in L\}$ is a line in $\mathbb{P}_2^*$). □

REMARK 2.2.7. In the proof above we used the following fact about σ-processes: let $\sigma\colon X \to Y$ be the σ-process for a 2-dimensional complex manifold Y in finitely many points. Then

$$H^q(X, \mathcal{O}_X) \cong H^q(Y, \mathcal{O}_Y) \quad \text{for} \quad q \geq 0.$$

This follows with the help of the Leray spectral sequence directly from the fact that

$$R^q\sigma_*\mathcal{O}_X = 0 \quad \text{for} \quad q > 0, \quad \sigma_*\mathcal{O}_X = \mathcal{O}_Y.$$

The vanishing of the image sheaf $R^q\sigma_*\mathcal{O}_X$ for $q > 0$ results from the following lemma: if $U \subset \mathbb{C}^2$ is an open Stein neighborood of 0 and $\sigma\colon V \to U$ the σ-process for U in 0, then $H^q(V, \mathcal{O}_V) = 0$ for $q > 0$. One proves this as follows: V is given as a submanifold of $U \times \mathbb{P}_1$. Let J be the sheaf of ideals of V in $U \times \mathbb{P}_1$. The exact sequence

$$0 \to J \to \mathcal{O}_{U\times\mathbb{P}_1} \to \mathcal{O}_V \to 0$$

gives

$$\cdots \to H^q(U \times \mathbb{P}_1, \mathcal{O}_{U\times\mathbb{P}_1}) \to H^q(V, \mathcal{O}_V) \to H^{q+1}(U \times \mathbb{P}_1, J) \to \cdots.$$

$H^q(U \times \mathbb{P}_1, \mathcal{O}_{U\times\mathbb{P}_1})$ vanishes for $q > 0$, because this is true for $H^q(\mathbb{P}_1, \mathcal{O}_{\mathbb{P}_1})$ and U is Stein. $H^{q+1}(U \times \mathbb{P}_1, J)$ vanishes for $q + 1 \geq 2$, i.e, for $q > 0$, since $U \times \mathbb{P}_1$ has a Stein covering

$$\{U \times U_0, U \times U_1\}, \quad U_i = \{(z_0 : z_1) \in \mathbb{P}_1 \mid z_i \neq 0\},$$

with two sets. Thus it follows that $H^q(V, \mathcal{O}_V) = 0$ for $q > 0$.

2.3. The splitting criterion of Horrocks. After considering these examples we return to the general theory. We give a cohomological criterion that a bundle E over $\mathbb{P}_n$ splits into a direct sum of line bundles.

THEOREM 2.3.1 (Horrocks). *A holomorphic bundle E over $\mathbb{P}_n$ splits precisely when*

$$H^i(\mathbb{P}_n, E(k)) = 0 \quad \textit{for} \quad i = 1, \dots, n-1 \quad \textit{and all} \quad k \in \mathbb{Z}.$$

PROOF. If $E = \bigoplus_i \mathcal{O}_{\mathbb{P}_n}(a_i)$ then the statement follows from the Bott formula. We prove the other direction by induction over n. For $n = 1$ the condition of the theorem is empty and every bundle splits by the theorem of Grothendieck. Thus the assertion is proved for $n = 1$. Suppose the criterion is correct for all $m < n$. If E is a holomorphic bundle over $\mathbb{P}_n$, then we tensor the sequence of sheaves

$$0 \to \mathcal{O}_{\mathbb{P}_n}(-1) \to \mathcal{O}_{\mathbb{P}_n} \to \mathcal{O}_{\mathbb{P}_{n-1}} \to 0$$

with $E(k)$ and consider the following segment of the associated long exact sequence of cohomology groups

$$\cdots \to H^i(\mathbb{P}_n, E(k)) \to H^i(\mathbb{P}_{n-1}, E|\mathbb{P}_{n-1}(k)) \\ \to H^{i+1}(\mathbb{P}_n, E(k-1)) \to \cdots.$$

From this we get $H^i(\mathbb{P}_{n-1}, E|\mathbb{P}_{n-1}(*)) = 0$ for $1 \leq i \leq n-2$ and thus $E|\mathbb{P}_{n-1}$ splits by the induction hypothesis:

$$E|\mathbb{P}_{n-1} \cong \bigoplus_{i=1}^{r} \mathcal{O}_{\mathbb{P}_{n-1}}(a_i)$$

Let $F = \bigoplus_{i=1}^r \mathcal{O}_{\mathbb{P}_n}(a_i)$; we want to show that F is isomorphic to E. Let $\phi\colon F|\mathbb{P}_{n-1} \to E|\mathbb{P}_{n-1}$ be an isomorphism.

Claim: ϕ can be extended to a homomorphism $\Phi\colon F \to E$.

Suppose this were proved; then Φ must be an isomorphism, for both bundles have the same rank and the same first Chern class and Φ induces a homomorphism

$$\det \Phi\colon \det F \to \det E$$

and thus a section

$$\begin{aligned} \det \Phi &\in H^0(\mathbb{P}_n, \det F^* \otimes \det E) \\ &= H^0(\mathbb{P}_n, \mathcal{O}_{\mathbb{P}_n}(-c_1(F)) \otimes \mathcal{O}_{\mathbb{P}_n}(c_1(E))) \\ &= H^0(\mathbb{P}_n, \mathcal{O}_{\mathbb{P}_n}). \end{aligned}$$

Since by construction $\det \Phi$ is non-zero on $\mathbb{P}_{n-1}$ and is constant on $\mathbb{P}_n$, it follows that $\det \Phi$ vanishes nowhere — that is, Φ is an isomorphism.

It thus remains to prove our claim. We tensor the sequence

$$0 \to \mathcal{O}_{\mathbb{P}_{n-1}}(-1) \to \mathcal{O}_{\mathbb{P}_n} \to \mathcal{O}_{\mathbb{P}_{n-1}} \to 0$$

with $\mathcal{H}om(F, E) = F^* \otimes E$ and obtain the cohomology sequence

$$\cdots \to H^0(\mathbb{P}_n, F^* \otimes E) \to H^0(\mathbb{P}_{n-1}, F^* \otimes E|\mathbb{P}_{n-1}) \to H^1(\mathbb{P}_n, F^* \otimes E(-1)) \to \cdots .$$

But the group $H^1(\mathbb{P}_n, F^* \otimes E(-1))$ vanishes since by assumption ($n > 1$!)

$$H^1(\mathbb{P}_n, F^* \otimes E(-1)) = \bigoplus_{i=1}^{r} H^1(\mathbb{P}_n, E(-a_i - 1)) = 0.$$

Thus every section in $(F^* \otimes E)|\mathbb{P}_{n-1} = \mathcal{H}om(F|\mathbb{P}_{n-1}, E|\mathbb{P}_{n-1})$ can be extended to a section in $\mathcal{H}om(F, E) = F^* \otimes E$ as claimed. □

This splitting criterion has a surprising consequence.

THEOREM 2.3.2. *A holomorphic bundle E over $\mathbb{P}_n$ splits precisely when its restriction to some plane $\mathbb{P}_2 \subset \mathbb{P}_n$ splits.*

PROOF. Let $n \geq 3$ and $\mathbb{P}_{n-1} \subset \mathbb{P}_n$ some hyperplane. It suffices to show that E over $\mathbb{P}_n$ splits if $E|\mathbb{P}_{n-1}$ splits. If $E|\mathbb{P}_{n-1}$ splits, then

$$H^i(\mathbb{P}_{n-1}, E|\mathbb{P}_{n-1}(*)) = 0 \quad \text{for} \quad i = 1, \ldots, n-2.$$

The cohomology sequence of

$$0 \to E(k-1) \to E(k) \to E|\mathbb{P}_{n-1}(k) \to 0$$

then gives

$$\cdots \to H^{i-1}(\mathbb{P}_{n-1}, E(k)|\mathbb{P}_{n-1}) \to H^i(\mathbb{P}_n, E(k-1)) \to H^i(\mathbb{P}_n, E(k)) \to H^i(\mathbb{P}_{n-1}, E(k)|\mathbb{P}_{n-1}) \to \cdots .$$

From this one deduces that the group

$$H^i(\mathbb{P}_n, E(k))$$

is independent of k for $i = 2, \ldots, n-2$ and for $i = n-1$

$$h^{n-1}(\mathbb{P}_n, E(k-1)) \leq h^{n-1}(\mathbb{P}_n, E(k))$$

for all k. With Theorem B we thus have

$$H^i(\mathbb{P}_n, E(k)) = 0 \quad \text{for } 2 \leq i \leq n-1 \text{ and for all } k \in \mathbb{Z}.$$

In order to see that also $H^1(\mathbb{P}_n, E(*)) = 0$ we apply the same considerations to E^*. With Serre duality we have

$$H^1(\mathbb{P}_n, E(k)) = H^{n-1}(\mathbb{P}_n, E^*(-k-n-1)) = 0. \qquad \square$$

EXAMPLE. From the Bott formula we have

$$H^i(\mathbb{P}_n, \Omega^p_{\mathbb{P}_n}(*)) = \mathbb{C}\delta_{ip}.$$

That means that the bundles $\Omega^p_{\mathbb{P}_n}$ do not split for $0 < p < n$. From the cohomological point of view however, they are the next simplest bundles.

2.4. Historical remarks. The splitting theorem of Grothendieck has a long history. It is in fact equivalent to a theorem on holomorphic invertible matrices on $\mathbb{C}^*$, as was noticed by Seshadri [114]. This theorem was proved by Birkhoff in 1913, [19], [20]. But in fact it was already known to Plemelj [96] in 1908 and to Hilbert [61] in 1905.

It was noticed by W. D. Geyer that Dedekind and Weber [25] proved it in an algebraic setting in 1892. In this form it can be found also in Hasse's Zahlentheorie as the Lemma of Witt. The present simple proof is due to Grauert and Remmert [46].

The notion of a uniform vector bundle appears first in a paper of Schwarzenberger [108]. He raised the question whether a uniform vector bundle on $\mathbb{P}_n$ is homogeneous. In §3 we will consider this question in some detail.

The construction of holomorphic bundles of rank 2 on $\mathbb{P}_2$ by blowing up points is due to Schwarzenberger [107]. Here we treat only the easier case of simple points.

Schwarzenberger gave in [108] another method of constructing vector bundles of rank 2 on a complex manifold X by considering two-fold branched coverings

$$\pi\colon \tilde{X} \to X, \quad \tilde{X} \text{ non-singular.}$$

For a holomorphic line bundle L on $\tilde{X}$ the direct image $\pi_*(L)$ is a holomorphic bundle of rank 2 on X. Schwarzenberger shows that any holomorphic vector bundle of rank 2 on a surface can be obtained by this procedure.

For example consider the standard covering

$$\pi\colon \mathbb{P}_1 \times \mathbb{P}_1 \to \mathbb{P}_2.$$

The line bundles on $\mathbb{P}_1 \times \mathbb{P}_1$ are all of the form

$$p_1^*\mathcal{O}_{\mathbb{P}_1}(a) \otimes p_2^*\mathcal{O}_{\mathbb{P}_1}(b), \quad a, b \in \mathbb{Z}.$$

Here the p_i are the projections onto the ith factor. Using Grothendieck's Riemann-Roch theorem [22] one gets the Chern classes of $\pi_*(p_1^*\mathcal{O}_{\mathbb{P}_1}(a) \otimes p_2^*\mathcal{O}_{\mathbb{P}_1}(b))$:

$$c_1 = a + b - 1$$

$$c_2 = \frac{a(a-1)}{2} + \frac{b(b-1)}{2}$$

The splitting theorem of Horrocks can be found in [65] in a more general setting. The present simple proof is due to Barth and Hulek [15]. Horrocks [27] also gives the following cohomological characterization of the p-forms:

If E is a holomorphic vector bundle on $\mathbb{P}_n$ with

$$H^q(\mathbb{P}_n, E(k)) = 0 \quad \text{for} \quad q \neq 0, n, p \quad \text{and all} \quad k \in \mathbb{Z}$$

and

$$H^p(\mathbb{P}_n, E(k)) = \delta_{k0}\mathbb{C},$$

then E is the direct sum of $\Omega^p_{\mathbb{P}_n}$ and some line bundles.

Finally, let us mention that theorem 2.3.2 has more recently been proved by Elencwajg and Forster [34] using different methods.

§3. Uniform bundles

In this paragraph we explain the "standard construction", which systematizes the study of a vector bundle over $\mathbb{P}_n$ by considering its restrictions to lines. As a first application we find that a bundle whose restriction to every line through some given point is trivial must itself be trivial. Then we show that uniform r-bundles over $\mathbb{P}_n$ always split if $r < n$. This is no longer true for $r \geq n$ (see the remarks at the end of this section). Finally we give an example of a uniform bundle which is not homogeneous.

3.1. The standard construction. Let G_n again denote the Grassmann manifold of lines in $\mathbb{P}_n$. Again we denote by $\ell \in G_n$ the point belonging to the line $L \subset \mathbb{P}_n$ and by E_ℓ the plane in $\mathbb{C}^{n+1}$ which is defined by L.

Over G_n one has the tautological 2-bundle

$$V = \{(\ell, v) \in G_n \times \mathbb{C}^{n+1} \mid v \in E_\ell\}.$$

The projective bundle $\mathbb{P}(V)$ of lines in V is the flag manifold

$$\mathbb{F}_n = \{(x, \ell) \in \mathbb{P}_n \times G_n \mid x \in L\}.$$

The projection

$$q\colon \mathbb{F}_n \to G_n$$

makes of $\mathbb{F}_n$ a holomorphic $\mathbb{P}_1$-bundle over G_n. On the other hand the holomorphic mapping

$$p\colon \mathbb{F}_n \to \mathbb{P}_n \quad (x,\ell) \mapsto x,$$

identifies $\mathbb{F}_n$ in a canonical fashion with the projective tangent bundle $\mathbb{P}(T_{\mathbb{P}_n})$ of $\mathbb{P}_n$

$$\mathbb{F}_n = \mathbb{P}(T_{\mathbb{P}_n}).$$

We call the diagram

$$\begin{array}{ccc} \mathbb{F}_n & \xrightarrow{q} & G_n \\ {\scriptstyle p}\downarrow & & \\ \mathbb{P}_n & & \end{array}$$

the *standard diagram.*

Let $\ell \in G_n$; the q-fibre over ℓ

$$\tilde{L} = q^{-1}(\ell) = \{(x,\ell) \mid x \in L\}$$

is mapped isomorphically under p to the line L in $\mathbb{P}_n$ determined by ℓ.

For $x \in \mathbb{P}_n$ we denote the p-fibre $p^{-1}(x) = \{(x,\ell) \mid x \in L\}$ by $\mathbb{F}(x)$. Let $G(x) = \{\ell \in G_n \mid x \in L\}$ be the submanifold of lines through the point x; $G(x)$ is isomorphic to $\mathbb{P}_{n-1}$. The projection q induces an isomorphism

$$q|\mathbb{F}(x)\colon \mathbb{F}(x) \to G(x).$$

Thus we have two commutative diagrams:

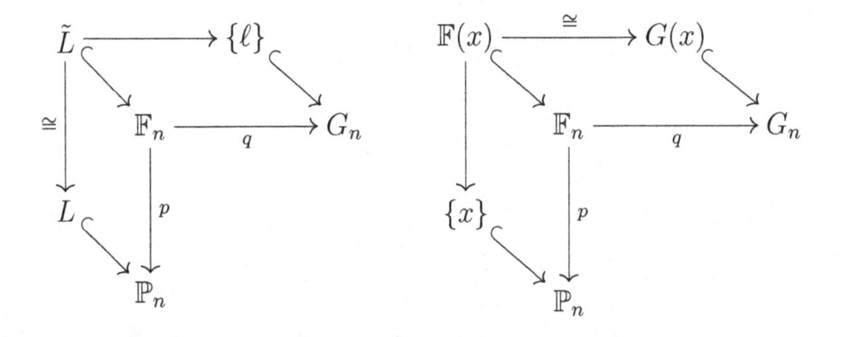

For $n = 2$ one can illustrate the situation as follows:

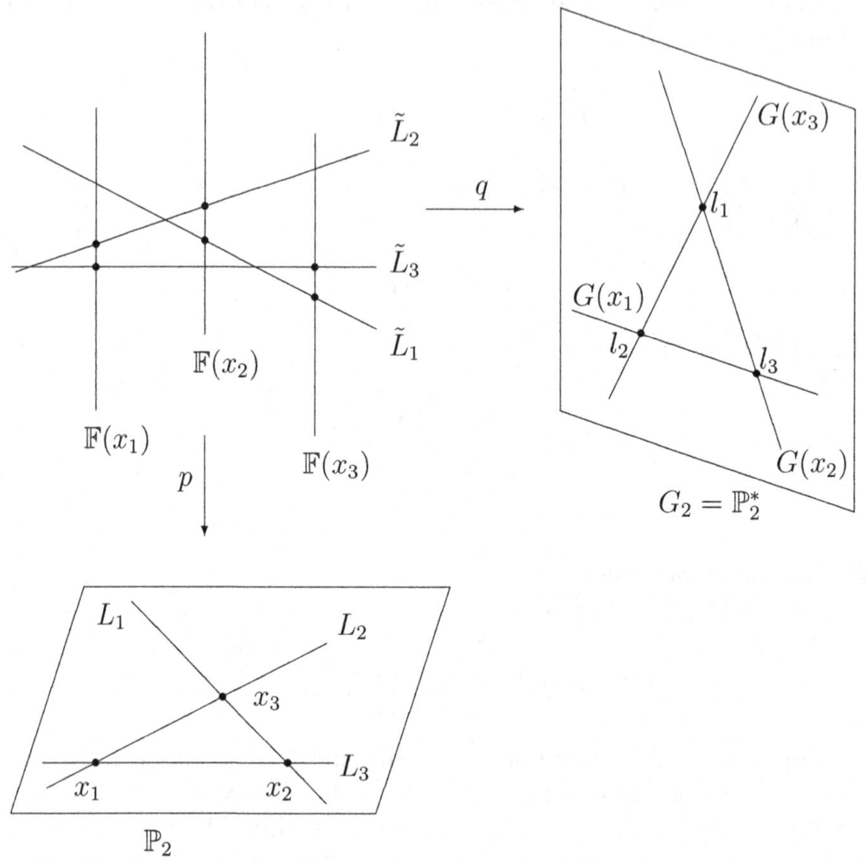

Finally for each $x \in \mathbb{P}_n$ we consider the following n-dimensional submanifold of the flag manifold:

$$\mathbb{B}(x) = \{(y, \ell) \mid x, y \in L\} = q^{-1}(G(x))$$

Let $f\colon \mathbb{B}(x) \to G(x)$ and $\sigma\colon \mathbb{B}(x) \to \mathbb{P}_n$ be the restrictions to $\mathbb{B}(x)$ of q resp. p. We then have the following commutative diagram

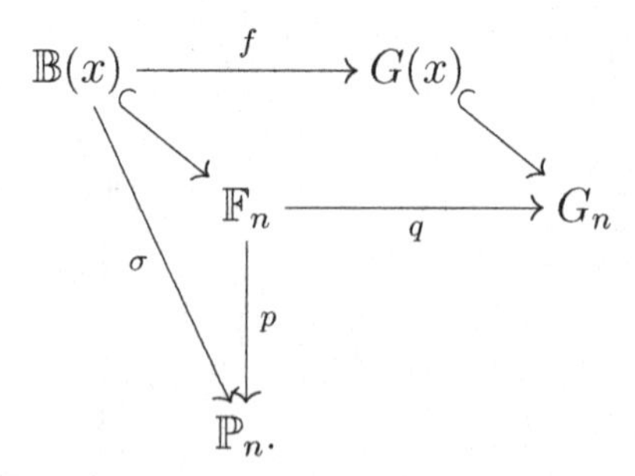

$\sigma\colon \mathbb{B}(x) \to \mathbb{P}_n$ is just the σ-process for $\mathbb{P}_n$ at the point x. $\mathbb{F}(x) \subset \mathbb{B}(x)$ is the exceptional set $\sigma^{-1}(x)$.

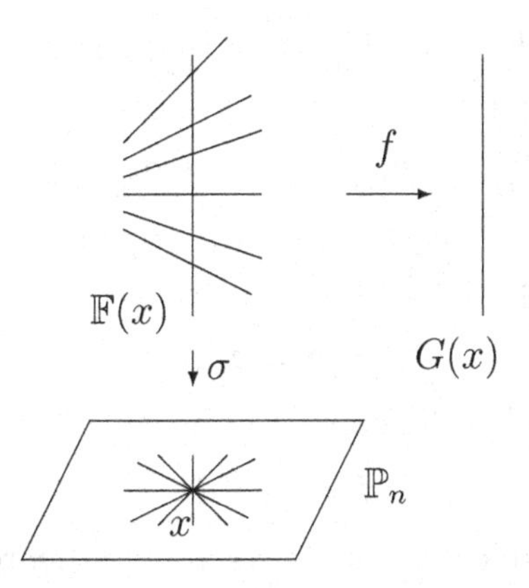

If $x, y \in \mathbb{P}_n$ are two different points, $L = \overline{xy} \subset \mathbb{P}_n$ the line through x and y, then the σ-processes $\mathbb{B}(x), \mathbb{B}(y) \subset \mathbb{F}_n$ meet precisely in the line $\tilde{L} = q^{-1}(\ell)$.

3.2. Uniform r-bundles over $\mathbb{P}_n$, $r < n$. We can now proceed to a first application of the standard construction.

THEOREM 3.2.1. *Let E be a holomorphic vector bundle of rank r over $\mathbb{P}_n$, $x \in \mathbb{P}_n$ a point, $E|L = \mathcal{O}_L^{\oplus r}$ for every line L through x. Then E is trivial.*

PROOF. We blow up $\mathbb{P}_n$ in the point x and consider the diagram

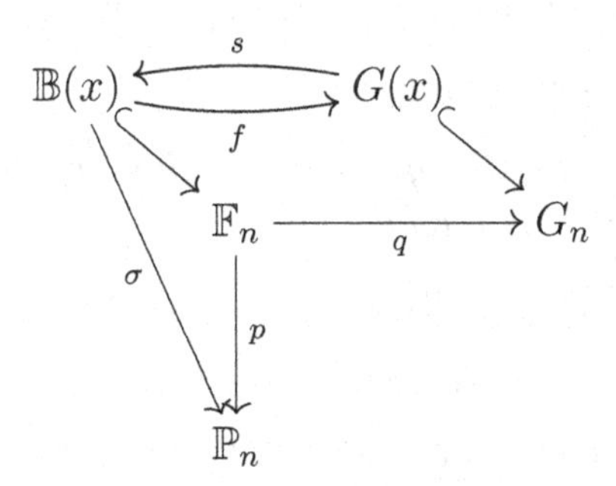

in which a section of f is given by $s(\ell) = (x, \ell)$. The bundle $\sigma^* E$ is trivial over every f-fibre $\tilde{L} = f^{-1}(\ell)$, $\ell \in G(x)$, for p induces an isomorphism

$$\sigma^* E|\tilde{L} = p^* E|\tilde{L} \cong E|L$$

and $E|L$ is trivial for all $\ell \in G(x)$ by assumption.

Claim. There is an r-bundle F over $G(x)$ with $\sigma^* E \cong f^* F$. If we have proved this, then $\sigma^* E$ is trivial, for from $\sigma \circ s = \text{const.}$ follows

$$\mathcal{O}_{G(x)}^{\oplus r} \cong s^* \sigma^* E \cong s^* f^* F \cong (f \circ s)^* F \cong F$$

and thus $\sigma^*E \cong f^*F \cong f^*\mathcal{O}_{G(x)}^{\oplus r} \cong \mathcal{O}_{\mathbb{B}(x)}^{\oplus r}$. If however σ^*E is trivial, then so is E, as the equation

$$\sigma_*\mathcal{O}_{\mathbb{B}(x)}^{\oplus r} \cong \sigma_*\sigma^*E \cong E \otimes \sigma_*\mathcal{O}_{\mathbb{B}(x)} \cong E$$

shows ($\sigma_*\mathcal{O}_{\mathbb{B}(x)} \cong \mathcal{O}_{\mathbb{P}_n}$, because σ is the σ-process of $\mathbb{P}_n$ at x). It thus remains to show that σ^*E is of the form

$$\sigma^*E \cong f^*F.$$

To this end we consider the coherent sheaf

$$F = f_*\sigma^*E.$$

F is locally free of rank r by the base-change theorem, for f is flat and

$$h^0(f^{-1}(\ell), \sigma^*E|f^{-1}(\ell)) = h^0(L, E|L) = h^0(L, \mathcal{O}_L^{\oplus r})$$

is equal to r for all lines L through x. The canonical homomorphism of sheaves

$$f^*f_*\sigma^*E \to \sigma^*E$$

is given on each f-fibre $\tilde{L} = f^{-1}(\ell)$ by the evaluation map

$$\begin{array}{ccc} f^*f_*\sigma^*E|\tilde{L} & \longrightarrow & \sigma^*E|\tilde{L} \\ \| \wr & \nearrow_{\text{ev.}} & \\ H^0(\tilde{L}, \sigma^*E|\tilde{L}) \otimes_{\mathbb{C}} \mathcal{O}_{\tilde{L}} & & \end{array}$$

and is thus an isomorphism, i.e., $\sigma^*E \cong f^*F$ with $F = f_*\sigma^*E$. □

We obtain the following important

Corollary. *A globally generated holomorphic vector bundle E over $\mathbb{P}_n$ with $c_1(E) = 0$ is trivial.*

Proof. Since E is globally generated we have an exact sequence

$$0 \to K \to \mathcal{O}_{\mathbb{P}_n}^{\oplus N} \to E \to 0.$$

Restricting this sequence to a line $L \subset \mathbb{P}_n$ we get

$$0 \to K|L \to \mathcal{O}_L^{\oplus N} \to E|L \to 0.$$

If we have $E|L = \bigoplus_{i=1}^r \mathcal{O}_L(a_i)$, then together with $E|L$ all line bundles $\mathcal{O}_L(a_i)$ are globally generated, i.e., $a_i \geq 0$. If the first Chern class of E vanishes, then it follows that

$$0 = \sum_{i=1}^r a_i,$$

and since $a_i \geq 0$ we must have $a_i = 0$ for $i = 1, \ldots, r$. Thus E is trivial on every line and hence trivial. □

In §2 the following lemma was left to be proved.

LEMMA 3.2.2. *Let E be a holomorphic r-bundle over $\mathbb{P}_n$. For each $a = (a_1, \ldots, a_r) \in \mathbb{Z}^r$ the set*

$$M_a = \{\ell \in G_n \mid a_E(\ell) > a\}$$

is a closed analytic subset of the Grassmann manifold G_n.

PROOF. For $M_k(a_1, \ldots, a_k) =$

$$\{\ell \in G_n \mid (a_E(\ell)_1, \ldots, a_E(\ell)_k) > (a_1, \ldots, a_k)\}$$

we have $M_1(a_1) \subset M_2(a_1, a_2) \subset \cdots \subset M_r(a_1, \ldots, a_r) = M_a$. Let

$$M'_k(a_1, \ldots, a_k) = \{\ell \in G_n \mid h^0(L, E(-a_k - 1)|L) > \sum_{i=1}^{k-1}(a_i - a_k)\}.$$

One checks that $M_k(a_1, \ldots, a_k)$ is given by

$$(*) \quad M_k(a_1, \ldots, a_k) = M_{k-1}(a_1, \ldots, a_{k-1} - 1) \cap \\ (M_{k-1}(a_1, \ldots, a_{k-1}) \cup M'_k(a_1, \ldots, a_k)).$$

Because of the semicontinuity theorem the sets $M'_k(a_1, \ldots, a_k)$ and $M_1(a_1) = \{\ell \in G_n \mid h^0(L, E(-a_1 - 1)|L) > 0\}$ are closed and analytic in G_n. By induction over k we see that each of the sets $M_k(a_1, \ldots, a_k) \subset G_n$ is closed and analytic and thus also

$$M_a = M_r(a_1, \ldots, a_r).$$

□

$S_E = M_{\underline{a}_E}$ is the set of jump lines. An r-bundle E over $\mathbb{P}_n$ which has no jump line does not necessarily have to split. For example the cotangent bundle $\Omega^1_{\mathbb{P}_n}$ is homogeneous and thus uniform but it does not split, since $H^1(\mathbb{P}_n, \Omega^1_{\mathbb{P}_n}) \neq 0$ (see 2.3.).

However, if the rank r is smaller than the dimension n of the base space, then we have

THEOREM 3.2.3. *For $r < n$ every uniform r-bundle over $\mathbb{P}_n$ splits as a direct sum of line bundles.*

PROOF. We prove the theorem by induction over r. For $r = 1$ there is nothing to prove. Suppose the assertion is true for all uniform r'-bundles with $1 \leq r' < r$, $r < n$. If E is a uniform r-bundle, then we can without restriction assume that E has the splitting type

$$\underline{a}_E = (a_1, \ldots, a_r), \quad a_1 \geq \cdots \geq a_r$$

with $a_1 = \cdots = a_k = 0$, $a_{k+1} < 0$. If $k = r$, then E is trivial by the previous theorem. Therefore let $k < r$, i.e.,

$$\underline{a}_E = (0, \ldots, 0, a_{k+1}, \ldots, a_r), \quad a_{k+i} < 0 \quad \text{for} \quad i = 1, \ldots, r - k.$$

In order to apply the induction hypothesis we try to write E as an extension of uniform bundles. If in fact

$$(*) \qquad 0 \to F \to E \to Q \to 0$$

is an exact sequence of vector bundles with F, Q uniform of rank smaller than r — and thus by the induction hypothesis splitting — then it follows from the Bott formula that $H^1(\mathbb{P}_n, Q^* \otimes F) = 0$. Thus (*) splits and hence also E.

To obtain an extension (*) we consider the standard diagram

$$\begin{array}{ccc} \mathbb{F}_n & \xrightarrow{q} & G_n \\ {\scriptstyle p}\downarrow & & \\ \mathbb{P}_n & & \end{array}$$

For $\tilde{L} = q^{-1}(\ell)$ we have (by means of $p\colon \tilde{L} \to L$)

$$p^*E|\tilde{L} \cong E|L$$

and thus because

$$E|L \cong \mathcal{O}_L^{\oplus k} \oplus \bigoplus_{i=1}^{r-k} \mathcal{O}_L(a_{k+i}), \quad a_{k+i} < 0$$

$$h^0(q^{-1}(\ell)), p^*E|q^{-1}(\ell)) = k \quad \text{for all} \quad \ell \in G_n.$$

Thus the direct image q_*p^*E is a vector bundle of rank k over G_n.

The canonical homomorphism of sheaves

$$q^*q_*p^*E \to p^*E$$

makes $\tilde{F} := q^*q_*p^*E$ into a subbundle of p^*E. For over each q-fibre $\tilde{L} = q^{-1}(\ell)$ the evaluation map

$$\tilde{F}|\tilde{L} = H^0(\tilde{L}, p^*E|\tilde{L}) \otimes_{\mathbb{C}} \mathcal{O}_{\tilde{L}} \to p^*E|\tilde{L}$$

identifies $\tilde{F}|\tilde{L}$ with $\mathcal{O}_L^{\oplus k} \subset \mathcal{O}_L^{\oplus k} \oplus \bigoplus_{i=1}^{r-k} \mathcal{O}_L(a_i) = E|L$.

Over $\mathbb{F}_n$ we thus obtain an exact sequence

$$(**) \qquad 0 \to \tilde{F} \to p^*E \to \tilde{Q} \to 0$$

of vector bundles, whose restriction to q-fibres $\tilde{L} = q^{-1}(\ell)$ looks as follows:

$$\begin{array}{ccccccc} 0 \to H^0(\tilde{L}, p^*E|\tilde{L}) \otimes \mathcal{O}_{\tilde{L}} & \longrightarrow & p^*E|\tilde{L} & \longrightarrow & \tilde{Q}|\tilde{L} & \longrightarrow & 0 \\ \| \wr & & \| \wr & & \| \wr & & \\ 0 \longrightarrow \mathcal{O}_L^{\oplus k} & \longrightarrow & \mathcal{O}_L^{\oplus k} \oplus \bigoplus_{i=1}^{r-k} \mathcal{O}_L(a_{k+i}) & \to & \bigoplus_{i=1}^{r-k} \mathcal{O}_L(a_{k+i}) & \to & 0 \end{array}$$

Thus over every line $\tilde{L} = q^{-1}(\ell)$ $\tilde{F}$ and $\tilde{Q}$ have the same splitting type.

Claim. There are bundles F, Q over $\mathbb{P}_n$ with

$$\tilde{F} = p^*F, \quad \tilde{Q} = p^*Q.$$

These bundles are then necessarily uniform and we obtain by projecting the bundle sequence

$$0 \to p^*F \to p^*E \to p^*Q \to 0$$

onto $\mathbb{P}_n$ the exact sequence

$$(*) \qquad 0 \to F \to E \to Q \to 0.$$

To prove our claim it suffices to show that $\tilde{F}$ and $\tilde{Q}$ are trivial on all p-fibres $\mathbb{F}(x)$ (the canonical morphisms $p^*p_*\tilde{F} \to \tilde{F}$, $p^*p_*\tilde{Q} \to \tilde{Q}$ are then isomorphisms). Let then $\mathbb{F}(x) = \mathbb{P}_{n-1}$ be any p-fibre. From

$$0 \to \tilde{F} \to p^*E \to \tilde{Q} \to 0$$

we obtain the exact sequence

$$0 \to \tilde{F}|\mathbb{F}(x) \to \mathcal{O}_{\mathbb{F}(x)}^{\oplus r} \to \tilde{Q}|\mathbb{F}(x) \to 0$$

and for the Chern classes of $\tilde{F}|\mathbb{F}(x)$ and $\tilde{Q}|\mathbb{F}(x)$ this means

$$c(\tilde{F}|\mathbb{F}(x)) \cdot c(\tilde{Q}|\mathbb{F}(x)) = 1.$$

If $r < n$ then this must imply

$$c(\tilde{F}|\mathbb{F}(x)) = 1 \quad \text{and} \quad c(\tilde{Q}|\mathbb{F}(x)) = 1$$

and in particular

$$c_1(\tilde{F}|\mathbb{F}(x)) = 0 \quad \text{and} \quad c_1(\tilde{Q}|\mathbb{F}(x)) = 0.$$

Since $\tilde{Q}|\mathbb{F}(x)$ is also globally generated, it follows from the previous corollary that $\tilde{Q}|\mathbb{F}(x)$ is trivial. Similarly it follows that $\tilde{F}^*|\mathbb{F}(x)$ and thus also $\tilde{F}|\mathbb{F}(x)$ is trivial. □

In II, §2 we will prove the following result of Van de Ven: The uniform 2-bundles on $\mathbb{P}_2$ are precisely the bundles $\mathcal{O}_{\mathbb{P}_2}(a) \oplus \mathcal{O}_{\mathbb{P}_2}(b)$ and $T_{\mathbb{P}_2}(a)$, $a, b \in \mathbb{Z}$. In particular uniform 2-bundles on $\mathbb{P}_n$ are homogeneous.

REMARK 3.2.4. As we hinted at the beginning of the proof just given uniform bundles can be characterized as follows: let

$$\begin{array}{ccc} \mathbb{F}_n & \xrightarrow{q} & G_n \\ {\scriptstyle p}\downarrow & & \\ \mathbb{P}_n & & \end{array}$$

be the standard diagram and E a holomorphic vector bundle over $\mathbb{P}_n$. E is uniform with

$$E|L = \mathcal{O}_L(a_1)^{r_1} \oplus \cdots \oplus \mathcal{O}_L(a_k)^{r_k},$$
$$a_1 > a_2 > \cdots > a_k,$$

if and only if there is a filtration

$$0 = F^0 \subset F^1 \subset \cdots \subset F^k = p^*E$$

of p^*E by subbundles F^i such that

$$F^i/F^{i-1} \simeq q^*(G_i) \otimes p^*\mathcal{O}_{\mathbb{P}_n}(a_i),$$

where G_i is a holomorphic vector bundle of rank r_i over G_n. This filtration is the relative Harder–Narasimhan filtration of p^*E (cf. [40]). It is constructed as follows:
$F^1 = q^*(q_*p^*E(-a_1)) \otimes p^*\mathcal{O}_{\mathbb{P}_n}(a_1)$ is a subbundle of p^*E of rank r_1 by the semicontinuity theorem. If one sets

$$Q = p^*E/F^1,$$

then for the same reason

$$q^*q_*(Q \otimes p^*\mathcal{O}_{\mathbb{P}_n}(-a_2)) \otimes p^*\mathcal{O}_{\mathbb{P}_n}(a_2)$$

is a subbundle of Q of rank r_2. If

$$\pi\colon p^*E \to Q$$

is the quotient map, then we set

$$F^2 = \pi^{-1}(q^*q_*(Q \otimes p^*\mathcal{O}_{\mathbb{P}_n}(-a_2)) \otimes p^*\mathcal{O}_{\mathbb{P}_n}(a_2)).$$

By continuing in this fashion we get the desired filtration of p^*E.

Conversely suppose we have such a filtration of p^*E. If one restricts it to $q^{-1}(\ell)$, $\ell \in G_n$, then because

$$a_1 > a_2 > \cdots > a_k$$

one sees that

$$F_i|q^{-1}(\ell) \simeq (F_{i-1}|q^{-1}(\ell)) \oplus (F_i/F_{i-1})|q^{-1}(\ell).$$

From this it follows that

$$E|L \simeq (p^*E)|q^{-1}(\ell) \simeq \mathcal{O}_L(a_1)^{\oplus r_1} \oplus \cdots \oplus \mathcal{O}_L(a_k)^{\oplus r_k}$$

for every line $L \subset \mathbb{P}_n$.

3.3. A non-homogeneous uniform $(3n-1)$-bundle over $\mathbb{P}_n$. Before we come to the construction of a non-homogeneous uniform bundle E over $\mathbb{P}_n$ we make the following definition.

DEFINITION 3.3.1. A holomorphic vector bundle E over $\mathbb{P}_n$ is k-*homogeneous* if for all linear embeddings

$$\phi_1, \phi_2 \colon \mathbb{P}_k \hookrightarrow \mathbb{P}_n$$

we have $\phi_1^* E \cong \phi_2^* E$.

An n-homogeneous bundle is homogeneous in the usual sense. A bundle is 1-homogeneous precisely when it is uniform. Let $\mathbb{P}_k \subset \mathbb{P}_{k+1}$ be the embedding $(z_0 : \cdots : z_k) \mapsto (z_0 : \cdots : z_k : 0)$. Since every linear embedding

$$\phi \colon \mathbb{P}_k \hookrightarrow \mathbb{P}_n$$

with $k < n$ can be extended to a linear embedding

$$\Phi \colon \mathbb{P}_{k+1} \hookrightarrow \mathbb{P}_n,$$

we see that $(k+1)$-homogeneous bundles are k-homogeneous. We call

$$h(E) = \max\{k \mid 0 \le k \le n;\ E \text{ is } k\text{-homogeneous}\}$$

the *degree of homogeneity* of E.

In 3.2 we saw that a uniform r-bundle E over $\mathbb{P}_n$ with $r < n$ splits and is thus homogeneous. For a bundle E of rank $r < n$ over $\mathbb{P}_n$ therefore only the values

$$h(E) = 0 \quad \text{or} \quad h(E) = n$$

are possible.

If however we place no condition on the rank then we have:

THEOREM 3.3.2. *Let $n \ge 2$. For every integer m with $1 \le m < n$ there is a holomorphic vector bundle E over $\mathbb{P}_n$ with degree of homogeneity*

$$h(E) = m - 1.$$

PROOF. We begin with the Euler sequence

$$0 \to \mathcal{O}_{\mathbb{P}_n}(-1) \to \mathcal{O}_{\mathbb{P}_n}^{\oplus(n+1)} \to T_{\mathbb{P}_n}(-1) \to 0. \tag{1}$$

The vector bundle fibre of $T_{\mathbb{P}_n}(-1)$ in a point

$$x = \mathbb{P}(\mathbb{C}v), \quad v \in \mathbb{C}^{n+1} \setminus 0,$$

is given by

$$T_{\mathbb{P}_n}(-1)(x) = \mathbb{C}^{n+1}/\mathbb{C}v.$$

Let us choose $m+1$ linearly independent vectors

$$w_0, \ldots, w_m \in \mathbb{C}^{n+1};$$

they determine sections

$$s_{w_i} \in H^0(\mathbb{P}_n, T_{\mathbb{P}_n}(-1))$$

in $T_{\mathbb{P}_n}(-1)$ which at the point $x = \mathbb{P}(\mathbb{C}v)$ are given by

$$s_{w_i}(x) = w_i/\mathbb{C}v \in T_{\mathbb{P}_n}(-1)(x).$$

Because $w_0, \ldots, w_m$ are linearly independent, the sections s_{w_i} have no common zeros. Thus they define a trivial 1-dimensional subbundle in $T_{\mathbb{P}_n}(-1)^{\oplus(m+1)}$

$$\mathcal{O}_{\mathbb{P}_n} \xrightarrow{(s_{w_0},\ldots,s_{w_m})} T_{\mathbb{P}_n}(-1)^{\oplus(m+1)}.$$

Let E be the quotient, i.e.,

$$0 \to \mathcal{O}_{\mathbb{P}_n} \to T_{\mathbb{P}_n}^{\oplus(m+1)} \to E \to 0 \tag{2}$$

is exact.

In order to determine the degree of homogeneity of E, we investigate the restrictions of this $(n(m+1)-1)$-bundle E to k-dimensional projective subspaces

$$\mathbb{P}(W) \subset \mathbb{P}_n.$$

Claim: Let $W_0 = \mathbb{C}w_0 + \cdots + \mathbb{C}w_m \subset \mathbb{C}^{n+1}$ be the subspace spanned by the vectors $w_0, \ldots, w_m$ and let $W \subset \mathbb{C}^{n+1}$ be a $(k+1)$-dimensional subspace.

i) If $W_0 \not\subset W$, then

$$E|\mathbb{P}(W) \cong T_{\mathbb{P}(W)}(-1)^{\oplus(m+1)} \oplus \mathcal{O}_{\mathbb{P}(W)}^{\oplus[(n-k)(m+1)-1]}$$

ii) If $W_0 \subset W$, then

$$E|\mathbb{P}(W) \cong E' \oplus \mathcal{O}_{\mathbb{P}(W)}^{\oplus(n-k)(m+1)}$$

with E' a bundle over $\mathbb{P}(W)$ such that $h^0(\mathbb{P}(W), E'^*) = 0$.

Proof.

i) If $W_0 \not\subset W$, then at least one of the vectors w_i — say w_0 — is not in W. The corresponding section

$$s_{w_0}|\mathbb{P}(W) \in H^0(\mathbb{P}(W), T_{\mathbb{P}_n}(-1)|\mathbb{P}(W))$$

is thus everywhere nonzero. It defines an exact sequence

$$0 \to \mathcal{O}_{\mathbb{P}(W)} \xrightarrow{s_{w_0}|\mathbb{P}(W)} T_{\mathbb{P}_n}(-1)|\mathbb{P}(W) \to Q \to 0$$

with an $(n-1)$-bundle Q over $\mathbb{P}(W)$. Together with the exact sequence

$$0 \to T_{\mathbb{P}(W)}(-1) \to T_{\mathbb{P}_n}(-1)|\mathbb{P}(W) \to \mathcal{O}_{\mathbb{P}(W)}^{\oplus(n-k)} \to 0$$

this gives a commutative diagram with exact rows and columns:

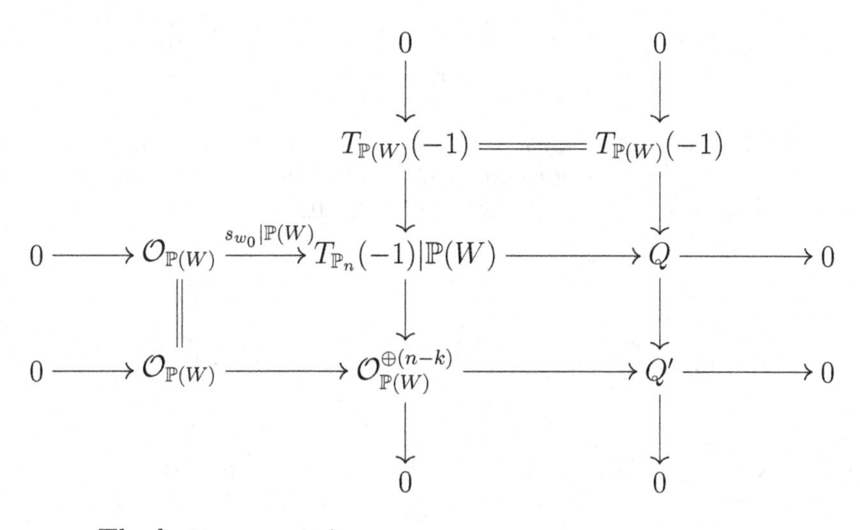

The bottom row gives

$$Q' = \mathcal{O}_{\mathbb{P}(W)}^{\oplus(n-k-1)}$$

and the right hand column gives

$$Q \cong T_{\mathbb{P}(W)}(-1) \oplus \mathcal{O}_{\mathbb{P}(W)}^{\oplus(n-k-1)}$$

since $H^1(\mathbb{P}(W), T_{\mathbb{P}(W)}(-1)) = 0$.

We now consider the diagram

$$\begin{array}{ccccccccc}
 & & & & 0 & & 0 & & \\
 & & & & \downarrow & & \downarrow & & \\
 & & & & T_{\mathbb{P}_n}(-1)^{\oplus m}|\mathbb{P}(W) & = & T_{\mathbb{P}_n}(-1)^{\oplus m}|\mathbb{P}(W) & & \\
 & & & & \downarrow & & \downarrow & & \\
0 & \longrightarrow & \mathcal{O}_{\mathbb{P}(W)} & \xrightarrow{(s_{w_i}|\mathbb{P}(W))} & T_{\mathbb{P}_n}(-1)^{\oplus(m+1)}|\mathbb{P}(W) & \longrightarrow & E|\mathbb{P}(W) & \longrightarrow & 0 \\
 & & \| & & \downarrow & & \downarrow & & \\
0 & \longrightarrow & \mathcal{O}_{\mathbb{P}(W)} & \xrightarrow{s_{w_0}|\mathbb{P}(W)} & T_{\mathbb{P}_n}(-1)|\mathbb{P}(W) & \longrightarrow & Q & \longrightarrow & 0 \\
 & & & & \downarrow & & \downarrow & & \\
 & & & & 0 & & 0 & &
\end{array}$$

The right hand column splits, because

$$H^1(\mathbb{P}(W), Q^* \otimes T_{\mathbb{P}_n}(-1)|\mathbb{P}(W)) = 0.$$

Altogether we then have

$$E|\mathbb{P}(W) \cong T_{\mathbb{P}_n}(-1)^{\oplus m}|\mathbb{P}(W) \oplus T_{\mathbb{P}(W)}(-1) \oplus \mathcal{O}_{\mathbb{P}(W)}^{\oplus(n-k-1)}$$
$$\cong T_{\mathbb{P}(W)}(-1)^{\oplus(m+1)} \oplus \mathcal{O}_{\mathbb{P}(W)}^{\oplus[(n-k)(m+1)-1]}$$

ii) Now let $W_0 = \mathbb{C}w_0 + \cdots + \mathbb{C}w_m \subset W$. We regard the sections s_{w_i} as sections in $T_{\mathbb{P}(W)}(-1)$. Then we get the following diagram with exact rows and columns

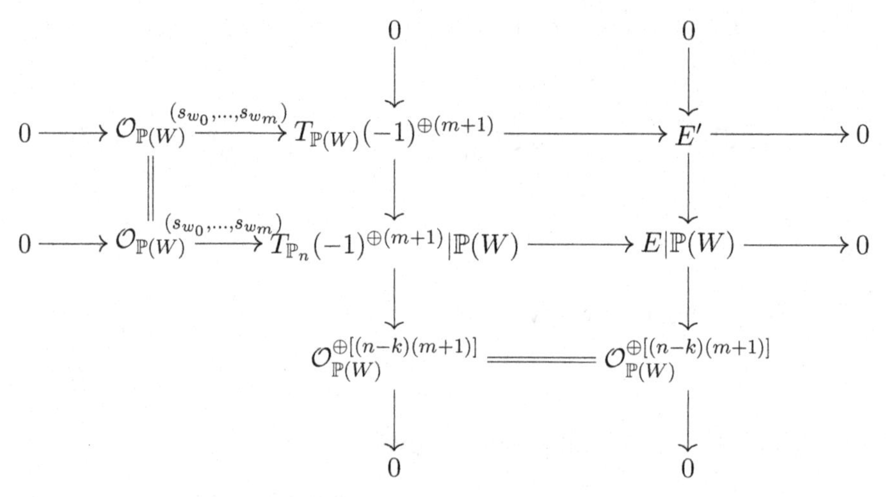

From the top row we get

$$h^0(\mathbb{P}(W), E'^*) = 0$$
$$h^1(\mathbb{P}(W), E') = 0.$$

The right hand column thus gives

$$E|\mathbb{P}(W) \cong E' \oplus \mathcal{O}_{\mathbb{P}(W)}^{\oplus[(n-k)(m+1)]}$$

with a bundle E' whose dual has no sections. Thus i) and ii) are proved.

For a k-dimensional subspace $\mathbb{P}(W) \subset \mathbb{P}_n$ we therefore have

$$E|\mathbb{P}(W) \cong \begin{cases} T_{\mathbb{P}(W)}(-1)^{\oplus(m+1)} \oplus \mathcal{O}_{\mathbb{P}(W)}^{\oplus[(n-k)(m+1)-1]} & \text{if } W_0 \not\subset W \\ E' \oplus \mathcal{O}_{\mathbb{P}(W)}^{\oplus[(n-k)(m+1)]} & \text{if } W_0 \subset W \end{cases}$$

and in particular

$$(3) \quad h^0(\mathbb{P}(W), E^*|\mathbb{P}(W)) = \begin{cases} (n-k)(m+1) - 1 & \text{if } W_0 \not\subset W \\ (n-k)(m+1) & \text{if } W_0 \subset W. \end{cases}$$

Since if $k = m-1$ we always have $W_0 \not\subset W$, we see that our bundle E is $(m-1)$-homogeneous. If $m < n$, one can find in addition to $\mathbb{P}(W_0)$

further m-dimensional projective subspaces $\mathbb{P}(W) \subset \mathbb{P}_n$; because of (3) our bundle E is not m-homogeneous.

Thus for every such $(m+1)$-tuple $(w_0, \dots, w_m)$, $1 \leq m < n$, we have constructed a bundle E over $\mathbb{P}_n$ with degree of homogeneity

$$h(E) = m - 1$$

and rank

$$\operatorname{rk}(E) = n(m+1) - 1. \qquad \square$$

If in particular we take $m = 2$, then for every $n \geq 3$ we get a nonhomogeneous uniform $(3n-1)$-bundle over $\mathbb{P}_n$.

3.4. Some historical remarks, further results, and open questions. The "standard construction" plays a very important rôle in the study of holomorphic vector bundles. It appears that Van de Ven was the first to use it in this context [134]. Theorem 3.2.1 is due to him — at least in the case of 2-bundles which are trivial on every line.

That uniform r-bundles over $\mathbb{P}_n$ split for $r < n$ was proved by Van de Ven [134] for $r = 2$ and by Sato [102] for $r > 2$. Sato uses a theorem of Tango [126] about holomorphic mappings from projective spaces to Grassmann manifolds. The proof given here was proposed by A. Hirschowitz.

In 1978 Elencwajg [31] extended the investigations of Van de Ven to show that uniform vector bundles of rank 3 over $\mathbb{P}_2$ are of the form

$$\mathcal{O}_{\mathbb{P}_2}(a) \oplus \mathcal{O}_{\mathbb{P}_2}(b) \oplus \mathcal{O}_{\mathbb{P}_2}(c) \quad \text{or} \quad T_{\mathbb{P}_2}(a) \oplus \mathcal{O}_{\mathbb{P}_2}(b) \quad \text{or} \quad S^2 T_{\mathbb{P}_2}(a).$$

In particular all uniform 3-bundles over $\mathbb{P}_2$ are homogeneous. Sato [102] had previously shown that uniform n-bundles over $\mathbb{P}_n$ for n odd are of the form

$$\mathcal{O}_{\mathbb{P}_n}(a_1) \oplus \cdots \oplus \mathcal{O}_{\mathbb{P}_n}(a_n) \qquad \text{or} \qquad T_{\mathbb{P}_n}(a) \qquad \text{or} \qquad \Omega^1_{\mathbb{P}_n}(a).$$

The results of Elencwajg and Sato thus yield a complete classification of uniform 3-bundles over $\mathbb{P}_n$. In particular all uniform 3-bundles are homogeneous. These results fortified the belief that the question of Schwarzenberger "Are uniform bundles homogeneous?" had a positive answer.

In 1979 Elencwajg [32] destroyed this belief. He constructed a 4-bundle over $\mathbb{P}_2$ which is uniform but not homogeneous. The examples in 3.3 of uniform nonhomogeneous bundles over $\mathbb{P}_n$, $n \geq 3$, are due to Hirschowitz.

PROBLEM 3.4.1. Determine the largest integer $k = k(n)$ such that uniform k-bundles over $\mathbb{P}_n$ are homogeneous.

From the examples of Hirschowitz we have in any case $k(n) \leq 3n-2$ for $n \geq 3$. In [35] it is shown that the result of Sato is still true for even n, i.e., uniform n-bundles over $\mathbb{P}_n$ are always of the form

$$\mathcal{O}_{\mathbb{P}_n}(a_1) \oplus \cdots \oplus \mathcal{O}_{\mathbb{P}_n}(a_n) \quad \text{or} \quad T_{\mathbb{P}_n}(a) \quad \text{or} \quad \Omega^1_{\mathbb{P}_n}(a).$$

Furthermore we conjecture that uniform $(n+1)$-bundles over $\mathbb{P}_n$ are of the form

$$\mathcal{O}_{\mathbb{P}_n}(a_1) \oplus \cdots \oplus \mathcal{O}_{\mathbb{P}_n}(a_{n+1}) \quad \text{or} \quad T_{\mathbb{P}_n}(a) \oplus \mathcal{O}_{\mathbb{P}_n}(b) \quad \text{or} \quad \Omega^1_{\mathbb{P}_n}(a) \oplus \mathcal{O}_{\mathbb{P}_n}(b).$$

This would give $k(n) \geq n+1$.

Finally we should mention that in [33] Elencwajg achieves a partial classification of uniform 4-bundles over $\mathbb{P}_2$. He shows: every uniform 4-bundle over $\mathbb{P}_2$ whose splitting type is not of the form $(a, a, a-1, a-1)$ or $(a, a-1, a-1, a-2)$ occurs in the following list:

$$\mathcal{O}_{\mathbb{P}_2}(a_1) \oplus \mathcal{O}_{\mathbb{P}_2}(a_2) \oplus \mathcal{O}_{\mathbb{P}_2}(a_3) \oplus \mathcal{O}_{\mathbb{P}_2}(a_4)$$
$$\mathcal{O}_{\mathbb{P}_2}(a_1) \oplus \mathcal{O}_{\mathbb{P}_2}(a_2) \oplus T_{\mathbb{P}_2}(a)$$
$$(S^2 T_{\mathbb{P}_2})(a) \oplus \mathcal{O}_{\mathbb{P}_2}(b)$$
$$(S^3 T_{\mathbb{P}_2})(a)$$
$$T_{\mathbb{P}_2}(a) \oplus T_{\mathbb{P}_2}(b)$$
$$T_{\mathbb{P}_2}(a) \otimes T_{\mathbb{P}_2}(b).$$

QUESTION 3.4.2. Are there nonhomogeneous uniform 4-bundles over $\mathbb{P}_2$ with splitting type $(1, 0, 0, -1)$?

§4. Examples of indecomposable $(n-1)$-bundles over $\mathbb{P}_n$

The bigger $n-r$ is the more difficult it becomes to find holomorphic r-bundles over $\mathbb{P}_n$ which cannot be decomposed as a direct sum of subbundles. The tangent bundles provide examples of indecomposable n-bundles over $\mathbb{P}_n$. In order to obtain indecomposable $(n-1)$-bundles over $\mathbb{P}_n$ we introduce for odd n the so-called null correlation bundle. A construction of Tango produces an indecomposable $(n-1)$-bundle for every n. All these examples are even simple bundles.

4.1. Simple bundles.

DEFINITION 4.1.1. A holomorphic r-bundle E over $\mathbb{P}_n$ is *indecomposable* if it is not the direct sum $E = F \oplus G$ of two proper subbundles $F, G \subset E$ and it is *simple* if $h^0(\mathbb{P}_n, E^* \otimes E) = 1$.

A bundle is thus simple if its only endomorphisms are the homotheties. A decomposable bundle $E = F \oplus G$, however, certainly has non-trivial endomorphisms, i.e., simple bundles are indecomposable.

A first example for a simple bundle is the tangent bundle on $\mathbb{P}_n$.

LEMMA 4.1.2. *$T_{\mathbb{P}_n}$ is simple and thus indecomposable.*

PROOF. We tensor the Euler sequence

$$0 \to \mathcal{O}_{\mathbb{P}_n}(-1) \to \mathcal{O}_{\mathbb{P}_n}^{\oplus(n+1)} \to T_{\mathbb{P}_n}(-1) \to 0$$

with $\Omega^1_{\mathbb{P}_n}(1)$ and obtain

$$0 \to \Omega^1_{\mathbb{P}_n} \to \Omega^1_{\mathbb{P}_n}(1)^{\oplus(n+1)} \to \Omega^1_{\mathbb{P}_n} \otimes T_{\mathbb{P}_n} \to 0.$$

The cohomology sequence looks as follows

$$\cdots \to H^0(\mathbb{P}_n, \Omega^1_{\mathbb{P}_n}(1)^{\oplus(n+1)}) \to H^0(\mathbb{P}_n, T^*_{\mathbb{P}_n} \otimes T_{\mathbb{P}_n}) \\ \to H^1(\mathbb{P}_n, \Omega^1_{\mathbb{P}_n}) \to H^1(\mathbb{P}_n, \Omega^1_{\mathbb{P}_n}(1)^{\oplus(n+1)}) \to \cdots .$$

From the Bott formula (1.1.) it follows that

$$H^0(\mathbb{P}_n, T^*_{\mathbb{P}_n} \otimes T_{\mathbb{P}_n}) \cong H^1(\mathbb{P}_n, \Omega^1_{\mathbb{P}_n}) \cong \mathbb{C}.$$

$T_{\mathbb{P}_n}$ is thus simple. Therefore also $\Omega^1_{\mathbb{P}_n}$ and all twisted bundles $\Omega^1_{\mathbb{P}_n}(k)$ are simple. □

Line bundles are naturally always simple. A useful necessary condition for bundles of higher rank is given by the following lemma.

LEMMA 4.1.3. *Let E be a holomorphic vector bundle of rank $r > 1$ over $\mathbb{P}_n$. If $h^0(\mathbb{P}_n, E)$ and $h^0(\mathbb{P}_n, E^*)$ are positive, then E is not simple.*

PROOF. Let $s \in H^0(\mathbb{P}_n, E)$ and $t \in H^0(\mathbb{P}_n, E^*)$ be non trivial sections. Then $s \otimes t \in H^0(\mathbb{P}_n, E \otimes E^*)$ is an endomorphism of E, which in each fibre has rank at most one and thus cannot be a homothety. □

We wish to employ this criterion to show that the 2-bundles which we constructed over $\mathbb{P}_2$ in §2 by blowing up points of $\mathbb{P}_2$ are indecomposable but not simple. These bundles E were constructed by an appropriate modification of extensions

$$(*) \qquad 0 \to [C] \to E' \to [-C] \to 0.$$

$\sigma\colon X \to \mathbb{P}_2$ was the modification with exceptional divisor C and $E' = \sigma^* E$. On a line L on which a of the blown up points $x_i \in \mathbb{P}_2$ lay $E|L$ had the splitting type

$$a_E(\ell) = (a, -a).$$

In particular E cannot be split. Since $c_1(E) = 0$, it follows that

$$E^* \cong E \otimes \det E^* \cong E.$$

To see that E is not simple it thus suffices (because of the lemma above) to show that

$$h^0(\mathbb{P}_2, E) > 0.$$

But from the construction of the extension (*) it follows that E has a section which vanishes only in the points x_i and thus $h^0(\mathbb{P}_2, E) > 0$.

4.2. The null correlation bundle. We shall construct the null correlation bundle N over $\mathbb{P}_n$ for n odd as the kernel of a bundle epimorphism

$$T_{\mathbb{P}_n}(-1) \to \mathcal{O}_{\mathbb{P}_n}(1).$$

Then we shall have the exact sequence

$$0 \to N \to T_{\mathbb{P}_n}(-1) \to \mathcal{O}_{\mathbb{P}_n}(1) \to 0. \tag{1}$$

Let us first convince ourselves that N will necessarily be simple. If one tensors (1) with N^*, one gets

$$h^0(\mathbb{P}_n, N^* \otimes N) \leq h^0(\mathbb{P}_n, N^* \otimes T_{\mathbb{P}_n}(-1)).$$

If one tensors the sequence dual to (1) with $T_{\mathbb{P}_n}(-1)$ one gets the sequence

$$0 \to T_{\mathbb{P}_n}(-2) \to T_{\mathbb{P}_n} \otimes \Omega^1_{\mathbb{P}_n} \to N^* \otimes T_{\mathbb{P}_n}(-1) \to 0$$

and thus the cohomology sequence

$$\cdots \to H^0(\mathbb{P}_n, T_{\mathbb{P}_n} \otimes \Omega^1_{\mathbb{P}_n}) \to H^0(\mathbb{P}_n, N^* \otimes T_{\mathbb{P}_n}(-1)) \\ \to H^1(\mathbb{P}_n, T_{\mathbb{P}_n}(-2)) \to \cdots.$$

The Bott formula shows that $H^1(\mathbb{P}_n, T_{\mathbb{P}_n}(-2)) = 0$; thus

$$h^0(\mathbb{P}_n, N^* \otimes T_{\mathbb{P}_n}(-1)) \leq h^0(\mathbb{P}_n, T_{\mathbb{P}_n} \otimes \Omega^1_{\mathbb{P}_n}) = 1.$$

Altogether we have

$$1 \leq h^0(\mathbb{P}_n, N^* \otimes N) \leq h^0(\mathbb{P}_n, N^* \otimes T_{\mathbb{P}_n}(-1)) \leq 1.$$

Thus every kernel of a bundle epimorphism

$$T_{\mathbb{P}_n}(-1) \to \mathcal{O}_{\mathbb{P}_n}(1)$$

must be simple.

In order to find an epimorphism $T_{\mathbb{P}_n}(-1) \to \mathcal{O}_{\mathbb{P}_n}(1)$ it suffices (dualizing) to construct a section without zeros in $\Omega^1_{\mathbb{P}_n}(2)$.

We consider $\mathbb{P}(\Omega^1_{\mathbb{P}_n})$ over $\mathbb{P}_n$, i.e., the hyperplanes in the tangent bundle of $\mathbb{P}_n$. We thus have

$$\mathbb{P}(\Omega^1_{\mathbb{P}_n}) \cong \{(x, H) \in \mathbb{P}_n \times \mathbb{P}_n^* \mid x \in H\},$$

where $\mathbb{P}_n^*$ is the projective space of hyperplanes in $\mathbb{P}_n$. Let

$$p \colon \mathbb{P}(\Omega^1_{\mathbb{P}_n}) \to \mathbb{P}_n$$

be the projection onto $\mathbb{P}_n$. For odd n — say $n+1 = 2m$ — the matrix

$$A = \begin{pmatrix} \begin{array}{cc|} 0 & -1 \\ 1 & 0 \\ \hline \end{array} & & 0 \\ & \ddots & \\ 0 & & \begin{array}{|cc} \hline 0 & -1 \\ 1 & 0 \end{array} \end{pmatrix} \qquad (m \text{ blocks})$$

is nonsingular and for all $x \in \mathbb{C}^{n+1}$ we have

$$\langle Ax, x \rangle = 0$$

$(\langle x, y \rangle = \sum x_i y_i)$.

If one chooses in $\mathbb{P}_n$ homogeneous coordinates $(x_0 : \cdots : x_n)$ and in $\mathbb{P}_n^*$ the dual homogeneous coordinates $(\xi_0 : \cdots : \xi_n)$, then with respect to these coordinates A defines an isomorphism

$$\Phi \colon \mathbb{P}_n \to \mathbb{P}_n^*.$$

$\langle Ax, x \rangle = 0$ then simply means that each point $x \in \mathbb{P}_n$ lies in the hyperplane $\Phi(x) \in \mathbb{P}_n^*$, i.e., $(x, \Phi(x)) \in \mathbb{P}_n(\Omega^1_{\mathbb{P}_n}) \subset \mathbb{P}_n \times \mathbb{P}_n^*$. Thus the graph of Φ defines a section

$$g \colon \mathbb{P}_n \hookrightarrow \mathbb{P}(\Omega^1_{\mathbb{P}_n}), \quad g(x) = (x, \Phi(x)).$$

We regard this section as a subbundle of $\Omega^1_{\mathbb{P}_n}$:

$$\mathcal{O}_{\mathbb{P}_n}(a) \hookrightarrow \Omega^1_{\mathbb{P}_n}$$

Claim: $a = -2$.

Proof. If $\mathcal{O}_{\mathbb{P}_n}(a)$ is a 1-dimensional subbundle of $\Omega^1_{\mathbb{P}_n}$, then $\Omega^1_{\mathbb{P}_n}(-a)$ has a non-vanishing section and hence

$$c_n(\Omega^1_{\mathbb{P}_n}(-a)) = 0.$$

From the formulae in (1.2.) then follows

$$0 = c_n(\Omega^1_{\mathbb{P}_n}(-a)) = -c_n(T_{\mathbb{P}_n}(a)) = -\sum_{i=0}^{n} c_i(T_{\mathbb{P}_n}) a^{n-i}$$
$$= -\sum_{i=0}^{n} \binom{n+1}{i} a^{n-i}.$$

Multiplication with a gives

$$0 = \sum_{i=0}^{n} \binom{n+1}{i} a^{n+1-i} = (1+a)^{n+1} - 1.$$

It follows that $1 + a = \pm 1$. The case $a + 1 = 1$, i.e., $a = 0$, cannot occur, because $c_n(T_{\mathbb{P}_n}) = n + 1 \neq 0$. Thus $a = -2$ as claimed.

The section g therefore defines a trivial one-dimensional subbundle $\mathcal{O}_{\mathbb{P}_n} \subset \Omega^1_{\mathbb{P}_n}(2)$ and thus an exact sequence (1). From this sequence together with the Euler sequence one easily computes the Chern classes of N:

$$c(N) = \frac{c(T_{\mathbb{P}_n}(-1))}{1+h} = \frac{1}{(1+h)(1-h)} = 1 + h^2 + h^4 + \cdots + h^{n-1}.$$

Thus we have

THEOREM 4.2.1. *For every odd n there is a simple $(n-1)$-bundle N over $\mathbb{P}_n$ with the Chern classes*

$$c(N) = 1 + h^2 + h^4 + \cdots + h^{n-1}.$$

Any bundle N with an exact sequence (1) will be called a *null correlation bundle* over $\mathbb{P}_n$ (n odd).

4.3. The example of Tango. We wish now to construct a simple $(n-1)$-bundle over $\mathbb{P}_n$ for arbitrary n. First we must do some preparatory work.

LEMMA 4.3.1 (Serre). *Let E be a globally generated r-bundle over $\mathbb{P}_n$, $r > n$. Then there is an exact sequence of vector bundles*

$$0 \to \mathcal{O}_{\mathbb{P}_n}^{r-n} \to E \to F \to 0$$

with a holomorphic n-bundle F.

PROOF. Since E is globally generated we have the exact sequence

$$0 \to K \to H^0(\mathbb{P}_n, E) \otimes \mathcal{O}_{\mathbb{P}_n} \xrightarrow{\text{ev.}} E \to 0.$$

The kernel K of the evaluation homomorphism is the bundle

$$K = \{(x, s) \in \mathbb{P}_n \times H^0(\mathbb{P}_n, E) \mid s(x) = 0\}.$$

By passing to the associated projective bundles we get a holomorphic mapping

$$f\colon \mathbb{P}(K) \to \mathbb{P}_n \times \mathbb{P}(H^0(\mathbb{P}_n, E)) \xrightarrow{\text{proj.}} \mathbb{P}(H^0(\mathbb{P}_n, E))$$
$$f(x, [s]) = [s].$$

For a point $[s] \in \mathbb{P}(H^0(\mathbb{P}_n, E))$ the fibre

$$f^{-1}([s]) \cong \{x \in \mathbb{P}_n \mid s(x) = 0\}$$

is isomorphic to the zero set of s.

Let $h^0(\mathbb{P}_n, E) = N + 1$. Then f is a mapping of the $(n + N - r)$-dimensional complex manifold $\mathbb{P}(K)$ to the N-dimensional projective space $\mathbb{P}(H^0(\mathbb{P}_n, E))$; thus

$$\operatorname{codim}\,(f(\mathbb{P}(K)), \mathbb{P}(H^0(\mathbb{P}_n, E))) \geq r - n.$$

Thus there is an $(r-n-1)$-dimensional projective subspace $\mathbb{P}(V)$ in $\mathbb{P}(H^0(\mathbb{P}_n, E))$ which does not meet $f(\mathbb{P}(K))$. This means that the sections $s \in V \subset H^0(\mathbb{P}_n, E)$ do not vanish. The composition

$$V \otimes \mathcal{O}_{\mathbb{P}_n} \hookrightarrow H^0(\mathbb{P}_n, E) \otimes \mathcal{O}_{\mathbb{P}_n} \xrightarrow{\text{ev.}} E$$

then provides the desired trivial subbundle $V \otimes \mathcal{O}_{\mathbb{P}_n} \subset E$ of rank $r-n$. □

Now from Theorem A and this lemma we get

COROLLARY. *If E is a holomorphic r-bundle over $\mathbb{P}_n$, $r > n$, then there is an exact sequence*

$$0 \to \mathcal{O}_{\mathbb{P}_n}(a)^{\oplus(r-n)} \to E \to F \to 0$$

with a holomorphic n-bundle F.

In order to find a non-vanishing holomorphic section in a globally generated r-bundle E over $\mathbb{P}_n$ one only need consider the top Chern class.

LEMMA 4.3.2. *If the top Chern class $c_r(E)$ of a globally generated r-bundle E over $\mathbb{P}_n$ vanishes, then E contains a trivial subbundle of rank* 1.

PROOF. Without restriction we may assume $r \leq n$. Then we consider again the sequence

$$0 \to K \to H^0(\mathbb{P}_n, E) \otimes \mathcal{O}_{\mathbb{P}_n} \xrightarrow{\text{ev.}} E \to 0.$$

Let $N+1 = h^0(\mathbb{P}_n, E)$. The mapping $f \colon \mathbb{P}(K) \to \mathbb{P}(H^0(\mathbb{P}_n, E))$ maps the $(N+(n-r))$-dimensional complex manifold $\mathbb{P}(K)$ into an N-dimensional projective space. Let $[s] \in \mathbb{P}(H^0(\mathbb{P}_n, E))$ be a regular value of f. $f^{-1}([s])$ is thus an $(n-r)$-dimensional complex submanifold Z of $\mathbb{P}(K)$.

$$Z = f^{-1}([s]) \cong \{x \in \mathbb{P}_n \mid s(x) = 0\}$$

is isomorphic to the zero set of s. We regard Z as submanifold of $\mathbb{P}_n$.

s is transverse regular on the zero section of E with zero set Z. $c_r(E)$ is thus identical with the dual class $d_{\mathbb{P}_n}(Z)$ of Z. By hypothesis we thus have

$$d_{\mathbb{P}_n}(Z) = 0, \quad \text{i.e., } \deg Z = 0.$$

This can only be the case if $Z = \varnothing$. s is thus without zeros. □

After these preparations we are in a position to explain the example of Tango. We start from the Euler sequence

$$0 \to \mathcal{O}_{\mathbb{P}_n}(-1) \to \mathcal{O}_{\mathbb{P}_n}^{\oplus(n+1)} \to T_{\mathbb{P}_n}(-1) \to 0.$$

The $(n-1)$ st exterior power of this sequence is

$$(*) \quad 0 \to \Lambda^{n-2} T_{\mathbb{P}_n}(-1) \otimes \mathcal{O}_{\mathbb{P}_n}(-1) \to \mathcal{O}_{\mathbb{P}_n}^{\oplus \binom{n+1}{n-1}} \to \Lambda^{n-1} T_{\mathbb{P}_n}(-1) \to 0.$$

Furthermore,

$$\Lambda^{n-1} T_{\mathbb{P}_n}(-1) \cong \Omega^1_{\mathbb{P}_n}(1) \otimes \det T_{\mathbb{P}_n}(-1) \cong \Omega^1_{\mathbb{P}_n}(2).$$

Let $E = ((\Lambda^{n-2} T_{\mathbb{P}_n}(-1)) \otimes \mathcal{O}_{\mathbb{P}_n}(-1))^*$. The sequence which is dual to (*) is

$$0 \to T_{\mathbb{P}_n}(-2) \to \mathcal{O}_{\mathbb{P}_n}^{\oplus \binom{n+1}{2}} \to E \to 0,$$

which shows that E is globally generated of rank

$$r = \binom{n+1}{2} - n = \binom{n}{2}.$$

For $n \geq 3$ we have $r \geq n$; thus there is an exact sequence

$$0 \to \mathcal{O}_{\mathbb{P}_n}^{\oplus (r-n)} \to E \to E' \to 0$$

with a holomorphic n-bundle E'. E' is again globally generated. The top Chern class of E' is

$$c_n(E') = c_n(E) = 0.$$

Thus E' contains a trivial subbundle of rank 1. Let F be the quotient, i.e.,

$$0 \to \mathcal{O}_{\mathbb{P}_n} \to E' \to F \to 0$$

is exact. Thus we have found an $(n-1)$-bundle F over $\mathbb{P}_n$ with

$$c(F) = c(E') = c(E) = \frac{1}{c(T_{\mathbb{P}_n}(-2))} = \frac{1-2h}{(1-h)^{n+1}}.$$

We now wish to show that F is simple. To this end we consider the defining sequences

$$\text{(i)} \qquad 0 \to T_{\mathbb{P}_n}(-2) \to \mathcal{O}_{\mathbb{P}_n}^{\oplus \binom{n+1}{2}} \to E \to 0$$

$$\text{(ii)} \qquad 0 \to \mathcal{O}_{\mathbb{P}_n}^{\oplus (r-n)} \to E \to E' \to 0$$

$$\text{(iii)} \qquad 0 \to \mathcal{O}_{\mathbb{P}_n} \to E' \to F \to 0.$$

If one tensors the dual of the last sequence with F and considers the associated cohomology sequence, one gets

$$h^0(\mathbb{P}_n, F^* \otimes F) \leq h^0(\mathbb{P}_n, E'^* \otimes F).$$

Doing the same with the second sequence gives

$$h^0(\mathbb{P}_n, E'^* \otimes F) \leq h^0(\mathbb{P}_n, E^* \otimes F).$$

It remains to show that

$$h^0(\mathbb{P}_n, E^* \otimes F) = 1.$$

If one tensors (iii) and (ii) with E^*, then one gets the exact sequences

$$0 \to E^* \to E^* \otimes E' \to E^* \otimes F \to 0$$

and

$$0 \to E^{*\oplus(r-n)} \to E^* \otimes E \to E^* \otimes E' \to 0.$$

Thus it suffices to show that

$$h^0(\mathbb{P}_n, E^*) = h^1(\mathbb{P}_n, E^*) = 0$$

and that E is simple. Because

$$\begin{aligned} E^* &= \Lambda^{n-2}T_{\mathbb{P}_n}(-1) \otimes \mathcal{O}_{\mathbb{P}_n}(-1) \cong \Lambda^2\Omega^1_{\mathbb{P}_n}(1) \otimes \det T_{\mathbb{P}_n}(-1) \otimes \mathcal{O}_{\mathbb{P}_n}(-1) \\ &\cong \Lambda^2\Omega^1_{\mathbb{P}_n}(1) = \Omega^2_{\mathbb{P}_n}(2) \end{aligned}$$

it follows from the Bott formula that

$$h^0(\mathbb{P}_n, E^*) = h^1(\mathbb{P}_n, E^*) = 0.$$

If we now tensor (i) with E^*, then the associated cohomology sequence gives us

$$h^0(\mathbb{P}_n, E^* \otimes E) = h^1(\mathbb{P}_n, E^* \otimes T_{\mathbb{P}_n}(-2)).$$

Finally we tensor the Euler sequence with $E^*(-1)$, getting

$$0 \to E^*(-2) \to E^*(-1)^{\oplus(n+1)} \to E^* \otimes T_{\mathbb{P}_n}(-2) \to 0,$$

whose cohomology sequence gives the desired result provided

$$h^1(\mathbb{P}_n, E^*(-1)) = 0 \quad \text{and} \quad h^2(\mathbb{P}_n, E^*(-2)) = 1.$$

This last statement, however, follows from the Bott formula together with the fact that

$$E^*(-1) = \Omega^2_{\mathbb{P}_n}(-1), \quad E^*(-2) = \Omega^2_{\mathbb{P}_n}.$$

Thus we have proved the

THEOREM 4.3.3 (Tango). *For every n there is a simple (and thus indecomposable) holomorphic $(n-1)$-bundle F over $\mathbb{P}_n$ with*

$$c(F) = \frac{1-2h}{(1-h)^{n+1}}.$$

4.4. Concluding remarks and open questions. In this paragraph we remarked that it is difficult to find indecomposable holomorphic vector bundles of small rank over $\mathbb{P}_n$. We have encountered simple n-bundles over $\mathbb{P}_n$ and it is comparatively simple to construct indecomposable (even simple) bundles of rank $r > n$ over $\mathbb{P}_n$ (cf. Maruyama [81]). The examples of Tango give simple $(n-1)$-bundles over $\mathbb{P}_n$. Indecomposable $(n-2)$-bundles over $\mathbb{P}_n$ are difficult to construct and are known only for $n = 4$ (examples of Horrocks and Mumford [68], cf. §5.2., example 6) and $n = 5$ (cf. Horrocks [69]).

The following theorem of Barth and Van de Ven [9] also sheds some light on this problem: a holomorphic 2-bundle over $\mathbb{P}_n$ which can be holomorphically extended over arbitrary $\mathbb{P}_N \supset \mathbb{P}_n$ is necessarily of the form

$$\mathcal{O}_{\mathbb{P}_n}(a) \oplus \mathcal{O}_{\mathbb{P}_n}(b).$$

This theorem has played an important psychological rôle in the development and has been written by Tjurin [128] and Sato [103] for bundles of arbitrary rank.

Hartshorne [55] conjectures that holomorphic 2-bundles over $\mathbb{P}_n$, $n \geq 7$, necessarily split. This would be equivalent (cf. §5) to the statement that every 2-codimensional analytic subvariety $Y \subset \mathbb{P}_n$ without singularities is a global complete intersection for $n \geq 7$, i.e., Y is the transversal intersection of two smooth hypersurfaces. Hartshorne even conjectures that every d-dimensional smooth subvariety of $\mathbb{P}_n$ for $d > \frac{2}{3}n$ is a global complete intersection.

For the sake of completeness we close this section with the following very difficult questions, which have to do with the conjecture of Hartshorne.

QUESTION 4.4.1. Are there indecomposable 2-bundles over $\mathbb{P}_5$?

It would be interesting to have contributions to the following.

QUESTION 4.4.2. For which n are there indecomposable n-bundles over $\mathbb{P}_{2n}$?

The example of Horrocks and Mumford gives one for $n = 2$.

§5. Holomorphic 2-bundles and codimension 2 locally complete intersections

Let $Y \subset \mathbb{P}_n$, $n \geq 3$, be a codimension 2 locally complete intersection. If one can extend the determinant bundle of the normal bundle of Y to a line bundle on all of $\mathbb{P}_n$, then even the normal bundle itself can be extended to a 2-bundle over $\mathbb{P}_n$. This bundle then has a section for which Y is precisely the set of zeros. The bundle is then decomposable precisely when Y is globally a complete intersection. In particular we get many examples of indecomposable 2-bundles over $\mathbb{P}_3$ with this construction.

5.1. Construction of 2-bundles associated to a locally complete intersection. Let E be a holomorphic 2-bundle over $\mathbb{P}_n$ with a section s whose zero set Y is of codimension 2. Let $U \subset \mathbb{P}_n$ be open and such that $E|U$ is trivial; let $s_1, s_2 \in H^0(U, E|U)$ be a local basis for E over U. Then

$$s|U = f_1 s_1 + f_2 s_2$$

for appropriate holomorphic functions $H^0(U, \mathcal{O})$. One obtains a global sheaf of ideals $J_Y \subset \mathcal{O}_{\mathbb{P}_n}$ with

$$J_Y|U = (f_1, f_2)\mathcal{O}_{\mathbb{P}_n}|U.$$

We have $\operatorname{Supp}(\mathcal{O}_{\mathbb{P}_n}/J_Y) = Y$ and since Y is by assumption of codimension 2, it follows that $Y = (Y, \mathcal{O}_{\mathbb{P}_n}/J_Y)$ is a codimension 2 locally complete intersection in $\mathbb{P}_n$ — the *zero locus* of the section $s \in H^0(\mathbb{P}_n, E)$. In general Y is not reduced.

J_Y/J_Y^2 is in a natural way a sheaf of $\mathcal{O}_{\mathbb{P}_n}/J_Y$-modules and as such locally free of rank 2, for if

$$s|U = f_1 s_1 + f_2 s_2,$$

then the germs $f_{1,x}, f_{2,x}$ for $x \in U \cap Y$ form a regular sequence and represent a $\mathcal{O}_{Y,x}$-module basis of $J_{Y,x}/J_{Y,x}^2$ $(\mathcal{O}_Y = \mathcal{O}_{\mathbb{P}_n}/J_Y)$. J_Y/J_Y^2 is the *conormal bundle* of Y in $\mathbb{P}_n$; $N_{Y/\mathbb{P}_n} = (\mathcal{O}_Y/J_Y^2)^*$.

Over U the sheaf J_Y has the free resolution

$$0 \to \mathcal{O}_U \xrightarrow{\alpha} \mathcal{O}_U \oplus \mathcal{O}_U \xrightarrow{\beta} J_Y|U \to 0$$

with

$$\alpha(g) = (-f_{2,x}g, f_{1,x}g),$$
$$\beta(g,h) = f_{1,x}g + f_{2,x}h$$

for $x \in U$, $g, h \in \mathcal{O}_{U,x}$, $f_{i,x} \in \mathcal{O}_{U,x}$ the germs of the f_i in the point x. The sequence is exact because $f_{1,x}, f_{2,x} \in \mathcal{O}_{U,x}$ is a regular sequence for every point $x \in U$.

These local sequences yield a global resolution

$$0 \to \det E^* \xrightarrow{\alpha} E^* \xrightarrow{\beta} J_Y \to 0 \tag{1}$$

with

$$\alpha(\phi_1 \wedge \phi_2) = \phi_1(s_x)\phi_2 - \phi_2(s_x)\phi_1,$$
$$\beta(\phi) = \phi(s_x)$$

for $x \in \mathbb{P}_n$, $\phi_1, \phi_2, \phi \in E_x^* = \operatorname{Hom}_{\mathcal{O}_{\mathbb{P}_n,x}}(E_x, \mathcal{O}_{\mathbb{P}_n,x})$. s_x denotes the germ of s in the point x.

The sequence (1) is the *Koszul complex* for s, and as we have just seen it is a locally free resolution of J_Y if s has a codimension 2 zero set.

If one restricts the Koszul complex to Y, one gets the exact sequence

$$\det E^* \otimes \mathcal{O}_Y \xrightarrow{\alpha\otimes 1_{\mathcal{O}_Y}} E^*|Y \to J_Y/J_Y^2 \to 0.$$

$\alpha \otimes 1_{\mathcal{O}_Y}$ is the zero homomorphism as one sees from the local description. Hence we get an isomorphism

$$E^*|Y \xrightarrow{\sim} J_Y/J_Y^2.$$

Thus E is an extension to all of $\mathbb{P}_n$ of the normal bundle $N_{Y/\mathbb{P}_n}$.

We now attempt to reverse this construction, and for a given locally complete intersection $(Y, \mathcal{O}_Y)$ we seek an extension of the normal bundle $N_{Y/\mathbb{P}_n}$ such that a holomorphic section of this extension has as zero set precisely Y with structure sheaf $\mathcal{O}_Y$. A necessary condition for the extendability of the normal bundle $N_{Y/\mathbb{P}_n}$ is the extendability of the determinant bundle $\det N_{Y/\mathbb{P}_n}$. If the determinant bundle is extendable, then E^* — if it exists — is an extension (in the group theoretic sense) of the sheaf of ideals J_Y by an extension over $\mathbb{P}_n$ of $\det N^*_{Y/\mathbb{P}_n}$ as the sequence (1) shows. The following theorem shows that the extendability of $\det N_{Y/\mathbb{P}_n}$ is also sufficient for the extendability of $N_{Y/\mathbb{P}_n}$.

THEOREM 5.1.1. *Let Y be a locally complete intersection of codimension 2 in $\mathbb{P}_n$, $n \geq 3$, with sheaf of ideals $J_Y \subset \mathcal{O}_{\mathbb{P}_n}$. Let the determinant bundle of the normal bundle be extendable:*

$$\det N_{Y/\mathbb{P}_n} = \mathcal{O}_{\mathbb{P}_n}(k)|Y \quad (k \in \mathbb{Z}).$$

Then there is a holomorphic 2-bundle E over $\mathbb{P}_n$ with a section s which has precisely $(Y, \mathcal{O}_Y)$ as zero set. s induces the exact sequence

$$0 \to \mathcal{O}_{\mathbb{P}_n} \xrightarrow{\cdot s} E \to J_Y(k) \to 0. \tag{2}$$

The Chern classes of E are given by

$$c_1(E) = k \qquad \text{and} \qquad c_2(E) = \deg Y.$$

PROOF. If there is a bundle E as claimed then

$$\det N^*_{Y/\mathbb{P}_n} = \det E^*|Y,$$

so we should choose $\det E^* = \mathcal{O}_{\mathbb{P}_n}(-k)$. Thus E^* is an extension

$$0 \to \mathcal{O}_{\mathbb{P}_n}(-k) \to E^* \to J_Y \to 0.$$

Thus we investigate the extensions of J_Y by $\mathcal{O}_{\mathbb{P}_n}(-k)$. These are classified by the global Ext-group (Griffiths and Harris, p. 725, [49])

$$\operatorname{Ext}^1_{\mathbb{P}_n}(J_Y, \mathcal{O}_{\mathbb{P}_n}(-k)).$$

For the calculation of this group we employ the lower term sequence of the spectral sequence (Griffiths and Harris, p. 706, [49])

$$E_2^{pq} = H^p(\mathbb{P}_n, \mathcal{E}xt^q_{\mathcal{O}_{\mathbb{P}_n}}(J_Y, \mathcal{O}_{\mathbb{P}_n}(-k))) \Rightarrow \mathrm{Ext}^{p+q}_{\mathbb{P}_n}(J_Y, \mathcal{O}_{\mathbb{P}_n}(-k)).$$

That is the following exact sequence

$$\begin{aligned} 0 &\to H^1(\mathbb{P}_n, \mathcal{H}om_{\mathcal{O}_{\mathbb{P}_n}}(J_Y, \mathcal{O}_{\mathbb{P}_n}(-k))) \to \mathrm{Ext}^1_{\mathbb{P}_n}(J_Y, \mathcal{O}_{\mathbb{P}_n}(-k)) \\ &\to H^0(\mathbb{P}_n, \mathcal{E}xt^1_{\mathcal{O}_{\mathbb{P}_n}}(J_Y, \mathcal{O}_{\mathbb{P}_n}(-k))) \to H^2(\mathbb{P}_n, \mathcal{H}om_{\mathcal{O}_{\mathbb{P}_n}}(J_Y, \mathcal{O}_{\mathbb{P}_n}(-k))). \end{aligned}$$

The sequence $0 \to J_Y \to \mathcal{O}_{\mathbb{P}_n} \to \mathcal{O}_Y \to 0$ gives rise to the long exact sequence

$$\begin{aligned} 0 \to \mathcal{H}om_{\mathcal{O}_{\mathbb{P}_n}}(\mathcal{O}_Y, \mathcal{O}_{\mathbb{P}_n}(-k)) &\to \mathcal{H}om_{\mathcal{O}_{\mathbb{P}_n}}(\mathcal{O}_{\mathbb{P}_n}, \mathcal{O}_{\mathbb{P}_n}(-k)) \to \\ &\to \mathcal{H}om_{\mathcal{O}_{\mathbb{P}_n}}(J_Y, \mathcal{O}_{\mathbb{P}_n}(-k)) \to \mathcal{E}xt^1_{\mathcal{O}_{\mathbb{P}_n}}(\mathcal{O}_Y, \mathcal{O}_{\mathbb{P}_n}(-k)) \to \cdots \end{aligned}$$

Because Y is a locally complete intersection of codimension 2 we have (Griffiths and Harris, p. 690, [49])

$$\mathcal{E}xt^i_{\mathcal{O}_{\mathbb{P}_n}}(\mathcal{O}_Y, \mathcal{O}_{\mathbb{P}_n}(-k)) = 0 \quad \text{for} \quad i = 0, 1,$$

and thus

$$\mathcal{H}om_{\mathcal{O}_{\mathbb{P}_n}}(J_Y, \mathcal{O}_{\mathbb{P}_n}(-k)) \cong \mathcal{H}om_{\mathcal{O}_{\mathbb{P}_n}}(\mathcal{O}_{\mathbb{P}_n}, \mathcal{O}_{\mathbb{P}_n}(-k)) = \mathcal{O}_{\mathbb{P}_n}(-k).$$

If we put this into the lower term sequence, we get the exact sequence

$$\begin{aligned} 0 \to H^1(\mathbb{P}_n, \mathcal{O}_{\mathbb{P}_n}(-k)) &\to \mathrm{Ext}^1_{\mathbb{P}_n}(J_Y, \mathcal{O}_{\mathbb{P}_n}(-k)) \to \\ &\to H^0(\mathbb{P}_n, \mathcal{E}xt^1_{\mathcal{O}_{\mathbb{P}_n}}(J_Y, \mathcal{O}_{\mathbb{P}_n}(-k))) \to H^2(\mathbb{P}_n, \mathcal{O}_{\mathbb{P}_n}(-k)) \to \dots. \end{aligned}$$

In particular for $n \geq 3$ we have

$$(*) \qquad \mathrm{Ext}^1_{\mathbb{P}_n}(J_Y, \mathcal{O}_{\mathbb{P}_n}(-k)) \simeq H^0(\mathbb{P}_n, \mathcal{E}xt^1_{\mathcal{O}_{\mathbb{P}_n}}(J_Y, \mathcal{O}_{\mathbb{P}_n}(-k))).$$

For $n = 2$ the group $H^2(\mathbb{P}_n, \mathcal{O}_{\mathbb{P}_n}(-k))$ is zero only for $k < 3$, so the equation (*) also holds in the case

$$n = 2, \quad k < 3.$$

In general for $n = 2$ we only have an exact sequence

$$\begin{aligned} 0 \to \mathrm{Ext}^1_{\mathbb{P}_2}(J_Y, \mathcal{O}_{\mathbb{P}_2}(-k)) \to H^0(\mathbb{P}_2, \mathcal{E}xt^1_{\mathcal{O}_{\mathbb{P}_2}}&(J_Y, \mathcal{O}_{\mathbb{P}_2}(-k))) \\ &\to H^2(\mathbb{P}_2, \mathcal{O}_{\mathbb{P}_2}(-k)). \end{aligned}$$

We now calculate $\mathcal{E}xt^1_{\mathcal{O}_{\mathbb{P}_n}}(J_Y, \mathcal{O}_{\mathbb{P}_n}(-k))$: from the $\mathcal{E}xt$-sequence associated to $0 \to J_Y \to \mathcal{O}_{\mathbb{P}_n} \to \mathcal{O}_Y \to 0$ we get

$$\mathcal{E}xt^1_{\mathcal{O}_{\mathbb{P}_n}}(J_Y, \mathcal{O}_{\mathbb{P}_n}(-k)) \xrightarrow{\sim} \mathcal{E}xt^2_{\mathcal{O}_{\mathbb{P}_n}}(\mathcal{O}_Y, \mathcal{O}_{\mathbb{P}_n}(-k)).$$

Since Y is a codimension 2 locally complete intersection, we have (Altman and Kleiman, [1], p. 12–14, Griffiths and Harris, [49], p. 690–692) the local fundamental isomorphism (LFI)

$$\mathcal{E}xt^2_{\mathcal{O}_{\mathbb{P}_n}}(\mathcal{O}_Y, \mathcal{O}_{\mathbb{P}_n}(-k)) \xrightarrow{\sim} \mathcal{H}om_{\mathcal{O}_Y}(\det J_Y/J_Y^2, \mathcal{O}_Y(-k)),$$

where $\mathcal{O}_Y(-k) = \mathcal{O}_{\mathbb{P}_n}(-k) \otimes \mathcal{O}_Y$.

However, by assumption $\det J_Y/J_Y^2 = \mathcal{O}_Y(-k)$, so

$$\mathcal{H}om_{\mathcal{O}_Y}(\det J_Y/J_Y^2, \mathcal{O}_Y(-k)) \simeq \mathcal{O}_Y.$$

Altogether we have a canonical isomorphism of sheaves

$$\mathcal{E}xt^1_{\mathcal{O}_{\mathbb{P}_n}}(J_Y, \mathcal{O}_{\mathbb{P}_n}(-k)) \simeq \mathcal{O}_Y$$

and thus

$$(*) \qquad \begin{aligned} \operatorname{Ext}^1_{\mathbb{P}_n}(J_Y, \mathcal{O}_{\mathbb{P}_n}(-k)) &\simeq H^0(\mathbb{P}_n, \mathcal{E}xt^1_{\mathcal{O}_{\mathbb{P}_n}}(J_Y, \mathcal{O}_{\mathbb{P}_n}(-k))) \\ &\simeq H^0(Y, \mathcal{O}_Y). \end{aligned}$$

We now consider the extension given by the $1 \in H^0(Y, \mathcal{O}_Y)$

$$0 \to \mathcal{O}_{\mathbb{P}_n}(-k) \to F \to J_Y \to 0$$

with a coherent sheaf F over $\mathbb{P}_n$.

Claim: F is locally free over $\mathbb{P}_n$.

PROOF. Let $x \in \mathbb{P}_n$. Because of (*) the germ 1_x of 1 in the point x is an element in

$$(\mathcal{E}xt^1_{\mathcal{O}_{\mathbb{P}_n}}(J_Y, \mathcal{O}_{\mathbb{P}_n}(-k)))_x = \operatorname{Ext}^1_{\mathcal{O}_{\mathbb{P}_n,x}}(J_{Y,x}, \mathcal{O}_{\mathbb{P}_n,x}(-k))$$

and defines the extension

$$0 \to \mathcal{O}_{\mathbb{P}_n,x}(-k) \to F_x \to J_{Y,x} \to 0.$$

Since 1_x naturally generates the $\mathcal{O}_{\mathbb{P}_n,x}$-module

$$\operatorname{Ext}^1_{\mathcal{O}_{\mathbb{P}_n,x}}(J_{Y,x}, \mathcal{O}_{\mathbb{P}_n,x}(-k)) \simeq \mathcal{O}_{Y,x},$$

F_x is a free $\mathcal{O}_{\mathbb{P}_n,x}$-module and thus F is locally free according to the following lemma of Serre.

LEMMA 5.1.2 (Serre). *Let A be a noetherian local ring, $I \subset A$ an ideal with a free resolution of length 1:*

$$0 \to A^p \to A^q \to I \to 0.$$

(e.g., the Koszul complex $0 \to A \to A^{\oplus 2} \to I \to 0$, if I is generated by a regular sequence (f_1, f_2)). Let $e \in \operatorname{Ext}^1_A(I, A)$ be represented by the extension

$$0 \to A \to M \to I \to 0.$$

Then M is a free A-module if and only if e generates the A-module $\operatorname{Ext}^1_A(I,A)$.

PROOF. The Ext-sequence associated to $0 \to A \to M \to I \to 0$ gives

$$\cdots \to \operatorname{Hom}_A(A,A) \xrightarrow{\delta} \operatorname{Ext}^1_A(I,A) \to \operatorname{Ext}^1_A(M,A) \to \operatorname{Ext}^1_A(A,A) = 0.$$

Thus $\operatorname{Ext}^1_A(M,A) = 0$ if and only if δ is surjective. Because $\delta(\mathrm{id}_A) = e$ this is the case precisely when e generates the A-module $\operatorname{Ext}^1_A(I,A)$.

It remains to show that

$$\operatorname{Ext}^1_A(M,A) = 0 \Rightarrow \ M \text{ is free.}$$

To this end we construct out of the exact sequences

$$0 \to A^p \to A^q \xrightarrow{\phi} I \to 0$$
$$0 \to A \underset{\alpha}{\to} M \underset{\beta}{\to} I \to 0$$

a free resolution of length 1 for M: let $\Phi\colon A^q \to M$ be a lifting of ϕ to M and $\psi\colon A \oplus A^q \to M$ be defined by $\psi(x,y) = \alpha(x) + \Phi(y)$. Then we have a commutative diagram

$$\begin{array}{ccccccccc}
0 & \longrightarrow & A & \longrightarrow & A \oplus A^q & \longrightarrow & A^q & \longrightarrow & 0 \\
 & & \| & & \downarrow{\scriptstyle\psi} & & \downarrow{\scriptstyle\phi} & & \\
0 & \longrightarrow & A & \xrightarrow[\alpha]{} & M & \xrightarrow[\beta]{} & I & \longrightarrow & 0.
\end{array}$$

It follows that $\ker\psi \cong \ker\phi \cong A^p$ and $\operatorname{coker}\psi = 0$. Thus we have an exact sequence

$$0 \to A^p \to A^r \to M \to 0 \quad (A^r = A \oplus A^q).$$

Since $\operatorname{Ext}^1_A(M,A) = 0$, this sequence splits. Hence M is a direct summand in A^r, hence projective and thus free. □

We now have an extension

$$(**) \qquad 0 \to \mathcal{O}_{\mathbb{P}_n}(-k) \to F \xrightarrow{\beta} J_Y \to 0$$

with a 2-bundle F over $\mathbb{P}_n$. $E = F^*$ is the bundle we want and (**) is the Koszul complex of a section $s \in H^0(\mathbb{P}_n, E)$. Multiplication with s, $\cdot s\colon \mathcal{O}_{\mathbb{P}_n} \to E$, is dual to the composition

$$E^* \overset{\beta}{\twoheadrightarrow} J_Y \hookrightarrow \mathcal{O}_{\mathbb{P}_n}.$$

Let $\beta'\colon E \twoheadrightarrow J_Y(k)$ be the composition

$$E \xrightarrow{\sim} E^* \otimes \det E = E^*(k) \xrightarrow{\beta(k)} J_Y(k),$$

where $E \xrightarrow{\sim} E^* \otimes \det E = \mathcal{H}om_{\mathcal{O}_{\mathbb{P}_n}}(E, \Lambda^2 E)$ is the canonical isomorphism given by $s \mapsto (t \mapsto s \wedge t)$. From this we get the exact sequence

$$(2) \qquad 0 \to \mathcal{O}_{\mathbb{P}_n} \xrightarrow{\cdot s} E \xrightarrow{\beta'} J_Y(k) \to 0,$$

and the proof of the theorem is complete. □

5.2. Examples. In the previous section we found a 2-bundle E with a section s so that Y is precisely the zero set of s, for every codimension 2 locally complete intersection Y in $\mathbb{P}_n$ for which the determinant bundle of the normal bundle is extendable. Before we proceed to concrete examples we prove the following

LEMMA 5.2.1. *Let E be a 2-bundle which belongs to a locally complete intersection $Y \subset \mathbb{P}_n$, $n \geq 3$. E splits if and only if Y is a global complete intersection.*

PROOF. Let Y be defined by the section $s \in H^0(\mathbb{P}_n, E)$. If E splits, say $E = \mathcal{O}_{\mathbb{P}_n}(a) \oplus \mathcal{O}_{\mathbb{P}_n}(b)$, then $s = (s_1, s_2)$ with $s_1 \in H^0(\mathbb{P}_n, \mathcal{O}_{\mathbb{P}_n}(a))$, $s_2 \in H^0(\mathbb{P}_n, \mathcal{O}_{\mathbb{P}_n}(b))$ and $Y = \{s = 0\}$ is the intersection of the hypersurfaces $\{s_1 = 0\}$ and $\{s_2 = 0\}$.

Conversely let Y be the intersection of two hypersurfaces V_a, V_b of degree a resp. b:

$$V_a = \{s_1 = 0\}, \quad V_b = \{s_2 = 0\}$$
$$\text{with} \quad s_1 \in H^0(\mathbb{P}_n, \mathcal{O}_{\mathbb{P}_n}(a)), \quad s_2 \in H^0(\mathbb{P}_n, \mathcal{O}_{\mathbb{P}_n}(b)).$$

The Koszul complex of the section $s = (s_1, s_2)$ in $\mathcal{O}_{\mathbb{P}_n}(a) \oplus \mathcal{O}_{\mathbb{P}_n}(b)$ gives the extension

$$0 \to \mathcal{O}_{\mathbb{P}_n}(-(a+b)) \to \mathcal{O}_{\mathbb{P}_n}(-a) \oplus \mathcal{O}_{\mathbb{P}_n}(-b) \to J_Y \to 0.$$

This extension defines a non-zero element in

$$\operatorname{Ext}^1_{\mathbb{P}_n}(J_Y, \mathcal{O}_{\mathbb{P}_n}(-(a+b))) \cong H^0(Y, \mathcal{O}_Y).$$

If we can show that $H^0(Y, \mathcal{O}_Y)$ is 1-dimensional, then every other nontrivial extension of J_Y by $\mathcal{O}_{\mathbb{P}_n}(-(a+b))$ must give the split bundle $\mathcal{O}_{\mathbb{P}_n}(-a) \oplus \mathcal{O}_{\mathbb{P}_n}(-b)$.

Claim: $h^0(Y, \mathcal{O}_Y) = 1$ for every global complete intersection Y of codimension 2 in $\mathbb{P}_n$, $n \geq 3$. (In particular Y is connected.)

PROOF. From the cohomology sequence belonging to

$$0 \to \mathcal{O}_{\mathbb{P}_n}(-(a+b)) \to \mathcal{O}_{\mathbb{P}_n}(-a) \oplus \mathcal{O}_{\mathbb{P}_n}(-b) \to J_Y \to 0$$

follows immediately (since $a, b > 0$, $n \geq 3$)

$$h^0(\mathbb{P}_n, J_Y) = h^1(\mathbb{P}_n, J_Y) = 0.$$

Thus

$$H^0(\mathbb{P}_n, \mathcal{O}_{\mathbb{P}_n}) \simeq H^0(Y, \mathcal{O}_Y). \qquad \square$$

This lemma shows that the existence of non-splitting 2-bundles over $\mathbb{P}_n$, $n \geq 3$, is closely connected with the question whether there are codimension 2 locally complete intersections $Y \subset \mathbb{P}_n$ which are not the intersection of two hypersurfaces.

On $\mathbb{P}_3$ we shall find sufficiently many examples of locally complete intersections which are disjoint unions of curves. On $\mathbb{P}_n$, $n \geq 4$ on the other hand all locally complete intersections of pure codimension 2 are connected and it is not so easy to find indecomposable 2-bundles as on $\mathbb{P}_3$. In fact until now no indecomposable 2-bundles on $\mathbb{P}_n$, $n \geq 5$, are known (cf. the remarks at the end of §4).

We now describe some concrete examples.

Example 1. Let Y consist of m simple points $x_1, \ldots, x_m$ in $\mathbb{P}_2$, $m > 1$. Since Y is 0-dimensional, each vector bundle over Y is trivial. In particular the bundle

$$\mathcal{O}_{\mathbb{P}_2}(k)|_Y = \mathcal{O}_Y(k) \quad \text{and} \quad \det N_{Y/\mathbb{P}_n}$$

are isomorphic for all k. Thus the hypotheses of the theorem in 5.1 are satisfied and for $k < 3$ one gets 2-bundles E over $\mathbb{P}_2$ with exact sequences

$$0 \to \mathcal{O}_{\mathbb{P}_2} \to E \to J_Y(k) \to 0.$$

E has the Chern classes

$$c_1(E) = k$$
$$c_2(E) = m.$$

For $k = 1$ the generic splitting type is $\underline{a}_E = (1, 0)$, for on any line L which does not meet Y the restriction $E|L$ is given by

$$0 \to \mathcal{O}_L \to E|L \to \mathcal{O}_L(1) \to 0,$$

and this sequence splits because $H^1(L, \mathcal{O}_L(-1)) = 0$. For $k = 2$ we have for any line $L \subset \mathbb{P}_2$ with $L \cap Y = \varnothing$ that $E|L$ is given by the extension

$$0 \to \mathcal{O}_L \to E|L \to \mathcal{O}_L(2) \to 0.$$

This however does not mean that the splitting type of E on these lines is $(2, 0)$, for not all extensions of $\mathcal{O}_L(2)$ by $\mathcal{O}_L$ split.

In fact the generic splitting type for $k = 2$ is

$$\underline{a}_E = (1, 1),$$

for $(1, 1)$ is the (lexicographically) smallest possible pair (a, b) with $a \geq b$ and $a + b = 2$ and on a line L' which meets exactly one point x_i

of Y, $E|L'$ has the section $s|L'$, which in x_i has its only zero (a simple zero). $E|L'$ is there given by

$$0 \to \mathcal{O}_{L'}(1) \to E|L' \to \mathcal{O}_{L'}(1) \to 0,$$

and this extension splits.

In the case $k = 1$ one easily sees that the set $S_E \subset \mathbb{P}_2^*$ of jump lines is precisely the set of the finitely many points ℓ which correspond to lines L in $\mathbb{P}_2$ with

$$\#(L \cap Y) \geq 2.$$

In the case $k = 2$ one cannot specify S_E so easily. Later (Ch. II) we shall see that $S_E \subset \mathbb{P}_2^*$ is a curve of degree $m - 1$.

Example 2. Let Y be a union of $d > 1$ disjoint lines $L_i \cong \mathbb{P}_1$ in $\mathbb{P}_3$. Since $L_i = H_i \cap H_i'$ is the intersection of two hyperplanes in $\mathbb{P}_3$, we have

$$\begin{aligned} N_{Y/\mathbb{P}_3}|L_i = N_{L_i/\mathbb{P}_3} &= N_{H_i/\mathbb{P}_3}|L_i \oplus N_{H_i'/\mathbb{P}_3}|L_i \\ &= \mathcal{O}_{L_i}(1) \oplus \mathcal{O}_{L_i}(1) \end{aligned}$$

and thus $\det N_{Y/\mathbb{P}_3}|L_i = \mathcal{O}_{L_i}(2)$. Hence $\det N_{Y/\mathbb{P}_3} = \mathcal{O}_Y(2)$. Thus the hypotheses of the theorem in 5.1 are satisfied. We obtain an extension

$$0 \to \mathcal{O}_{\mathbb{P}_3} \xrightarrow{\cdot s} E \to \mathcal{J}_Y(2) \to 0$$

with an indecomposable 2-bundle E with Chern classes

$$\begin{aligned} c_1(E) &= 2 \\ c_2(E) &= d. \end{aligned}$$

Restricting E to a line $L \subset \mathbb{P}_3$ which intersects precisely one of the lines L_i transversally one sees that again

$$\underline{a}_E = (1, 1).$$

Furthermore one easily sees that $S_E \subset G_3$ contains all points $\ell \in G_3$ whose associated lines L meet the set Y transversally with $\#(L \cap Y) \geq 2$. This set is purely 2-codimensional in the 4-dimensional complex manifold G_3. We shall see later on that S_E is a hypersurface.

Example 3. For Y we choose a disjoint union of r elliptic curves C_i of degree d_i in $\mathbb{P}_3$. The tangent bundle T_C of such a curve is trivial; hence from the exact sequence

$$0 \to T_C \to T_{\mathbb{P}_3}|C \to N_{C/\mathbb{P}_3} \to 0$$

one deduces

$$\det N_{C/\mathbb{P}_3} \cong \det T_{\mathbb{P}_3}|C = \mathcal{O}_C(4)$$

and thus

$$\det N_{Y/\mathbb{P}_3} = \mathcal{O}_Y(4).$$

To Y we can therefore associate a 2-bundle E with

$$c_1(E) = 4, \qquad c_2(E) = \sum_{i=1}^{r} d_i.$$

If in particular one chooses $r = 2$, $d_1 = d_2 = 3$, C_1, C_2 two plane elliptic curves in two different planes H_1, H_2, then $E(-2)$ is a 2-bundle over $\mathbb{P}_3$ with

$$c_1 = 0, \quad c_2 = 2$$

and generic splitting type $(0, 0)$.

Example 4. Let Y consist of r disjoint conics D_i, $\deg D_i = 2$, in $\mathbb{P}_3$. If for such a conic D one chooses a projective plane $H \subset \mathbb{P}_3$ with $D \subset H$, then from the normal bundle sequence

$$0 \to N_{D/H} \to N_{D/\mathbb{P}_3} \to N_{H/\mathbb{P}_3}|D \to 0$$

we deduce

$$\det N_{D/\mathbb{P}_3} \cong N_{D/H} \otimes N_{H/\mathbb{P}_3}|D \cong \mathcal{O}_D(2) \otimes \mathcal{O}_D(1) = \mathcal{O}_D(3)$$

and thus

$$\det N_{Y/\mathbb{P}_3} = \mathcal{O}_Y(3).$$

The associated 2-bundle E has the Chern classes

$$c_1(E) = 3, \quad c_2(E) = 2r$$

and the generic splitting type $(2, 1)$.

Example 5. We consider r pairs of positive integers

$$(a_i, b_i) \quad \text{with} \quad a_i + b_i = p \geq 2 \quad \text{for} \quad i = 1, \ldots, r.$$

To each pair (a_i, b_i) we choose polynomials

$$p_i \in H^0(\mathbb{P}_3, \mathcal{O}_{\mathbb{P}_3}(a_i)), \quad q_i \in H^0(\mathbb{P}_3, \mathcal{O}_{\mathbb{P}_3}(b_i)).$$

Let Y_i be the intersection of the hypersurfaces defined by p_i and q_i. By choosing the polynomials p_i and q_i appropriately one can achieve that the intersections Y_i are smooth and pairwise disjoint curves. Let Y be their union. The Koszul complex for Y_i

$$0 \to \mathcal{O}_{\mathbb{P}_3}(-(a_i + b_i)) \to \mathcal{O}_{\mathbb{P}_3}(-a_i) \oplus \mathcal{O}_{\mathbb{P}_3}(-b_i) \to J_{Y_i} \to 0$$

shows that the determinant bundle

$$\det N_{Y_i/\mathbb{P}_3}$$

is isomorphic to $\mathcal{O}_{Y_i}(a_i + b_i) = \mathcal{O}_{Y_i}(p)$ for all $i = 1, \ldots, r$. Thus we have

$$\det N_{Y/\mathbb{P}_3} = \mathcal{O}_Y(p).$$

To Y there is thus associated a 2-bundle E with Chern classes

$$c_1(E) = p, \quad c_2(E) = \deg Y = \sum_{i=1}^{r} a_i b_i.$$

Example 6. Horrocks and Mumford [68] have shown that in $\mathbb{P}_4$ there is a 2-dimensional complex torus

$$Y = \mathbb{C}^2/\Gamma$$

($\Gamma = \sum_{i=1}^{4} \mathbb{Z}\omega_i$, $\omega_1, \dots, \omega_4 \in \mathbb{C}^2$ linearly independent over $\mathbb{R}$). The tangent bundle to Y is trivial. From the exact sequence

$$0 \to T_Y \to T_{\mathbb{P}_4}|Y \to N_{Y/\mathbb{P}_4} \to 0$$

it follows that

$$\det N_{Y/\mathbb{P}_4} = \mathcal{O}_Y(5).$$

Thus to Y belongs a 2-bundle E over $\mathbb{P}_4$ with

$$c_1(E) = 5, \quad c_2(E) = 10.$$

Hence Y is of degree 10.

This bundle is essentially the only non-splitting (consider the Chern classes!) holomorphic 2-bundle over $\mathbb{P}_4$ which is known.

In Chapter II, §3 we construct the bundle E without using the existence of non-singular abelian surfaces in $\mathbb{P}_4$.

5.3. Historical remarks. The correspondence between locally complete intersections of codimension 2 in $\mathbb{P}_n$ and holomorphic 2-bundles over $\mathbb{P}_n$ is essentially due to Serre [112]. In its present form it has been repeatedly discovered (Horrocks [66], Barth–Van de Ven [9], Hartshorne [55], Grauert–Mülich [45]). Here we have followed more or less the presentation of Hartshorne. We remark that this correspondence also holds for manifolds; the attentive reader will easily discover which conditions must be required.

The question whether there is an analogous connection between bundles of higher rank over $\mathbb{P}_n$ and subspaces of $\mathbb{P}_n$ will be dealt with in the next paragraph.

§6. Existence of holomorphic structures on topological bundles

In this paragraph we shall first report on the topological classification of complex vector bundles over $\mathbb{P}_n$, $n \leq 6$. By refining the methods we have developed in previous sections we then show that every continuous bundle over $\mathbb{P}_n$, $n \leq 3$, has at least one holomorphic structure.

6.1. Topological classification of bundles over $\mathbb{P}_n$, $n \leq 6$. The topological classification of line bundles over $\mathbb{P}_n$ is simple (cf. §1). For every number $c_1 \in \mathbb{Z}$ there is exactly one complex line bundle over $\mathbb{P}_n$ with c_1 as first Chern class. Each of these line bundles possesses exactly one holomorphic structure $\mathcal{O}_{\mathbb{P}_n}(c_1)$.

a) Bundles over $\mathbb{P}_1$: In §2.1 we gave a complete description of bundles over the projective line. To given numbers $r > 0$, $c_1 \in \mathbb{Z}$ there is up to topological equivalence exactly one complex vector bundle of rank r with c_1 as first Chern class. The various analytically inequivalent holomorphic bundles which belong to a fixed pair

$$(r, c_1) \in \mathbb{N} \times \mathbb{Z}$$

are given by

$$\mathcal{O}_{\mathbb{P}_1}(a_1) \oplus \cdots \oplus \mathcal{O}_{\mathbb{P}_1}(a_r) \quad a_1 \geq \cdots \geq a_r$$
$$c_1 = a_1 + \cdots + a_r.$$

b) The Schwarzenberger condition:
Schwarzenberger [62] showed with the help of the Riemann–Roch theorem that the Chern classes of holomorphic r-bundles over $\mathbb{P}_n$ must satisfy certain number-theoretic conditions. By using the Atiyah–Singer index theorem (or with homotopy theoretic methods (cf. Switzer [121])) one can show that these so-called Schwarzenberger conditions must also hold for continuous r-bundles over $\mathbb{P}_n$.

Let E be a continuous r-bundle over $\mathbb{P}_n$,

$$c_t(E) = 1 + c_1(E)t + \cdots + c_r(E)t^r \in \mathbb{Z}[t]$$

the Chern polynomial of E,

$$c_t(E) = \prod_{i=1}^{r}(1 + x_i t)$$

its factorization over $\mathbb{C}$. The x_i must then satisfy the *Schwarzenberger condition:*

$$(S_n^r): \qquad \sum_{i=1}^{r} \binom{n + x_i + s}{s} \in \mathbb{Z} \quad \text{for all} \quad s \in \mathbb{Z}.$$

For example in the simplest cases these conditions — expressed in terms of the Chern classes of E — look as follows:

(S_n^1) : no condition

(S_2^2) : no condition

(S_3^2) : $c_1c_2 \equiv 0 \quad (2)$

(S_4^2) : $c_2(c_2 + 1 - 3c_1 - 2c_1^2) \equiv 0 \quad (12)$

(S_3^3) : $c_1c_2 \equiv c_3 \quad (2)$.

c) r-bundles over $\mathbb{P}_n$, $n \leq r$:

Let E be a topological r-bundle over $\mathbb{P}_n$, $n < r$. Then E has $r - n$ linearly independent continuous sections and thus a trivial subbundle of rank $r - n$. The latter is (topologically) a direct summand; i.e., as a continuous bundle E is equivalent to a bundle of the form

$$E' \oplus \mathcal{O}_{\mathbb{P}_n}^{\oplus(r-n)}$$

with E' an n-bundle. Topological n-bundles can be classified using a theorem of A. Thomas [127]. For every n-tuple $(c_1, c_2, \ldots, c_n) \in \mathbb{Z}^n$ which satisfies the Schwarzenberger condition (S_n^n) there is exactly one complex n-bundle (up to topological equivalence) with $c_1, \ldots, c_n$ as Chern classes. In particular for $n = 2, 3$ this means

$n = 2$: The topological 2-bundles over $\mathbb{P}_2$ correspond bijectively to pairs (c_1, c_2) of integers (the condition S_2^2 is empty).

$n = 3$: The topological 3-bundles over $\mathbb{P}_3$ are classified by triples (c_1, c_2, c_3) with
$c_1c_2 \equiv c_3 \quad (2)$.

d) 2-bundles over $\mathbb{P}_3$:

The Schwarzenberger condition (S_3^2) for 2-bundles over $\mathbb{P}_3$ is

$$c_1c_2 \equiv 0 \quad (2).$$

Atiyah and Rees [2] have shown that for every pair $(c_1, c_2) \in \mathbb{Z}^2$ with $c_1c_2 \equiv 0$ (2) there is a 2-bundle with these Chern classes. For $c_1 \equiv 1$ (2) there is exactly one such bundle, while for $c_1 \equiv 0$ (2) there are two topologically inequivalent bundles. These two inequivalent 2-bundles with even first Chern class are distinguished by means of the so-called α-invariant, which is defined as follows: if $c_1(E) = 2k$ then the first Chern class of $E(-k)$ is 0, i.e., the structure group of $E(-k)$ can be reduced to $\mathrm{Sp}(1) \subset U(2)$ [62]. Thus to classify 2-bundles E with

$c_1(E) \equiv 0\ (2)$ it suffices to determine the symplectic line bundles over $\mathbb{P}_3$. But symplectic line bundles over $\mathbb{P}_3$ are stable in the topological sense, that is they are classified by the group

$$\widetilde{\mathrm{KSp}}(\mathbb{P}_3) \cong \mathbb{Z} \oplus \mathbb{Z}/2\mathbb{Z}.$$

Let π be the projection onto the summand $\mathbb{Z}/2\mathbb{Z}$, i.e., the direct image associated to the map $\mathbb{P}_3 \to$ point [2]. The α-invariant of a complex 2-bundle E over $\mathbb{P}_3$ with $c_1(E) = 2k$ is then given by

$$\alpha(E) = \pi(E(-k)) \in \mathbb{Z}/2\mathbb{Z}.$$

In case E is a holomorphic 2-bundle over $\mathbb{P}_3$ with $c_1(E) = 2k$ the α-invariant can be computed analytically. According to Atiyah and Rees we then have

$$\alpha(E) \equiv h^0(\mathbb{P}_3, E(-k-2)) + h^1(\mathbb{P}_3, E(-k-2)) \bmod 2.$$

Thus for a holomorphic 2-bundle E over $\mathbb{P}_3$ with $c_1(E) = 2k$ the invariant $\alpha(E)$ is just the holomorphic semicharacteristic of $E(-k-2)$. (The holomorphic Euler characteristic $\chi(\mathbb{P}_3, E(-k-2))$ vanishes because of Serre duality!)

By calculating the Postnikov tower of the fibration

$$BU(2) \to BU$$

together with a few additional homotopy theoretic considerations one can push the topological classification of rank 2 bundles over $\mathbb{P}_n$ up to $n = 6$ (cf. Switzer [121]). The result is contained in the following diagrams.

c_1 **even:**

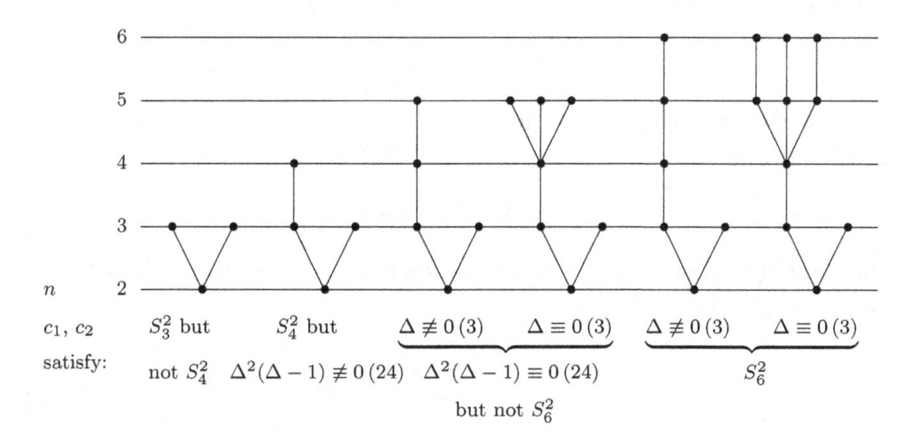

Here $\Delta = \frac{c_1^2-4c_2}{4}$. Of the two bundles over $\mathbb{P}_3$ the one which extends to $\mathbb{P}_4$ has $\alpha \equiv \frac{\Delta(\Delta-1)}{12}$ mod 2.

c_1 **odd:**

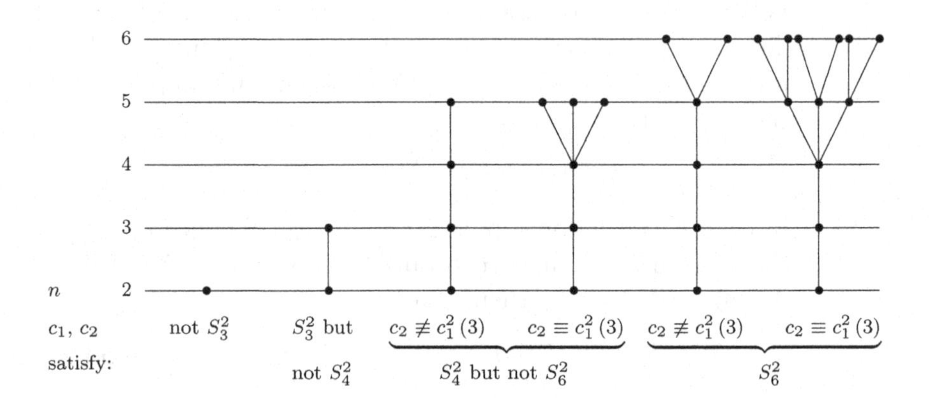

A dot • on a line n means an isomorphism class of rank 2 bundles over $\mathbb{P}_n$; a line-segment connecting a dot on line n with a dot on line $n-1$ means the given rank 2 bundle over $\mathbb{P}_n$ restricts to the given bundle over $\mathbb{P}_{n-1}$.

6.2. 2-bundles over $\mathbb{P}_2$. According to 6.1 for every pair $(c_1, c_2) \in \mathbb{Z}^2$ there is precisely one topological 2-bundle with these numbers as Chern classes. In order to see that every 2-bundle over $\mathbb{P}_2$ has at least one holomorphic structure we must for given $c_1, c_2 \in \mathbb{Z}$ construct a holomorphic 2-bundle with these Chern classes.

THEOREM 6.2.1 (Schwarzenberger). *For every pair $(c_1, c_2) \in \mathbb{Z}^2$ there is a holomorphic 2-bundle E over $\mathbb{P}_2$ with*

$$c_1(E) = c_1, \quad c_2(E) = c_2.$$

PROOF. We use the same methods as in the construction of the first examples in 2.2. Let $x_1, \dots, x_4 \in \mathbb{P}_2$ be four points,

$$\sigma : X \to \mathbb{P}_2$$

the blowing up of these points, $C_i = \sigma^{-1}(x_i)$ the exceptional curve over the point x_i. Let

$$L = [\sum_{i=1}^{4} k_i C_i]$$

be the line bundle associated to the divisor

$$\sum_{i=1}^{4} k_i C_i, \quad k_i \geq 0.$$

We consider extensions

$$(*) \qquad 0 \to L \otimes \sigma^* \mathcal{O}_{\mathbb{P}_2}(b) \to E' \to L^* \otimes \sigma^* \mathcal{O}_{\mathbb{P}_2}(a) \to 0$$

over X. If E' has the form

$$E' = \sigma^* E$$

for some 2-bundle E over $\mathbb{P}_2$, then this bundle has the Chern classes

$$\begin{aligned} c_1(E) &= a + b \\ c_2(E) &= c_2(\sigma^* E) = \\ &= (\sum_{i=1}^{4} k_i C_i + b\sigma^* H) \cdot (\sum_{i=1}^{4} (-k_i) C_i + a\sigma^* H) \\ &= \sum_{i=1}^{4} k_i^2 \; + ab. \end{aligned}$$

Here one must use that $C_i C_j = 0$ for $i \neq j$, $C_i^2 = -1$, $C_i \cdot \sigma^* H = 0$, $(\sigma^* H)^2 = 1$.

Because every positive integer is the sum of 4 squares, one can for all $c_1, c_2 \in \mathbb{Z}$ find integers $a, b, k_1, \dots, k_4$ with

$$\begin{aligned} c_1 &= a + b \\ c_2 &= \sum_{i=1}^{4} k_i^2 \; + ab, \quad k_i \geq 0, \quad a - b < 0. \end{aligned}$$

By proceeding in a manner similar to that in 2.2 (Lemma 2.2.6), one sees that a bundle E' from an extension

$$(*) \qquad 0 \to L \otimes \sigma^* \mathcal{O}_{\mathbb{P}_2}(b) \to E' \to L^* \otimes \sigma^* \mathcal{O}_{\mathbb{P}_2}(a) \to 0$$

is of the form $E' = \sigma^* E$ if and only if the restrictions to the exceptional curves C_i are of the form

$$(**) \qquad 0 \to \mathcal{O}_{C_i}(-k_i) \to \mathcal{O}_{C_i}^{\oplus 2} \to \mathcal{O}_{C_i}(k_i) \to 0.$$

(One must first extend (**) over a small neighborhood $\sigma^{-1}(U_i) = V_i$ of C_i to an extension

$$0 \to [k_i C_i] | V_i \to \mathcal{O}_{V_i}^{\oplus 2} \to [-k_i C_i] | V_i \to 0$$

and then show that this prolongation is uniquely determined up to equivalence.)

These considerations thus show that for the proof of the theorem the following suffices:

Claim: For $a, b \in \mathbb{Z}$, $a - b < 0$, $k_1, \ldots, k_4 \geq 0$ and $C = \sum_{i=1}^4 k_i C_i$ there is an extension

$$(*) \qquad 0 \to [C] \otimes \sigma^* \mathcal{O}_{\mathbb{P}_2}(b) \to E' \to [-C] \otimes \sigma^* \mathcal{O}_{\mathbb{P}_2}(a) \to 0$$

with $E'|C_i \cong \mathcal{O}_{C_i}^{\oplus 2}$ for $i = 1, \ldots, 4$.

PROOF. Without restriction we can take $b = 0$, $a < 0$. The extensions

$$(*) \qquad 0 \to [C] \to E' \to [-C] \otimes \sigma^* \mathcal{O}_{\mathbb{P}_2}(a) \to 0$$

are classified by

$$\operatorname{Ext}^1_X([-C] \otimes \sigma^* \mathcal{O}_{\mathbb{P}_2}(a), [C]) = H^1(X, [2C] \otimes \sigma^* \mathcal{O}_{\mathbb{P}_2}(-a)).$$

Let $C' = C_1 + \cdots + C_4$ be the exceptional divisor of the blowing up $\sigma \colon X \to \mathbb{P}_2$.

Since there are extensions of the form

$$0 \to [C]|C' \to \mathcal{O}_{C'}^{\oplus 2} \to [-C]|C' \to 0,$$

it suffices to show that the restriction mapping

$$(***) \qquad H^1(X, [2C] \otimes \sigma^* \mathcal{O}_{\mathbb{P}_2}(-a)) \to H^1(C', [2C]|C')$$

is surjective. Thus we only need to show that

$$H^2(X, [2C - C'] \otimes \sigma^* \mathcal{O}_{\mathbb{P}_2}(-a))$$

vanishes.

With Serre duality and because of the formula (Griffiths and Harris [49], p. 187)

$$\omega_X = \sigma^* \omega_{\mathbb{P}_2} \otimes [C']$$

one has

$$\begin{aligned} h^2(X, [2C - C'] \otimes \sigma^* \mathcal{O}_{\mathbb{P}_2}(-a)) &= h^0(X, [2C' - 2C] \otimes \sigma^* \mathcal{O}_{\mathbb{P}_2}(a - 3)) \\ &\leq h^0(X, [2C'] \otimes \sigma^* \mathcal{O}_{\mathbb{P}_2}(a - 3)). \end{aligned}$$

From the sequences

$$0 \to [(k-1)C'] \to [kC'] \to [kC']|C' \to 0, \quad k > 0,$$

follows immediately

$$h^0(X, [kC'] \otimes \sigma^* \mathcal{O}_{\mathbb{P}_2}(a - 3)) = h^0(X, \sigma^* \mathcal{O}_{\mathbb{P}_2}(a - 3)) \quad \text{for all} \quad k > 0.$$

Because $a < 0$ we have

$$h^0(X, \sigma^* \mathcal{O}_{\mathbb{P}_2}(a - 3)) = h^0(\mathbb{P}_2, \mathcal{O}_{\mathbb{P}_2}(a - 3)) = 0,$$

and thus altogether

$$H^2(X, [2C - C'] \otimes \sigma^* \mathcal{O}_{\mathbb{P}_2}(-a)) = 0. \qquad \square$$

COROLLARY. *Every topological bundle over $\mathbb{P}^2$ has at least one holomorphic structure.*

6.3. 2-bundles over $\mathbb{P}_3$. In order to show that every 2-bundle over $\mathbb{P}_3$ has a holomorphic structure we must for given integers c_1, c_2 with $c_1 c_2 \equiv 0\,(2)$ find a holomorphic 2-bundle E with $c_1(E) = c_1$, $c_2(E) = c_2$ and in case $c_1 \equiv 0\,(2)$ we must further realize both values of $\alpha \in \mathbb{Z}/2\mathbb{Z}$ by

$$\alpha(E) \equiv h^0(\mathbb{P}_3, E(-k-2)) + h^1(\mathbb{P}_3, E(-k-2)) \mod 2.$$

We obtain such bundles with the construction in 5.1. There for every codimension 2 locally complete intersection $(Y, \mathcal{O}_Y)$ in $\mathbb{P}_3$ (for which the determinant bundle of the normal bundle is extendable over $\mathbb{P}_3$) we had constructed a 2-bundle E with a section $s \in H^0(\mathbb{P}_3, E)$ whose zero locus was precisely Y.

If $c_1(E)$ is even, then the α-invariant can be computed from Y as follows.

LEMMA 6.3.1. *Let $Y \subset \mathbb{P}_3$ be the codimension 2 zero set of a section $s \in H^0(\mathbb{P}_3, E)$ in a 2-bundle E, $c_1(E) = 2k$. Then*

$$\alpha(E) \equiv \left\{ \begin{array}{ll} h^0(Y, \mathcal{O}_Y(k-2)) + 1 & \text{for } \; k-2 \equiv 0 \mod 4 \\ h^0(Y, \mathcal{O}_Y(k-2)) & \text{otherwise} \end{array} \right\} \mod 2.$$

PROOF. We proceed from the Koszul complex of the section s

$$0 \to \mathcal{O}_{\mathbb{P}_3}(-2k) \to E^* \to J_Y \to 0.$$

By twisting with $\mathcal{O}_{\mathbb{P}_3}(k-2)$ we obtain

$$(*) \qquad 0 \to \mathcal{O}_{\mathbb{P}_3}(-k-2) \to E^*(k-2) \to J_Y(k-2) \to 0.$$

We have

$$E^*(k-2) \cong E \otimes \det E^* \otimes \mathcal{O}_{\mathbb{P}_3}(k-2) \cong E(-k-2).$$

The exact cohomology sequence of (*) thus yields

$$\text{(i)} \quad 0 \to H^0(\mathbb{P}_3, \mathcal{O}_{\mathbb{P}_3}(-k-2)) \to H^0(\mathbb{P}_3, E(-k-2)) \to H^0(\mathbb{P}_3, J_Y(k-2)) \to 0$$

and

$$\text{(ii)} \qquad 0 \to H^1(\mathbb{P}_3, E(-k-2)) \to H^1(\mathbb{P}_3, J_Y(k-2)) \to 0.$$

From the exact sequence

$$0 \to J_Y(k-2) \to \mathcal{O}_{\mathbb{P}_3}(k-2) \to \mathcal{O}_Y(k-2) \to 0$$

one further obtains

$$\text{(iii)}\quad 0 \to H^0(\mathbb{P}_3, J_Y(k-2)) \to H^0(\mathbb{P}_3, \mathcal{O}_{\mathbb{P}_3}(k-2)) \\ \to H^0(Y, \mathcal{O}_Y(k-2)) \to H^1(\mathbb{P}_3, J_Y(k-2)) \to 0.$$

(i)–(iii) together give (calculating modulo 2)

$$\begin{aligned}\alpha(E) &= h^0(\mathcal{O}_{\mathbb{P}_3}(-k-2)) + h^0(J_Y(k-2)) + h^1(J_Y(k-2)) \\ &\equiv h^0(\mathcal{O}_{\mathbb{P}_3}(-k-2)) + h^0(\mathcal{O}_{\mathbb{P}_3}(k-2)) + h^0(\mathcal{O}_Y(k-2)) \\ &\equiv \begin{cases} h^0(\mathcal{O}_Y(k-2)) + 1 & \text{for} \quad k-2 \equiv 0 \mod 4 \\ h^0(\mathcal{O}_Y(k-2)) & \text{otherwise,} \end{cases}\end{aligned}$$

for we have

$$h^0(\mathbb{P}_3, \mathcal{O}_{\mathbb{P}_3}(m)) \equiv \begin{cases} 1 \mod 2 & \text{for} \quad m \geq 0 \quad \text{and} \quad m \equiv 0\ (4) \\ 0 \mod 2 & \text{otherwise.} \end{cases}$$

□

Example: Let Y be the intersection of two hypersurfaces

$$Y = V_a \cap V_b$$

of degree a resp. b, $a + b = 2k$. The associated 2-bundle splits $E = \mathcal{O}_{\mathbb{P}_3}(a) \oplus \mathcal{O}_{\mathbb{P}_3}(b)$.

On sees immediately that

$$\begin{aligned} h^0(\mathbb{P}_3, E(-k-2)) &= h^0(\mathbb{P}_3, \mathcal{O}_{\mathbb{P}_3}(\frac{|a-b|}{2} - 2)) \\ h^1(\mathbb{P}_3, E(-k-2)) &= 0 \end{aligned}$$

and thus

$$\alpha(E) \equiv h^0(\mathbb{P}_3, \mathcal{O}_{\mathbb{P}_3}(\frac{|a-b|}{2} - 2)).$$

We set

$$d(m) = \begin{cases} 1 & \text{for} \quad m \geq 0 \quad \text{and} \quad m \equiv 0 \mod 4 \\ 0 & \text{otherwise} \end{cases}$$

Then $d(m) \equiv h^0(\mathbb{P}_3, \mathcal{O}_{\mathbb{P}_3}(m)) \mod 2$ for all $m \in \mathbb{Z}$.

The α-invariant is additive in the following sense.

LEMMA 6.3.2. *Let s be a section in a 2-bundle E with zero set Y, $c_1(E) = 2k$. If Y is the disjoint union of r complete intersections $Y_i = V_{a_i} \cap V_{b_i}$, $i = 1, \ldots, r$, with $a_i \leq b_i$ and $a_i + b_i = 2k$, then*

$$\alpha(E) \equiv (r-1)d(k-2) + \sum_{i=1}^{r} d(k - a_i - 2) \mod 2.$$

PROOF. Above we saw that

$$\alpha(E) + d(k-2) \equiv h^0(Y, \mathcal{O}_Y(k-2)) \mod 2,$$

i.e.,

$$\alpha(E) + d(k-2) \equiv \sum_{i=1}^{r} h^0(Y_i, \mathcal{O}_{Y_i}(k-2)) \mod 2.$$

Let $E_i = \mathcal{O}_{\mathbb{P}_3}(a_i) \oplus \mathcal{O}_{\mathbb{P}_3}(b_i)$ be the 2-bundle belonging to Y_i. Then

$$\alpha(E_i) + d(k-2) \equiv h^0(Y_i, \mathcal{O}_{Y_i}(k-2)) \mod 2.$$

Hence we obtain

$$\alpha(E) \oplus d(k-2) \equiv \sum_{i=1}^{r} (\alpha(E_i) + d(k-2)) \mod 2$$

or in other words

$$\alpha(E) \equiv (r-1)d(k-2) + \sum_{i=1}^{r} \alpha(E_i).$$

From the above example, however, one sees that

$$\alpha(E_i) \equiv d(k - a_i - 2).$$ □

For integers c_1, c_2 with $c_1 c_2 \equiv 0 \mod 2$ we let

$$\Delta(c_1, c_2) = c_1^2 - 4c_2,$$

respectively for a 2-bundle E over $\mathbb{P}_3$

$$\Delta(E) = \Delta(c_1(E), c_2(E)).$$

Then we evidently have

$$\Delta(E) = \Delta(E(k))$$

and

$$\Delta(c_1, c_2) \equiv \begin{cases} 0 \mod 4 \text{ for } c_1 \text{ even} \\ 1 \mod 8 \text{ for } c_1 \text{ odd.} \end{cases}$$

We can now show

THEOREM 6.3.3 (Atiyah, Rees, Horrocks). *Every topological 2-bundle over $\mathbb{P}_3$ has a holomorphic structure.*

PROOF. We first show that for every pair $(c_1, c_2) \in \mathbb{Z}^2$ with $c_1 c_2 \equiv 0 \mod 2$ there is a holomorphic 2-bundle E with

$$c_1(E) = c_1, \qquad c_2(E) = c_2.$$

To do this we again employ Example 5 in 5.2. There we considered a locally complete intersection

$$Y = \bigcup_{i=1}^{r} Y_i$$

in $\mathbb{P}_3$ which consisted of r disjoint complete intersections Y_i. Y_i was the intersection of two hypersurfaces $Y_i = V_{a_i} \cap V'_{b_i}$ of degree a_i and b_i with $a_i \leq b_i$, $a_i + b_i = p \geq 2$ for $i = 1, \ldots, r$. The associated 2-bundle had the Chern classes

$$c_1(E) = p, \qquad c_2(E) = \sum_{i=1}^{r} a_i b_i.$$

Now suppose $c_1, c_2 \in \mathbb{Z}$ are given with $c_1 c_2 \equiv 0 \bmod 2$. By a simple calculation one sees that one can choose the numbers

$$p, \quad r, \quad a_i$$

in more than one way so that

$$a_i \leq p/2$$

and $\Delta(c_1, c_2) = p^2 - 4\sum_{i=1}^{r} a_i(p - a_i)$.

Furthermore for c_1 odd one can also choose p odd (for c_1 even p is necessarily also even). To p, r, a_i, $b_i = p - a_i$ then there belongs a 2-bundle E with

$$c_1(E) = p, \qquad c_2(E) = \sum_{i=1}^{r} a_i b_i.$$

The twisted bundle $E' = E(\frac{c_1 - p}{2})$ then has the desired Chern classes

$$c_1(E') = c_1(E) + 2\frac{c_1 - p}{2} = c_1$$

$$c_2(E') = c_2(E) + \frac{c_1 - p}{2} c_1(E) + \left(\frac{c_1 - p}{2}\right)^2 = c_2.$$

Finally we must see that with these bundles both values of the α-invariant can be realized. Let c_1 be even then. By the previous lemma the α-invariant of the bundle $E' = E(\frac{c_1 - p}{2})$ is given by

$$\alpha(E') \equiv \alpha(E) \equiv (r-1)d(\tfrac{p}{2} - 2) + \textstyle\sum_{i=1}^{r} d(\tfrac{p}{2} - a_i - 2) \mod 2.$$

Now some elementary — if somewhat complicated — number theoretic considerations (Atiyah, Rees [2] §6) show that by appropriate choice of p, r, a_i one can achieve both values of the α-invariant. □

6.4. 3-bundles over $\mathbb{P}_3$. To every triple $(c_1, c_2, c_3) \in \mathbb{Z}^3$ with

$$(S_3^3): \qquad c_1 c_2 \equiv c_3 \mod 2$$

there is exactly one topological 3-bundle E with Chern classes

$$c_1(E) = c_1, \qquad c_2(E) = c_2, \qquad c_3(E) = c_3.$$

Then we have the following

THEOREM 6.4.1 (Vogelaar). *Every topological 3-bundle over $\mathbb{P}_3$ has a holomorphic structure.*

From this theorem together with the preceding considerations one has immediately the

COROLLARY. *Every topological bundle over $\mathbb{P}_3$ has a holomorphic structure.*

For the proof of the theorem we use once again the method of construction in 5.1 but in a somewhat more general form. In 5.1 we had for a codimension 2 locally complete intersection $Y \subset \mathbb{P}_3$ for which the bundle

$$\det N_{Y/\mathbb{P}_3} \otimes \mathcal{O}_Y(-k)$$

was trivial for appropriate k a 2-bundle E as extension

$$0 \to \mathcal{O}_{\mathbb{P}_3} \to E \to J_Y(k) \to 0.$$

The hypothesis on $\det N_{Y/\mathbb{P}_3} \otimes \mathcal{O}_Y(-k)$ was used essentially to show that E is locally free.

For constructing 3-bundles it would then be natural to consider extensions

$$0 \to \mathcal{O}_{\mathbb{P}_3} \oplus \mathcal{O}_{\mathbb{P}_3} \to E \to J_Y(k) \to 0$$

and to investigate under what conditions the sheaf E is locally free. We formulate the result in a theorem.

THEOREM 6.4.2 (Vogelaar). *Let $Y \subset \mathbb{P}_3$ be a codimension 2 locally complete intersection. Let the bundle*

$$\det N_{Y/\mathbb{P}_3} \otimes \mathcal{O}_Y(-k)$$

be generated by two global sections

$$t_1, t_2 \in H^0(Y; \det N_{Y/\mathbb{P}_3} \otimes \mathcal{O}_Y(-k)).$$

Then there is a 3-bundle E over $\mathbb{P}_3$ with two sections s_1, s_2 which are linearly dependent precisely over Y.

PROOF. The extensions

$$0 \to \mathcal{O}_{\mathbb{P}_3} \oplus \mathcal{O}_{\mathbb{P}_3} \to E \to \mathcal{J}_Y(k) \to 0$$

are classified by

$$\begin{aligned}\operatorname{Ext}^1_{\mathbb{P}_3}(\mathcal{J}_Y(k), \mathcal{O}_{\mathbb{P}_3} \oplus \mathcal{O}_{\mathbb{P}_3}) &\cong H^0(\mathbb{P}_3, \mathcal{E}xt^1_{\mathcal{O}_{\mathbb{P}_3}}(\mathcal{J}_Y(k), \mathcal{O}_{\mathbb{P}_3}))^{\oplus 2} \\ &\cong H^0(\mathbb{P}_3, \mathcal{E}xt^2_{\mathcal{O}_{\mathbb{P}_3}}(\mathcal{O}_Y(k), \mathcal{O}_{\mathbb{P}_3}))^{\oplus 2} \\ &\underset{\text{(LFI)}}{\cong} H^0(Y, \det N_{Y/\mathbb{P}_3} \otimes \mathcal{O}_Y(-k))^{\oplus 2}\end{aligned}$$

as we saw in 5.1.

Let

$$0 \to \mathcal{O}_{\mathbb{P}_3} \oplus \mathcal{O}_{\mathbb{P}_3} \to E \to \mathcal{J}_Y(k) \to 0$$

be the extension defined by

$$(t_1, t_2) \in H^0(Y, \det N_{Y/\mathbb{P}_3} \otimes \mathcal{O}_Y(-k))^{\oplus 2}.$$

Since by hypothesis $t_{1,x}, t_{2,x}$ generate the $\mathcal{O}_{\mathbb{P}_3,x}$-module $(\det N_{Y/\mathbb{P}_3} \otimes \mathcal{O}_Y(-k))_x$, E_x is a free $\mathcal{O}_{\mathbb{P}_3,x}$-module and thus E is a 3-bundle over $\mathbb{P}_3$. For in fact we have the

LEMMA 6.4.3. *Let A be a local noetherian ring, $I \subset A$ an ideal with a free resolution of length* 1. *An element $e = (e_1, e_2) \in \operatorname{Ext}^1_A(I, A^{\oplus 2})$ which defines the extension*

$$(*) \qquad 0 \to A^{\oplus 2} \to M \to I \to 0$$

leads to a free A-module M if and only if (e_1, e_2) is a system of generators for the A-module $\operatorname{Ext}^1_A(I, A)$.

PROOF. Let $p_i \colon A \oplus A \twoheadrightarrow A$ be the ith projection $i = 1, 2$. ¿From (*) we get the following commutative diagram with exact rows:

$$\begin{array}{ccccccc} \cdots \longrightarrow \operatorname{Hom}_A(A^{\oplus 2}, A^{\oplus 2}) & \xrightarrow{\delta} & \operatorname{Ext}^1_A(I, A^{\oplus 2}) & \longrightarrow & \operatorname{Ext}^1_A(M, A^{\oplus 2}) & \longrightarrow & 0 \\ \downarrow p_{i*} & & \downarrow p_{i*} & & \downarrow p_{i*} & & \\ \cdots \longrightarrow \operatorname{Hom}_A(A^{\oplus 2}, A) & \xrightarrow[\delta]{} & \operatorname{Ext}^1_A(I, A) & \longrightarrow & \operatorname{Ext}^1_A(M, A) & \longrightarrow & 0 \end{array}$$

Now $\operatorname{Ext}^1_A(M, A) = 0$ if and only if the δ in the lower row is surjective. Since however $\delta(\mathrm{id}_{A^{\oplus 2}}) = e = (e_1, e_2)$ and the elements $\phi_1 = p_{1*}(\mathrm{id}_{A^{\oplus 2}})$, $\phi_2 = p_{2*}(\mathrm{id}_{A^{\oplus 2}})$ generate the A-module $\operatorname{Hom}_A(A^{\oplus 2}, A)$ and further $\delta(\phi_i) = e_i$, $i = 1, 2$, we see that δ is surjective precisely when e_1, e_2 generate the A-module $\operatorname{Ext}^1_A(I, A)$. The rest of the proof goes as in the proof of the analogous lemma 5.1.2. □

The monomorphism $\mathcal{O}_{\mathbb{P}_3} \oplus \mathcal{O}_{\mathbb{P}_3} \hookrightarrow E$ in the extension

$$0 \to \mathcal{O}_{\mathbb{P}_3} \oplus \mathcal{O}_{\mathbb{P}_3} \to E \to J_Y(k) \to 0$$

thus obtained gives us two sections $s_1, s_2 \in H^0(\mathbb{P}_3, E)$ which are linearly dependent exactly over Y. J_Y is the sheaf of ideals of the zero set of the section

$$s_1 \wedge s_2 \in H^0(\mathbb{P}_3, \Lambda^2 E)$$

and the epimorphism $E \twoheadrightarrow J_Y(k)$ in the extension can be identified with the composition

$$E \xrightarrow{\sim} (\Lambda^2 E)^* \otimes \det E \xrightarrow{(s_1 \wedge s_2)^t \otimes \mathrm{id}} J_Y \otimes \det E = J_Y(k),$$

where $E \xrightarrow{\sim} (\Lambda^2 E)^* \otimes \det E = \mathcal{H}om_{\mathcal{O}_{\mathbb{P}_3}}(\Lambda^2 E, \Lambda^3 E)$ denotes the canonical isomorphism

$$s \mapsto (s_1 \wedge s_2 \mapsto s \wedge s_1 \wedge s_2). \qquad \square$$

We have the following addendum about the Chern classes.

Addendum.

(a) $c_1(E) = k$

(b) $c_2(E) = \deg Y$ if Y is smooth

(c) $c_3(E) = \deg(s = 0)$ for a general section $s \in \mathbb{C}s_1 + \mathbb{C}s_2 \subset H^0(\mathbb{P}_3, E)$.

For (b) cf. Kleiman, [75].

We come now to the proof of the theorem of Vogelaar. Let $Y = \bigcup_{i=1}^r Y_i$ be a disjoint union of r smooth complete intersections Y_i,

$$Y_i = V_{a_i} \cap V'_{b_i}$$

of degree $a_i b_i$. Then $\deg Y = \sum_{i=1}^r a_i b_i$ and

$$\det N_{Y/\mathbb{P}_3}|Y_i = \mathcal{O}_{Y_i}(a_i + b_i).$$

We choose $k \in \mathbb{Z}$ so that

$$a_i + b_i - k \geq 0 \qquad \text{for} \quad i = 1, \dots, r.$$

One then finds hypersurfaces H_i and H'_i of degree $a_i + b_i - k$ which meet Y_i transversally and such that

$$(Y_i \cap H_i) \cap (Y_i \cap H'_i) = \varnothing.$$

These hypersurfaces then define sections

$$t_1, t_2 \in H^0(Y, \det N_{Y/\mathbb{P}_3} \otimes \mathcal{O}_Y(-k))$$

without common zeros. Thus they generate the bundle

$$\det N_{Y/\mathbb{P}_3} \otimes \mathcal{O}_Y(-k).$$

By the previous theorem we obtain a 3-bundle E over $\mathbb{P}_3$ with

$$(*)\qquad \begin{cases} c_1(E) = k & k \in \mathbb{Z} \\ c_2(E) = \sum_{i=1}^r a_i b_i & a_i, b_i > 0 \\ c_3(E) = \sum_{i=1}^r (a_i + b_i - k) a_i b_i & a_i + b_i - k \geq 0. \end{cases}$$

Using (rather difficult) methods of elementary number theory one can again demonstrate that (*) gives all triples

$$(c_1, c_2, c_3) \in \mathbb{Z}^3$$

with $c_1 c_2 \equiv c_3 \bmod 2$ (Vogelaar [137]).

6.5. Concluding remarks. In this paragraph we showed that every topological complex vector bundle over $\mathbb{P}_n$, $n \leq 3$, admits a holomorphic structure. For $n \geq 4$ this is an unsolved problem.

There is a conjecture (cf. [48], [105]) that unstable (cf. Ch. II, §1) holomorphic 2-bundles over $\mathbb{P}_n$, $n \geq 5$, split. In particular every holomorphic 2-bundle E with $c_1^2 - 4c_2 \leq 0$ would split. Rees [101] and Smith [118] have constructed topological 2-bundles over $\mathbb{P}_n$, $n \geq 5$, which are topologically nontrivial but have vanishing Chern classes (for $n = 5, 6$ this follows from the diagrams in §6.1.). These would then have no holomorphic structure.

Barth and Elencwajg [14] have shown that on $\mathbb{P}_4$ there is at any rate no stable 2-bundle with $c_1 = 0$, $c_2 = 3$. This would be the first topologically possible case.

Chapter II is to a large extent concerned with the question how many holomorphic structures a fixed topological bundle can carry. This moduli problem can be satisfactorily treated for stable bundles. We shall investigate moduli spaces for stable 2-bundles over $\mathbb{P}_2$.

The method of Vogelaar presented in Theorem 6.4.2 for constructing holomorphic vector bundles of rank 3 over $\mathbb{P}_3$ can be extended to higher rank and higher base dimension. The statement and proofs are almost word for word the same.

CHAPTER 2

Stability and moduli spaces

§1. Stable bundles

In this section we introduce the crucial concept of *stability* of holomorphic vector bundles over $\mathbb{P}_n$. We begin by collecting together several theorems about torsion-free, normal and reflexive sheaves which we shall need later. Then we define stable and semistable torsion-free sheaves in the sense of Mumford and Takemoto and compare this concept of stability with that of $\mathcal{O}_{\mathbb{P}_n}(1)$-stability as introduced by Gieseker and Maruyama. In a final section we investigate the stability of a number of bundles with which we became acquainted in earlier sections.

1.1. Some useful results from sheaf theory. We begin by reminding the reader of several definitions from homological algebra.

Let F be a coherent sheaf over the n-dimensional complex manifold X, $x \in X$ a point. The stalk F_x is a finitely generated module over the n-dimensional regular local noetherian ring $\mathcal{O}_{X,x}$. The *homological dimension* $\operatorname{dh} F_x$ of F_x over $\mathcal{O}_{X,x}$ is the minimal length of a free resolution of F_x. $\operatorname{dh} F_x$ is the smallest integer k so that for all finitely generated $\mathcal{O}_{X,x}$-modules and all $i > k$ we have

$$\operatorname{Ext}^i_{\mathcal{O}_{X,x}}(F_x, M) = 0$$

(see Serre, [113], IV–27). The *homological codimension* (or *depth*) $\operatorname{codh} F_x$ is defined as the maximal length of an F_x-sequence in $\mathcal{O}_{X,x}$ (Serre, [113], IV–14). Since $\mathcal{O}_{X,x}$ is an n-dimensional regular local noetherian ring, we have the

Syzygy Theorem: $\operatorname{dh} F_x + \operatorname{codh} F_x = n$.

The homological dimension can also be characterized as follows:

LEMMA 1.1.1. $\operatorname{dh} F_x \leq q$ *precisely when for all* $i > q$ *we have*

$$(\mathcal{E}xt^i_{\mathcal{O}_X}(F, \mathcal{O}_X))_x = 0.$$

PROOF. If $\operatorname{dh} F_x \leq q$, then

$$(\mathcal{E}xt^i_{\mathcal{O}_X}(F, \mathcal{O}_X))_x = \operatorname{Ext}^i_{\mathcal{O}_{X,x}}(F_x, \mathcal{O}_{X,x})$$

vanishes for $i > q$.

We prove the opposite implication by descending induction over q. For $q \geq n$ there is nothing to prove because of the Syzygy theorem. Suppose then that

$$(\mathcal{E}xt^i_{\mathcal{O}_X}(F, \mathcal{O}_X))_x = 0 \quad \text{for} \quad i > q.$$

By the induction hypothesis it follows that

$$\operatorname{dh} F_x \leq q + 1.$$

If M is any finitely generated $\mathcal{O}_{X,x}$-module with an exact sequence

$$0 \to K \to \mathcal{O}_{X,x}^{\oplus m} \to M \to 0,$$

then we have from the Ext-sequence

$$\cdots \to \operatorname{Ext}^{q+1}_{\mathcal{O}_{X,x}}(F_x, K) \to \operatorname{Ext}^{q+1}_{\mathcal{O}_{X,x}}(F_x, \mathcal{O}_{X,x}^{\oplus m}) \\ \to \operatorname{Ext}^{q+1}_{\mathcal{O}_{X,x}}(F_x, M) \to \operatorname{Ext}^{q+2}_{\mathcal{O}_{X,x}}(F_x, K) \to \cdots$$

the equation

$$\operatorname{Ext}^{q+1}_{\mathcal{O}_{X,x}}(F_x, M) = 0,$$

i.e., $\operatorname{dh} F_x \leq q$. □

In what follows we shall need an estimate for the dimension of the support of the Ext-sheaves.

LEMMA 1.1.2. *For a coherent analytic sheaf F over an n-dimensional complex manifold X we have*

$$\dim(\operatorname{supp} \mathcal{E}xt^i_{\mathcal{O}_X}(F, \mathcal{O}_X)) \leq n - i$$

for all i.

PROOF. Let us suppose there were a point $x \in X$ with

$$\dim_x(\operatorname{supp} \mathcal{E}xt^i_{\mathcal{O}_X}(F, \mathcal{O}_X)) > n - i.$$

Then there is a prime ideal

$$p \subset \mathcal{O}_{X,x}$$

with

$$\dim(\mathcal{O}_{X,x}/p) > n - i$$

so that the germ of an analytic set defined by p lies in the support of the sheaf

$$\mathcal{E}xt^i_{\mathcal{O}_X}(F, \mathcal{O}_X).$$

The localization

$$(\mathcal{O}_{X,x})_p$$

is a regular local noetherian ring, and from

$$n = \dim(\mathcal{O}_{X,x}/p) + \dim(\mathcal{O}_{X,x})_p$$

it follows that

$$\dim(\mathcal{O}_{X,x})_p < i.$$

From this we deduce with the help of the Syzygy theorem that

$$\mathrm{dh}\,(F_x)_p < i$$

and thus

$$(*) \qquad (\mathrm{Ext}^i_{\mathcal{O}_{X,x}}(F_x, \mathcal{O}_{X,x}))_p = \mathrm{Ext}^i_{(\mathcal{O}_{X,x})_p}((F_x)_p, (\mathcal{O}_{X,x})_p)) = 0.$$

But because the germ of an analytic set defined by p lies in the support of $\mathcal{E}xt^i_{\mathcal{O}_X}(F, \mathcal{O}_X)$, the localization

$$(\mathrm{Ext}^i_{\mathcal{O}_{X,x}}(F_x, \mathcal{O}_{X,x}))_p$$

must be non-vanishing in contradiction to $(*)$. □

Let F again be a coherent analytic sheaf over the n-dimensional complex manifold X.

DEFINITION 1.1.3. The *m th singularity set* of the homological codimension of F is

$$S_m(F) = \{x \in X \mid \mathrm{codh}\, F_x \leq m\}.$$

From the syzygy theorem follows

$$S_m(F) = \{x \in X \mid \mathrm{dh}\, F_x \geq n - m\}$$

and thus from Lemma 1.1.1 the equation

$$S_m(F) = \bigcup_{i=n-m}^{n} \mathrm{supp}\,(\mathcal{E}xt^i_{\mathcal{O}_X}(F, \mathcal{O}_X)).$$

From Lemma 1.1.2 we then obtain

LEMMA 1.1.4. *The sets $S_m(F)$ are closed analytic in X of codimension $\geq n - m$.*

The stalk F_x of F at x is free if and only if the homological dimension of F_x is zero. Thus the *singularity set*

$$S(F) = \{x \in X \mid F_x \text{ is not free over } \mathcal{O}_{X,x}\}$$

coincides with $S_{n-1}(F)$.

COROLLARY. *The singularity set $S(F)$ of a coherent analytic sheaf F over X is a closed analytic subset of X of codimension at least* 1.

Over $X \setminus S(F)$ the sheaf is locally free. Let us assume that X is irreducible, i.e., connected. The *rank of F* is defined by

$$\mathrm{rk}\, F = \mathrm{rk}\,(F|\,X \backslash S(F)).$$

DEFINITION 1.1.5. A coherent sheaf F over X is a *k th syzygy sheaf* if there is an exact sequence

$$0 \to F \to \mathcal{O}_X^{\oplus p_1} \to \mathcal{O}_X^{\oplus p_2} \to \cdots \to \mathcal{O}_X^{\oplus p_k}.$$

For the singularity sets of such sheaves we have the following estimate of the codimension.

THEOREM 1.1.6. *The codimension of the singularity set $S(F)$ in X of a coherent sheaf F which is locally a k th syzygy sheaf is greater than k:*

$$\operatorname{codim}(S(F), X) > k$$

PROOF. Let $U \subset X$ be an open neighborhood of $x \in S(F)$ over which there is an exact sequence

$$0 \to F|U \xrightarrow{\phi_1} \mathcal{O}_U^{\oplus p_1} \xrightarrow{\phi_2} \mathcal{O}_U^{\oplus p_2} \to \ldots \xrightarrow{\phi_k} \mathcal{O}_U^{\oplus p_k}.$$

Let

$$F_i = \operatorname{coker} \phi_i, \qquad i = 1, \ldots, k$$
$$F_0 = F|U.$$

Then we have exact sequences

$$0 \to F_{i-1} \to \mathcal{O}_U^{\oplus p_i} \to F_i \to 0, \qquad i = 1, \ldots, k$$

and hence (cf. Serre [113], IV–28)

$$\operatorname{dh} F_{i,x} = \begin{cases} 0 & \text{if } F_{i,x} \text{ free over } \mathcal{O}_{X,x} \\ \operatorname{dh} F_{i-1,x} + 1 & \text{otherwise.} \end{cases}$$

Thus it follows that

$$S(F) \cap U = S(F_0) = S_{n-1}(F_0) \subset S_{n-2}(F_1) \subset \cdots \subset S_{n-k-1}(F_k)$$

and thus by lemma 1.1.4

$$\dim S(F) \leq n - k - 1.$$ □

DEFINITION 1.1.7. A coherent sheaf F over X is *torsion-free* if every stalk F_x is a torsion free $\mathcal{O}_{X,x}$-module; that is $fa = 0$ for $f \in \mathcal{O}_{X,x}$, $a \in F_x$ always implies $a = 0$ or $f = 0$.

Locally free sheaves are torsion-free. Sub-sheaves of torsion-free sheaves are again torsion-free. In particular coherent sheaves which are locally 1st syzygy sheaves are torsion-free. The converse is also true. One can in fact even arrange that a torsion-free sheaf of rank r is locally a subsheaf of a free sheaf of the same rank:

LEMMA 1.1.8. *Let F be a torsion-free coherent sheaf of rank r. For every $x \in X$ there is an open neighborhood U of x and a monomorphism of sheaves*

$$F|U \hookrightarrow \mathcal{O}_U^{\oplus r}.$$

PROOF. Because F is coherent it suffices to find a monomorphism of the stalks

$$\phi_x \colon F_x \hookrightarrow \mathcal{O}_{X,x}^{\oplus r}$$

for each $x \in X$. Let K be the quotient field of $\mathcal{O}_{X,x}$. Since F_x is torsion-free, the canonical mapping

$$F_x \to F_x \otimes_{\mathcal{O}_{X,x}} K \cong K^{\oplus r}$$

is injective. By multiplying with a suitable element $h \in \mathcal{O}_{X,x}$ one gets the desired monomorphism

$$\phi_x \colon F_x \overset{\cdot h}{\hookrightarrow} h \cdot F_x \subset \mathcal{O}_{X,x}^{\oplus r}. \qquad \square$$

Thus the torsion-free sheaves are precisely the locally 1st syzygy sheaves. With Theorem 1.1.6 we deduce:

COROLLARY. *The singularity set of a torsion-free coherent sheaf is at least 2-codimensional.*

In particular all torsion-free coherent sheaves over a Riemann surface are already locally free.

Let $F^* = \mathcal{H}om_{\mathcal{O}_X}(F, \mathcal{O}_X)$ be the dual sheaf of F and

$$\mu \colon F \to F^{**}$$

the canonical sheaf homomorphism of F into its bidual sheaf F^{**}. Then $\ker \mu$ is precisely the torsion subsheaf $T(F)$ of F, i.e.,

$$\ker \mu_x = \{a \in F_x \mid fa = 0 \quad \text{for some} \quad f \in \mathcal{O}_{X,x} \setminus \{0\}\}, \qquad x \in X$$

(cf. Grauert/Remmert [44] p. 233). Therefore μ is a monomorphism if and only if F is torsion-free.

DEFINITION 1.1.9. A coherent sheaf F for which $\mu \colon F \to F^{**}$ is an isomorphism is called *reflexive.*

LEMMA 1.1.10. *The singularity set of a reflexive coherent sheaf is at least 3-codimensional.*

PROOF. By Theorem 1.1.6 it suffices to show that reflexive sheaves are locally 2nd syzygy sheaves. To show this we begin with a presentation

$$\mathcal{O}_U^{\oplus q} \to \mathcal{O}_U^{\oplus p} \to F^*|U \to 0$$

over an open set U. We have the dual exact sequence

$$0 \to F^{**}|U \to \mathcal{O}_U^{\oplus p} \to \mathcal{O}_U^{\oplus q}$$

with $F|U \cong F^{**}|U$ — i.e., the desired sequence for $F|U$. □

In particular all reflexive sheaves over surfaces are locally free.

We summarize our results on the codimension of singularity sets in the following table:

F	coherent	torsion-free	reflexive
$\operatorname{codim}(S(F), X)$	≥ 1	≥ 2	≥ 3

DEFINITION 1.1.11. A sheaf F over X is *normal* if for every open set $U \subset X$ and every analytic set $A \subset U$ of codimension at least 2 the restriction map

$$F(U) \to F(U \setminus A)$$

is an isomorphism.

Reflexive sheaves can be characterized as follows.

LEMMA 1.1.12. *A coherent sheaf F is reflexive if and only if it is torsion-free and normal.*

PROOF. 1) If F is reflexive, then $F \cong \mathcal{H}om_{\mathcal{O}_X}(F^*, \mathcal{O}_X)$. But the dual sheaf of any coherent sheaf is torsion-free and normal, because $\mathcal{O}_X$ is torsion-free and normal (second Riemann extension theorem).

2) Suppose conversely F is torsion-free and normal; then the canonical mapping

$$\mu\colon F \to F^{**}$$

is a monomorphism and outside of $S(F)$ an isomorphism. Since by the corollary to Lemma 1.1.8 the singularity set $S(F)$ is at least 2-codimensional, one has the following commutative diagram

$$\begin{array}{ccc} F(U \setminus S(F)) & \xrightarrow{\sim\,\mu} & F^{**}(U \setminus S(F)) \\ \uparrow \wr\ \text{res} & & \uparrow \text{res}\ \wr \\ F(U) & \xrightarrow{\quad\mu\quad} & F^{**}(U) \end{array}$$

for every open set $U \subset X$. Thus

$$\mu\colon F(U) \to F^{**}(U)$$

is an isomorphism, i.e., F is reflexive. □

The sheaves of ideals of analytic sets in X having codimension at least 2 give examples of torsion-free non-normal sheaves.

We now wish to give a simple example to show that reflexive sheaves F of rank $r \geq 2$ over complex manifolds of dimension $n \geq 3$ are not in general locally free:

EXAMPLE 1.1.13. We consider the Euler sequence over $\mathbb{P}_3$

$$0 \to \mathcal{O}_{\mathbb{P}_3}(-1) \to \mathcal{O}_{\mathbb{P}_3}^{\oplus 4} \xrightarrow{p} T_{\mathbb{P}_3}(-1) \to 0.$$

Let $v \in \mathbb{C}^4 \setminus \{0\}$ and $s \in H^0(\mathbb{P}_3, T_{\mathbb{P}_3}(-1))$ be the section defined by v:

$$s(x) = p(x, v), \qquad x \in \mathbb{P}_3 \qquad \text{(cf. Ch. I, §3.3)}.$$

s has as its only zero the point $x_0 \in \mathbb{P}_3$ determined by v. We have an exact sequence

$$(*) \qquad 0 \to \mathcal{O}_{\mathbb{P}_3} \xrightarrow{s} T_{\mathbb{P}_3}(-1) \to F \to 0$$

with a coherent sheaf F of rank 2 over $\mathbb{P}_3$. By construction

$$S(F) = \{x_0\}.$$

In order to show that F is reflexive we dualize the sequence (*) and get the exact sequence

$$(**) \qquad 0 \to F^* \to T_{\mathbb{P}_3}^*(1) \xrightarrow{s^t} \mathcal{O}_{\mathbb{P}_3} \to \mathcal{O}_{\mathbb{P}_3}/J_{x_0} \to 0,$$

where J_{x_0} is the sheaf of ideals of the 3-codimensional locally complete intersection $\{x_0\}$ in $\mathbb{P}_3$.

The sequence (**) can be further split up in the diagram:

$$\begin{array}{ccccccccc} & & & & & & 0 & & \\ & & & & & & \downarrow & & \\ 0 & \longrightarrow & F^* & \longrightarrow & T_{\mathbb{P}_3}^*(1) & \longrightarrow & J_{x_0} & \longrightarrow & 0 \\ & & & & & \overset{s^t}{\searrow} & \downarrow & & \\ & & & & & & \mathcal{O}_{\mathbb{P}_3} & & \\ & & & & & & \downarrow & & \\ & & & & & & \mathcal{O}_{\mathbb{P}_3}/J_{x_0} & & \\ & & & & & & \downarrow & & \\ & & & & & & 0 & & \end{array}$$

If we now dualize this diagram, then because

$$\mathcal{E}xt^i_{\mathcal{O}_{\mathbb{P}_3}}(\mathcal{O}_{\mathbb{P}_3}/J_{x_0}, \mathcal{O}_{\mathbb{P}_3}) = 0 \qquad \text{for} \quad 0 \leq i < 3$$

(cf. Griffiths and Harris [49] p. 690) we get

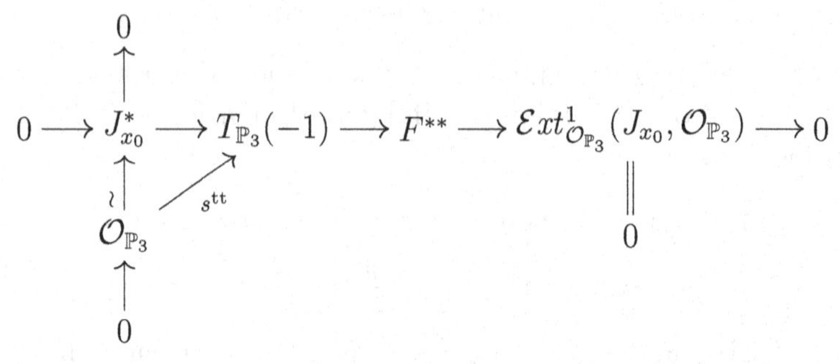

Thus we get the following canonical diagram with exact rows

$$\begin{array}{ccccccccc} 0 & \longrightarrow & \mathcal{O}_{\mathbb{P}_3} & \xrightarrow{s} & T_{\mathbb{P}_3}(-1) & \longrightarrow & F & \longrightarrow & 0 \\ & & \downarrow & & \downarrow & & \downarrow{\scriptstyle \mu} & & \\ 0 & \longrightarrow & \mathcal{O}^{**}_{\mathbb{P}_3} & \xrightarrow{s^{tt}} & T^{**}_{\mathbb{P}_3}(-1) & \longrightarrow & F^{**} & \longrightarrow & 0. \end{array}$$

F is thus reflexive.

Obviously one has in general: let E be a 3-bundle over an n-dimensional manifold ($n \geq 3$) with a section s whose zero set Y has codimension 3; then the exact sequence

$$0 \to \mathcal{O} \xrightarrow{s} E \to F \to 0$$

yields a reflexive sheaf F of rank 2 with $S(F) = Y$.

DEFINITION 1.1.14. Let F be a torsion-free sheaf of rank r over X.

$$\det F = (\Lambda^r F)^{**}$$

is called the *determinant bundle* of F.

$\det F$ is in fact a line bundle over X, for we have

LEMMA 1.1.15. *A reflexive sheaf of rank* 1 *is a line bundle.*

PROOF. By Lemma 1.1.8 and Lemma 1.1.12 we only need show that every normal sheaf of ideals $0 \neq J \subset \mathcal{O}_X$ is invertible — i.e., a sheaf of principal ideals. To this end let $x \in X$ and

$$J_x = q_1 \cap \cdots \cap q_m$$

be an (irreducible) primary decomposition of J_x in $\mathcal{O}_{X,x}$, U a Stein open neighborhood of x such that there exist sheaves of ideals $Q_1, \ldots, Q_m \subset \mathcal{O}_U$ with

$$Q_{i,x} = q_i \qquad \text{for} \quad i = 1, \ldots, m$$

and

$$J|U = Q_1 \cap \cdots \cap Q_m.$$

Claim: $\operatorname{codim}_x(\operatorname{supp}(\mathcal{O}_U/Q_i), U) = 1$ for $i = 1, \ldots, m$.

PROOF. We assume that $A = \operatorname{supp} \mathcal{O}_U/Q_1$ has codimension at least 2 in U. Since J is normal and $Q_1(U \setminus A) = \mathcal{O}(U \setminus A)$, we get a commutative diagram:

$$\begin{array}{ccc} J(U) = Q_1(U) \cap \cdots \cap Q_m(U) & \hookrightarrow & Q_2(U) \cap \cdots \cap Q_m(U) \\ \downarrow \wr \, \text{res} & & \downarrow \text{res} \\ J(U \setminus A) = Q_1(U \setminus A) \cap \cdots \cap Q_m(U \setminus A) & = & Q_2(U \setminus A) \cap \cdots \cap Q_m(U \setminus A) \end{array}$$

The second restriction map must therefore also be an isomorphism and thus we have

$$Q_1(U) \cap \cdots \cap Q_m(U) = Q_2(U) \cap \cdots \cap Q_m(U).$$

Since U is Stein, it follows by Theorem A that

$$q_1 \cap \cdots \cap q_m = q_2 \cap \cdots \cap q_m$$

in contradiction to the assumption that the decomposition $J_x = q_1 \cap \cdots \cap q_m$ be irreducible.

Thus for the prime ideals $p_i := \operatorname{rad} q_i \subset \mathcal{O}_{X,x}$ we have

$$\dim \mathcal{O}_{X,x}/p_i = n - 1 \qquad (n = \dim X)$$

and therefore ($\mathcal{O}_{X,x}$ is factorial!)

$$p_i = (f_i)$$

is a principal ideal. It follows that $q_i = (f_i^{k_i})$ for a suitable $k_i \geq 1$ and

$$J_x = q_1 \cap \cdots \cap q_m = (f_1^{k_1} \ldots f_m^{k_m}). \qquad \square$$

In the next section we shall often have to do with subsheaves

$$F \subset E$$

and quotient sheaves

$$E \twoheadrightarrow Q$$

of reflexive sheaves E.

Let $F \subset E$ be a coherent subsheaf of the reflexive sheaf E of rank $s = \operatorname{rk} F$. With $Q = E/F$ we get the exact sequence of sheaves

$$(*) \qquad 0 \to F \to E \to Q \to 0.$$

We consider coherent sheaves F' with

i) $F \subset F' \subset E$
ii) $\operatorname{rk} F' = \operatorname{rk} F$.

We call such an F' an *extension of F in E*. If in addition

iii) F' is normal,

then we call F' a *normal extension of* F *in* E. Every extension F' of F in E leads to a canonical diagram as follows

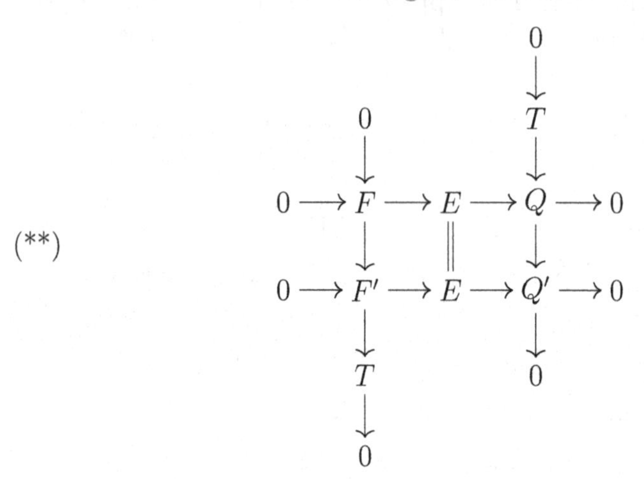

where T is a torsion sheaf. Since $T \subset Q$, we have $T \subset T(Q)$ and thus

$$\hat{F}_E := \ker(E \twoheadrightarrow Q/T(Q))$$

is the largest extension of F in E.

Claim: $\hat{F}_E$ is normal.

This follows from the following simple lemma.

LEMMA 1.1.16. *Let* $0 \to F \to E \to Q \to 0$ *be an exact sequence of sheaves,* E *reflexive. Then* F *is normal, if* Q *is torsion-free.*

PROOF. Let $U \subset X$ be open and $A \subset U$ an analytic set in U of codimension at least 2. Since Q, F are torsion-free, the restriction mappings

$$F(U) \hookrightarrow F(U \setminus A), \qquad Q(U) \hookrightarrow Q(U \setminus A)$$

are injective. From the diagram

$$\begin{array}{ccccccc} 0 \longrightarrow & F(U) & \longrightarrow & E(U) & \longrightarrow & Q(U) \\ & \downarrow & & \downarrow \wr & & \downarrow \\ 0 \longrightarrow & F(U \setminus A) & \longrightarrow & E(U \setminus A) & \longrightarrow & Q(U \setminus A) \end{array}$$

follows that $F(U) \to F(U \setminus A)$ is an isomorphism. □

We call $\hat{F}_E$ the *maximal normal extension of* F *in* E.

REMARK. If $F \subset E$ is a subsheaf of the reflexive sheaf E, then $\hat{F} = F^{**}$ is the smallest normal subsheaf of E with $F \subset \hat{F}$. In general

$\hat{F} \subset \hat{F}_E$ is a proper inclusion. If $\hat{F} \neq \hat{F}_E$, then $\operatorname{supp}(\hat{F}_E/\hat{F})$ is purely 1-codimensional.

We further show

LEMMA 1.1.17. *A monomorphism $F \hookrightarrow F'$ between torsion-free sheaves of the same rank induces a monomorphism* $\det F \hookrightarrow \det F'$ *of the determinant bundles.*

PROOF. Outside of the analytic set $A = S(F') \cup S(F'/F)$ the map $F \to F'$ is an isomorphism and thus also $\det F \to \det F'$. Over X therefore $\ker(\det F \to \det F')$ is a torsion-sheaf and as subsheaf of a torsion-free sheaf it must be zero. □

1.2. Stability: definitions and elementary properties. Let F be a torsion-free coherent sheaf of rank r over $\mathbb{P}_n$. We define the first Chern class of F by

$$c_1(F) = c_1(\det F),$$

where $\det F = (\Lambda^r F)^{**}$ denotes the determinant bundle of F (cf. 1.1.14). Since the singularity set $S(F)$ of F has codimension at least 2, there are lines $L \subset \mathbb{P}_n$ which do not meet $S(F)$. Let

$$F|L \cong \mathcal{O}_L(a_1) \oplus \cdots \oplus \mathcal{O}_L(a_r).$$

Then

$$c_1(F) = a_1 + \cdots + a_r.$$

We set

$$\mu(F) = \frac{c_1(F)}{\operatorname{rk}(F)}.$$

DEFINITION 1.2.1 (Mumford/Takemoto). A torsion-free coherent sheaf E over $\mathbb{P}_n$ is *semistable* if for every coherent subsheaf $0 \neq F \subset E$ we have

$$\mu(F) \leq \mu(E).$$

If moreover for all coherent subsheaves $F \subset E$ with $0 < \operatorname{rk} F < \operatorname{rk} E$ we have

$$\mu(F) < \mu(E),$$

then E is *stable.*

We wish to show that to demonstrate the stability of a torsion-free sheaf E it suffices to consider only such subsheaves F in E whose quotient E/F is torsion-free.

THEOREM 1.2.2. *Let E be a torsion-free sheaf over $\mathbb{P}_n$. The following statements are equivalent:*

i) *E is stable (semistable).*
ii) *$\mu(F) < \mu(E)$ ($\mu(F) \leq \mu(E)$) for all coherent subsheaves $F \subset E$ with $0 < \operatorname{rk} F < \operatorname{rk} E$ whose quotient E/F is torsion-free.*
iii) *$\mu(Q) > \mu(E)$ ($\mu(Q) \geq \mu(E)$) for all torsion-free quotients $E \twoheadrightarrow Q$ of E with $0 < \operatorname{rk} Q < \operatorname{rk} E$.*

PROOF. 1) We first prove:
if $F \hookrightarrow F'$ is a monomorphism of torsion-free sheaves of the same rank, then $c_1(F) \leq c_1(F')$. $F \hookrightarrow F'$ induces namely by Lemma 1.1.17 a monomorphism $\det F \hookrightarrow \det F'$ and thus we have

$$c_1(F) = c_1(\det F) \leq c_1(\det F') = c_1(F').$$

2) The equivalence of i) and ii) follows immediately from 1) if for a subsheaf $F \subset E$ with quotient $Q = E/F$ we consider the maximal extension

$$\hat{F}_E = \ker(E \twoheadrightarrow Q/T(Q)).$$

$\hat{F}_E$ has the same rank as F and $E/\hat{F}_E$ is torsion-free with

$$\mu(F) \leq \mu(\hat{F}_E).$$

To see the equivalence of ii) and iii) we consider an exact sequence of sheaves

$$0 \to F \to E \to Q \to 0$$

with torsion-free sheaves F and Q of ranks s and t. E has rank $s+t$ and since $c_1(E) = c_1(F) + c_1(Q)$ the condition

$$\mu(F) < \mu(E)$$

is equivalent to

$$\frac{s+t}{s} c_1(F) < c_1(F) + c_1(Q),$$

i.e.,

$$c_1(F) < \frac{s}{t} c_1(Q).$$

This inequality holds if and only if

$$c_1(Q) + c_1(F) < \frac{s+t}{t} c_1(Q),$$

i.e.,

$$\mu(Q) > \mu(E).$$

The calculation is analogous for the case of semistability. □

The following useful remark follows directly from the definition of stability:

REMARK 1.2.3. If E is a torsion-free sheaf of rank 2 over $\mathbb{P}_n$ with odd first Chern class, then we have: E is stable if and only if it is semistable.

If namely for a subsheaf $F \subset E$ of rank 1 the condition

$$c_1(F) = \mu(F) \leq \mu(E) = c_1(E)/2$$

holds, then we also have

$$c_1(F) < c_1(E)/2,$$

because $c_1(E)$ is odd. More generally we have: for an r-bundle E with $(c_1(E), r) = 1$ the concepts of stability and semistability coincide, as the reader can easily convince himself.

We summarize in the following lemma some simple properties of stability and semistability.

LEMMA 1.2.4. i) *Line bundles are stable.*
 ii) *The sum $E_1 \oplus E_2$ of two semistable sheaves is semistable if and only if $\mu(E_1) = \mu(E_2)$.*
 iii) *E is (semi)stable if and only if E^* is.*
 iv) *E is (semi)stable if and only if $E(k)$ is.*

PROOF. The first statement is trivial.

If E_1 and E_2 are semistable with $\mu = \mu(E_1) = \mu(E_2)$, then $\mu(E_1 \oplus E_2) = \mu$ and for every subsheaf $0 \neq F \subset E_1 \oplus E_2$ one has the following diagram

$$\begin{array}{ccccccccc} 0 & \longrightarrow & F_1 & \longrightarrow & F & \longrightarrow & F_2 & \longrightarrow & 0 \\ & & \downarrow & & \downarrow & & \downarrow & & \\ 0 & \longrightarrow & E_1 & \longrightarrow & E_1 \oplus E_2 & \longrightarrow & E_2 & \longrightarrow & 0 \end{array}$$

with $F_1 = F \cap (E_1 \oplus 0)$, $F_2 = F/F_1$. Since E_i is semistable it follows that $c_1(F_i) \leq \text{rk}\,(F_i)\mu$ and thus

$$\mu(F) = \frac{c_1(F_1) + c_1(F_2)}{\text{rk}\, F_1 + \text{rk}\, F_2} \leq \mu = \mu(E_1 \oplus E_2).$$

Thus $E_1 \oplus E_2$ is semistable.

Conversely if $E_1 \oplus E_2$ is semistable, then $\mu(E_i) = \mu(E_1 \oplus E_2)$ because E_i occurs both as subsheaf and as quotient of $E_1 \oplus E_2$.

The third statement follows from the equivalence of the conditions ii) and iii) in Theorem 1.2.2. The fourth statement is trivial. □

For reflexive sheaves of rank 2 over $\mathbb{P}_n$ there is a stability criterion which is often very useful. If E is a torsion-free sheaf of rank 2 over

$\mathbb{P}_n$, then there is a uniquely determined integer k_E such that

$$c_1(E(k_E)) \in \{0, -1\},$$

namely

$$k_E = \begin{cases} -\frac{c_1(E)}{2} & \text{for } c_1(E) \text{ even} \\ -\frac{c_1(E)+1}{2} & \text{for } c_1(E) \text{ odd.} \end{cases}$$

We set

$$E_{\text{norm}} := E(k_E)$$

and call E *normalized* if $E = E_{\text{norm}}$. We then have the following criterion.

LEMMA 1.2.5. *A reflexive sheaf E of rank 2 over $\mathbb{P}_n$ is stable if and only if E_{norm} has no sections:*

$$H^0(\mathbb{P}_n, E_{\text{norm}}) = 0.$$

If $c_1(E)$ is even, then E is semistable if and only if

$$H^0(\mathbb{P}_n, E_{\text{norm}}(-1)) = 0.$$

PROOF. If $H^0(\mathbb{P}_n, E_{\text{norm}}) \neq 0$, then there is a monomorphism

$$\mathcal{O}_{\mathbb{P}_n} \hookrightarrow E_{\text{norm}}.$$

Since however $\mu(\mathcal{O}_{\mathbb{P}_n}) = 0 \geq \mu(E_{\text{norm}})$, we see that E_{norm} and thus also E is not stable.

If conversely $H^0(\mathbb{P}_n, E_{\text{norm}}) = 0$ and $F \subset E$ is a coherent subsheaf of rank 1 with torsion-free quotient $Q = E/F$, then by Lemma 1.1.16 it follows that F is reflexive and by Lemma 1.1.15 that F is a line bundle and hence of the form

$$F = \mathcal{O}_{\mathbb{P}_n}(k), \qquad k = c_1(F) = \mu(F).$$

The inclusion $F \subset E$ defines a non-zero section s in $E(-k)$. Thus we must have

$$-k > k_E$$

because we assumed $H^0(\mathbb{P}_n, E_{\text{norm}}) = 0$. Thus we have

$$k < -k_E, \qquad \text{i.e., } \mu(F) < \mu(E).$$

In exactly the same way one shows that E is semistable if and only if $h^0(\mathbb{P}_n, E_{\text{norm}}(-1)) = 0$, provided $c_1(E)$ is even. □

In the case $c_1(E)$ odd the concepts "stable" and "semistable" coincide (Remark 1.2.3).

REMARK 1.2.6. a) For a torsion-free sheaf E of rank r over $\mathbb{P}_n$ which is normalized — i.e., $c_1(E) \in \{0, -1, \dots, -r+1\}$ — we have: if E is stable then $H^0(\mathbb{P}_n, E) = 0$ and

$$H^0(\mathbb{P}_n, E^*) = 0 \qquad \text{for } c_1 = 0$$
$$H^0(\mathbb{P}_n, E^*(-1)) = 0 \qquad \text{for } c_1 < 0.$$

b) For normalized reflexive sheaves E of rank 3 over $\mathbb{P}_n$ one has the criterion:
E is stable if and only if

$$H^0(\mathbb{P}_n, E) = H^0(\mathbb{P}_n, E^*) = 0 \qquad \text{for } c_1 = 0$$
$$H^0(\mathbb{P}_n, E) = H^0(\mathbb{P}_n, E^*(-1)) = 0 \qquad \text{for } c_1 = -1, -2.$$

In the case $c_1 = -1, -2$ the concepts "stable" and "semistable" are again equivalent. If $c_1 = 0$, then E is semistable if and only if

$$H^0(\mathbb{P}_n, E(-1)) = H^0(\mathbb{P}_n, E^*(-1)) = 0.$$

The proofs are analogous to that of Lemma 1.2.5.

With the help of the semistability criterion 1.2.5 and the Riemann–Roch theorem one gets an inequality for the Chern classes of stable 2-bundles over $\mathbb{P}_2$:

LEMMA 1.2.7 (Schwarzenberger). *For the Chern classes c_1, c_2 of a stable 2-bundle E over $\mathbb{P}_2$ we have*

$$c_1^2 - 4c_2 < 0.$$

If E is semistable we have

$$c_1^2 - 4c_2 \leq 0.$$

PROOF. The discriminant $\Delta_E = c_1^2 - 4c_2$ of E is invariant with respect to tensoring with $\mathcal{O}_{\mathbb{P}_2}(k)$, i.e.,

$$\Delta_E = \Delta_{E(k)}.$$

Thus we may as well assume E is normalized. If E is stable then

$$H^0(\mathbb{P}_2, E) = 0$$

and for reasons of duality also

$$H^2(\mathbb{P}_2, E) = 0.$$

Thus we have

$$(*) \qquad 0 \leq h^1(\mathbb{P}_2, E) = -\chi(\mathbb{P}_2, E).$$

The Riemann–Roch formula for a 2-bundle E over $\mathbb{P}_2$ is

$$(1) \qquad \chi(\mathbb{P}_2, E) = \tfrac{1}{2}(c_1(E)^2 - 2c_2(E) + 3c_1(E) + 4).$$

This is easily seen as follows: $\chi(\mathbb{P}_2, E)$ is some polynomial in the Chern classes of E with rational coefficients. This universal polynomial can be determined by calculating it for suitable special cases.

Let $E = \mathcal{O}_{\mathbb{P}_2}(a) \oplus \mathcal{O}_{\mathbb{P}_2}(b)$, $a, b > 0$. Then

$$\begin{aligned}\chi(\mathbb{P}_2, E) &= \tbinom{a+2}{2} + \tbinom{b+2}{2} = \tfrac{1}{2}((a+b)^2 - 2ab + 3(a+b) + 4)\\ &= \tfrac{1}{2}(c_1(E)^2 - 2c_2(E) + 3c_1(E) + 4).\end{aligned}$$

This is the formula (1).

From (*) and (1) follows that for a normalized stable 2-bundle E over $\mathbb{P}_2$

$$(**) \qquad c_1(E)^2 - 2c_2(E) + 3c_1(E) + 4 \leq 0.$$

If we now set $c_1(E) = 0, -1$ we get the desired inequality $\Delta_E < 0$.

If E is normalized and semistable but not stable, then $c_1(E) = 0$ and an analogous calculation shows that

$$0 \leq h^1(\mathbb{P}_2, E(-1)) = -\chi(\mathbb{P}_2, E(-1)) = c_2(E),$$

i.e., $c_2(E) \geq 0$, which means $\Delta_E \leq 0$. $\square$

From the inequality (**) one can further deduce that there is no stable 2-bundle over $\mathbb{P}_2$ with

$$c_1 = 0, \quad c_2 = 1 \qquad (\Delta = -4).$$

We shall later show that to every other choice c_1, c_2 with $\Delta = c_1^2 - 4c_2 < 0$, $\Delta \neq -4$, there is at least one stable 2-bundle over $\mathbb{P}_2$ with these Chern classes. In the next section we shall see that over $\mathbb{P}_3$ there are indeed stable 2-bundles with $c_1 = 0$, $c_2 = 1$.

We now want to show that stable bundles are simple, i.e., have only homotheties as endomorphisms. As a preparation we prove

LEMMA 1.2.8. *Let $\phi\colon E_1 \to E_2$ be a nontrivial sheaf homomorphism between semistable sheaves E_1, E_2. If at least one of the sheaves is stable and $\mu(E_1) = \mu(E_2)$, then ϕ is either a monomorphism or generically an epimorphism.*

PROOF. The image $I = \operatorname{Im} \phi \subset E_2$ is a torsion-free quotient of E_1 with $\operatorname{rk} I > 0$, since ϕ is nontrivial. If we had

$$\operatorname{rk} I < \operatorname{rk} E_2 \qquad \text{and} \qquad \operatorname{rk} I < \operatorname{rk} E_1,$$

then we would have

$$\mu(I) \leq \mu(E_2) = \mu(E_1) < \mu(I) \qquad \text{if } E_1 \text{ is stable}$$

or

$$\mu(I) < \mu(E_2) = \mu(E_1) \leq \mu(I) \qquad \text{if } E_2 \text{ is stable.}$$

Both are impossible. Therefore we must have

$$\operatorname{rk} I = \operatorname{rk} E_1 \qquad \text{or} \qquad \operatorname{rk} I = \operatorname{rk} E_2.$$

In the former case ϕ is a monomorphism, in the latter an epimorphism outside of $S(\operatorname{coker}\phi)$. □

As a corollary we have:

COROLLARY. *Let $\phi\colon E_1 \to E_2$ be a nontrivial sheaf homomorphism between two semistable vector bundles E_1, E_2 with $\operatorname{rk} E_1 = \operatorname{rk} E_2$ and $c_1(E_1) = c_1(E_2)$. Let at least one of the bundles be stable. Then ϕ is an isomorphism.*

PROOF. By Lemma 1.2.8 we see that ϕ is a monomorphism and thus also $\det\phi\colon \det E_1 \hookrightarrow \det E_2$. Since $c_1(E_1) = c_1(E_2)$, $\det\phi$ is in fact an isomorphism and thus also ϕ. □

As an application of this corollary we have

THEOREM 1.2.9. *Stable bundles are simple.*

PROOF. Let $\phi\colon E \to E$ be an arbitrary endomorphism of E, $x \in \mathbb{P}_n$ a point and $c \in \mathbb{C}$ some eigenvalue of $\phi(x)\colon E(x) \to E(x)$. The endomorphism

$$\phi - c \cdot \mathrm{id}_E$$

is then not an isomorphism. By the corollary to Lemma 1.2.8 it follows that

$$\phi - c \cdot \mathrm{id}_E = 0,$$

i.e., ϕ is a homothety. □

For holomorphic vector bundles of rank 2 over $\mathbb{P}_n$ we also have the converse of the theorem:

THEOREM 1.2.10. *Every simple 2-bundle over $\mathbb{P}_n$ is stable.*

PROOF. Without loss of generality we may assume E is normalized. If E were not stable, we would have $H^0(\mathbb{P}_n, E) \neq 0$ and because $E^* \cong E \otimes \det E^*$ also $H^0(\mathbb{P}_n, E^*) \neq 0$. Two nontrivial sections

$$0 \neq s \in H^0(\mathbb{P}_n, E), \qquad 0 \neq t \in H^0(\mathbb{P}_n, E^*)$$

then define an endomorphism

$$t \otimes s \in H^0(\mathbb{P}_n, E^* \otimes E),$$

which is not a homothety (cf. Ch. I, 4.1.3) in contradiction to the simplicity of E. □

At this point we wish to mention a further stability concept due to Gieseker and Maruyama.

DEFINITION 1.2.11 (Gieseker). A torsion-free coherent sheaf E over $\mathbb{P}_n$ is *Gieseker-stable (Gieseker-semistable)* if and only if for all coherent subsheaves $F \subset E$ with $0 < \operatorname{rk} F < \operatorname{rk} E$ we have

$$p_F(k) < p_E(k) \qquad (p_F(k) \leq p_E(k))$$

for all sufficiently large integers $k \in \mathbb{Z}$. Here

$$p_F(k) = \frac{\chi(F(k))}{\operatorname{rk} F}$$

and $\chi(F(k)) = \sum(-1)^i h^i(\mathbb{P}_n, F(k))$ is the Euler characteristic of $F(k)$.

By the Riemann–Roch theorem (cf. Borel, Serre [22], p. 113) the Euler characteristic $\chi(E(k))$ is a polynomial in the Chern classes of $E(k)$ with rational coefficients. If one fixes E, then $\chi(E(k))$ becomes a polynomial in the variable k. The coefficients of this polynomial can again be determined by computing $\chi(E(k))$ for suitable E. For $E = \mathcal{O}_{\mathbb{P}_n}(a_1) \oplus \cdots \oplus \mathcal{O}_{\mathbb{P}_n}(a_r)$, $a_i > 0$, $k > 0$, $c_1 = \sum a_i$, we have

$$\begin{aligned}\chi(E(k)) &= \sum_{i=1}^{r} \binom{a_i + k + n}{n} \\ &= r\frac{k^n}{n!} + \left(c_1 + r \cdot \frac{n+1}{2}\right)\frac{k^{n-1}}{(n-1)!} + \text{terms of lower order.}\end{aligned}$$

Thus for every torsion-free coherent sheaf E over $\mathbb{P}_n$:

$$p_E(k) = \frac{k^n}{n!} + \left(\mu(E) + \frac{n+1}{2}\right)\frac{k^{n-1}}{(n-1)!} + \text{terms of lower order.}$$

If $F \subset E$ is a coherent subsheaf with $0 < \operatorname{rk} F < \operatorname{rk} E$, then

$$(*) \quad p_E(k) - p_F(k) = (\mu(E) - \mu(F))\frac{k^{n-1}}{(n-1)!} + \text{terms of lower order,}$$

i.e., for sufficiently large $k \in \mathbb{Z}$

$$p_E(k) - p_F(k)$$

has the same sign as $\mu(E) - \mu(F)$. We have thus proved

LEMMA 1.2.12. *Stable torsion-free coherent sheaves over $\mathbb{P}_n$ are also Gieseker-stable. Gieseker-semistable sheaves over $\mathbb{P}_n$ are also semistable.*

As in Theorem 1.2.2 one can show that the following statements are equivalent for a torsion-free sheaf E over $\mathbb{P}_n$:

i) E is Gieseker-stable (Gieseker-semistable)

ii) For every coherent subsheaf $F \subset E$, $0 < \operatorname{rk} F < \operatorname{rk} E$, with torsion-free quotient E/F

$$p_F(k) < p_E(k) \qquad (p_F(k) \leq p_E(k)) \qquad \text{for } k \gg 0.$$

iii) For every torsion-free quotient $E \twoheadrightarrow Q$, $0 < \operatorname{rk} Q < \operatorname{rk} E$

$$p_Q(k) > p_E(k) \qquad (p_Q(k) \geq p_E(k)) \qquad \text{for } k \gg 0.$$

Further one can prove the following analog to Theorem 1.2.9.

LEMMA 1.2.13. *Gieseker-stable bundles over $\mathbb{P}_n$ are simple.*

PROOF. Let E be a Gieseker-stable vector bundle of rank r over $\mathbb{P}_n$, $\phi\colon E \to E$ an arbitrary endomorphism, $I = \operatorname{Im} \phi \subset E$ its image. If $0 < \operatorname{rk} I < r$, then since I is both a subsheaf and a quotient of E we have

$$p_E(k) < p_I(k) < p_E(k) \qquad \text{for } k \gg 0.$$

From this contradiction follows: $\operatorname{rk} I = 0$ or $\operatorname{rk} I = r$ — i.e., if $\phi \neq 0$, then ϕ is an isomorphism. As in the proof of 1.2.9 the simplicity of E follows. □

We get the following

COROLLARY. *For holomorphic 2-bundles over $\mathbb{P}_n$ the concepts "stable", "Gieseker-stable" and "simple" are equivalent.*

If E is a non-stable but Gieseker-semistable 2-bundle over $\mathbb{P}_n$, then $E \cong \mathcal{O}_{\mathbb{P}_n}(a) \oplus \mathcal{O}_{\mathbb{P}_n}(a)$ for some $a \in \mathbb{Z}$. We see this as follows: because $c_1(E)$ must necessarily be even, we may assume $c_1(E) = 0$. Let $s \in H^0(\mathbb{P}_n, E)$ be a non-zero section. Since $H^0(\mathbb{P}_n, E(-1)) = 0$ the zero locus Y of s is 2-codimensional or empty. Thus we get an exact sequence

$$0 \to \mathcal{O}_{\mathbb{P}_n} \to E \to J_Y \to 0.$$

Since E is Gieseker-semistable, it follows for $k \gg 0$:

$$\begin{aligned} \chi(\mathcal{O}_{\mathbb{P}_n}(k)) = p_{\mathcal{O}_{\mathbb{P}_n}}(k) \leq p_E(k) \leq p_{J_Y}(k) \\ = \chi(J_Y(k)) = \chi(\mathcal{O}_{\mathbb{P}_n}(k)) - \chi(\mathcal{O}_Y(k)), \end{aligned}$$

i.e.,

$$\chi(\mathcal{O}_Y(k)) \leq 0.$$

If $Y \neq \varnothing$, then we get a contradiction to Theorems A and B. Therefore $Y = \varnothing$ and we have $E = \mathcal{O}_{\mathbb{P}_n} \oplus \mathcal{O}_{\mathbb{P}_n}$.

We close this section with an example of a 3-bundle over $\mathbb{P}_2$ which is Gieseker-stable but not stable.

EXAMPLE (Maruyama). Let E be a stable 2-bundle over $\mathbb{P}_2$ with $c_1(E) = 0$ and $H^1(\mathbb{P}_2, E) \neq 0$. A non-zero element $t \in H^1(\mathbb{P}_2, E)$ defines a nontrivial extension

$$0 \to E \to F \to \mathcal{O}_{\mathbb{P}_2} \to 0$$

with a 3-bundle F over $\mathbb{P}_2$ which has $c_1(F) = 0$. In the associated cohomology sequence

$$0 = H^0(\mathbb{P}_2, E) \to H^0(\mathbb{P}_2, F) \to H^0(\mathbb{P}_2, \mathcal{O}_{\mathbb{P}_2}) \xrightarrow{\delta} H^1(\mathbb{P}_2, E) \to \cdots$$

we have $\delta(1) = t \neq 0$ and thus

$$H^0(\mathbb{P}_2, F) = 0.$$

F is not stable since $H^0(\mathbb{P}_2, F^*) \neq 0$.

Claim. F is Gieseker-stable.

PROOF. 1) Let $F' \subset F$ be a coherent subsheaf of rank 1 with torsion-free quotient F/F'. Then $F' = \mathcal{O}_{\mathbb{P}_2}(a)$ for some $a \in \mathbb{Z}$. Since $H^0(\mathbb{P}_2, F) = 0$, we have $a < 0$. From the formula (*) we have

$$\begin{aligned} p_F(k) - p_{F'}(k) &= (\mu(F) - \mu(F'))k + \text{const} \\ &= -ak + \text{const}, \end{aligned}$$

i.e., $p_{F'}(k) < p_F(k)$ for $k \gg 0$.

2) It remains to consider the case of a torsion-free quotient $F \twoheadrightarrow Q$ of rank 1. We want to show

$$p_F(k) < p_Q(k) \qquad \text{for} \quad k \gg 0.$$

Let $a = c_1(Q)$; then we have

$$(**) \quad p_F(k) - p_Q(k) = -ak + \text{const} = -ak + \left(\frac{1 - h^1(\mathbb{P}_2, E)}{3} - \chi(Q)\right).$$

Now $Q^* \subset F^*$ is a subsheaf of the form $Q^* = \mathcal{O}_{\mathbb{P}_2}(-a)$. Since

$$h^0(\mathbb{P}_2, F^*(-1)) = 0,$$

it follows that $a \leq 0$.

In the case that $a < 0$ it follows from (**) that $p_F(k) < p_Q(k)$ for $k \gg 0$. If $a = 0$, then $Q^* = \mathcal{O}_{\mathbb{P}_2}$. Since $H^0(\mathbb{P}_2, E^*) = 0$, the composition $Q^* \to F^* \twoheadrightarrow E^*$ must vanish. Q^* thus lies in the kernel of the epimorphism $F^* \twoheadrightarrow E^*$. Since this kernel is also trivial, we have $Q^* = \ker(F^* \twoheadrightarrow E^*)$, i.e., the sequence

$$0 \to Q^* \to F^* \to E^* \to 0$$

is exact. By dualizing this sequence one gets $Q = Q^{**}$ and hence $Q = \mathcal{O}_{\mathbb{P}_2}$. If one puts this into formula (**), one has

$$p_F(k) - p_Q(k) = \frac{1 - h^1(E)}{3} - 1 < 0 \qquad (\text{since } h^1(E) \geq 1).$$

Thus we have proved the Gieseker-stability of F! □

REMARK. The dual bundle F^* of F is not Gieseker-stable. Hence Gieseker-stability is not invariant with respect to dualizing, in contrast to stability.

1.3. Examples of stable bundles. In this section we investigate the null-correlation bundle over $\mathbb{P}_3$ and its restriction to planes, the tangent bundle over $\mathbb{P}_n$ and finally 2-bundles which come from 2-codimensional locally complete intersections.

Example A. Let N be a null correlation bundle over $\mathbb{P}_3$, i.e., the kernel of a bundle morphism

$$T_{\mathbb{P}_3}(-1) \to \mathcal{O}_{\mathbb{P}_3}(1).$$

From the exact sequence

$$(*) \qquad 0 \to N \to T_{\mathbb{P}_3}(-1) \to \mathcal{O}_{\mathbb{P}_3}(1) \to 0$$

one gets the Chern classes of N:

$$c_1(N) = 0$$
$$c_2(N) = 1.$$

In the following theorem we summarize the properties of these bundles.

THEOREM 1.3.1. *Let N be a null-correlation bundle over $\mathbb{P}_3$.*

i) *N is stable.*
ii) *The restriction $N|H$ of N to any plane $H \subset \mathbb{P}_3$ is semistable but not stable.*

PROOF. i) In order to prove that N is stable it suffices to show that $h^0(\mathbb{P}_3, N) = 0$. From the cohomology sequence associated with (*) we get

$$0 \to H^0(\mathbb{P}_3, N) \to H^0(\mathbb{P}_3, T_{\mathbb{P}_3}(-1)) \to H^0(\mathbb{P}_3, \mathcal{O}_{\mathbb{P}_3}(1)) \to H^1(\mathbb{P}_3, N) \to \cdots.$$

The sequence dual to (*)

$$0 \to \mathcal{O}_{\mathbb{P}_3}(-1) \to \Omega^1_{\mathbb{P}_3}(1) \to N^* \to 0$$

gives

$$H^1(\mathbb{P}_3, N^*) \cong H^1(\mathbb{P}_3, \Omega^1_{\mathbb{P}_3}(1)) = 0.$$

Now $N \cong N^*$, so we also have $h^1(\mathbb{P}_3, N) = 0$ and from the long exact sequence it then follows that

$$h^0(\mathbb{P}_3, N) = h^0(\mathbb{P}_3, T_{\mathbb{P}_3}(-1)) - h^0(\mathbb{P}_3, \mathcal{O}_{\mathbb{P}_3}(1)) = 0.$$

Thus null-correlation bundles are stable.

ii) Let $H \subset \mathbb{P}_3$ be a plane; $E = N|H$ has Chern classes

$$c_1(E) = 0 \qquad c_2(E) = 1,$$

and thus by Lemma 1.2.7 it cannot be stable. E is however semistable, for the sequence

$$0 \to E(-1) \to T_{\mathbb{P}_3}(-2)|H \to \mathcal{O}_H \to 0$$

gives

$$h^0(H, E(-1)) \leq h^0(H, T_{\mathbb{P}_3}(-2)|H) = h^0(H, T_H(-2) \oplus \mathcal{O}_H(-1)) = 0.$$

We remark that E is not Gieseker-semistable (cf. remarks following the Corollary to Lemma 1.2.13). □

In what follows we shall encounter the null-correlation bundles several times. In particular we shall see that all stable 2-bundles over $\mathbb{P}_3$ with Chern classes $c_1 = 0$, $c_2 = 1$ are null-correlation bundles.

In the next section we shall investigate more generally what happens when semistable 2-bundles are restricted to hyperplanes. A theorem of Barth [12] says: the null-correlation bundles are the only stable 2 -bundles which do not remain stable on some hyperplane.

Example B. Next we investigate the tangent bundle $T_{\mathbb{P}_n}$ over $\mathbb{P}_n$.

THEOREM 1.3.2. *The tangent bundle $T_{\mathbb{P}_n}$ is stable.*

PROOF. It suffices to show that $\Omega^1_{\mathbb{P}_n}(1)$ is stable. Let then $F \subset \Omega^1_{\mathbb{P}_n}(1)$ be a coherent subsheaf with torsion-free quotient and $0 < \operatorname{rk} F < n$. We consider the dual Euler sequence

$$0 \to \Omega^1_{\mathbb{P}_n}(1) \to \mathcal{O}_{\mathbb{P}_n}^{\oplus(n+1)} \to \mathcal{O}_{\mathbb{P}_n}(1) \to 0.$$

Because $\Omega^1_{\mathbb{P}_n}(1)/F$ is torsion-free, there is a line L on which $F|L$ is a vector subbundle of $\Omega^1_{\mathbb{P}_n}(1)|L$ and so that $c_1(F) = c_1(F|L)$. Then $F|L$ is also a subbundle of $\mathcal{O}_L^{\oplus(n+1)}$ and thus we must have $c_1(F) \leq 0$.

If in fact $c_1(F) < 0$ then

$$\mu(F) = \frac{c_1(F)}{\operatorname{rk}(F)} < -\frac{1}{n} = \frac{c_1(\Omega^1_{\mathbb{P}_n}(1))}{n} = \mu(\Omega^1_{\mathbb{P}_n}(1)).$$

Hence if we can show that the case

$$F \subset \Omega^1_{\mathbb{P}_n}(1), \qquad c_1(F) = 0$$

cannot occur, then the stability of $\Omega^1_{\mathbb{P}_n}(1)$ is proved.

We proceed as follows: $\Omega^1_{\mathbb{P}_n}(1)/F$ torsion-free implies $\mathcal{O}_{\mathbb{P}_n}^{\oplus(n+1)}/F$ torsion-free; thus F defines an exact sequence

$$0 \to F \to \mathcal{O}_{\mathbb{P}_n}^{\oplus(n+1)} \to Q \to 0$$

with a torsion-free sheaf Q with $c_1(Q) = -c_1(F)$. If $c_1(F)$ were 0, then Q would be a globally generated torsion-free sheaf with $c_1(Q) = 0$. We show that such sheaves must be trivial. The inclusion $F \subset \Omega^1_{\mathbb{P}_n}(1)$ would then lead to nontrivial sections of $\Omega^1_{\mathbb{P}_n}(1)$, which is impossible. Thus the theorem is proved if we show the following.

LEMMA 1.3.3. *A torsion-free globally generated sheaf Q over $\mathbb{P}_n$ with $c_1(Q) = 0$ is trivial.*

PROOF. We blow up $\mathbb{P}_n$ in a point $x \in \mathbb{P}_n$ in which Q is free. We then get a commutative diagram (notation as in Ch. I, §3.1):

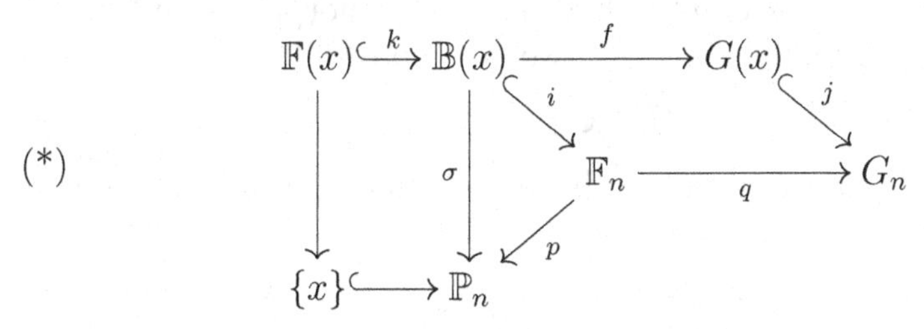

The epimorphism

$$\mathcal{O}_{\mathbb{P}_n}^{\oplus N} \to Q$$

induces a commutative diagram

$$\begin{array}{ccc} \mathcal{O}_{\mathbb{F}_n}^{\oplus N} & \xrightarrow{\;g\;} & p^*Q \\ \| & & \uparrow h \\ q^*q_*\mathcal{O}_{\mathbb{F}_n}^{\oplus N} & \longrightarrow & q^*q_*p^*Q. \end{array}$$

h is an epimorphism since g is. Since Q is torsion-free, $S(Q)$ has codimension at least 2 in $\mathbb{P}_n$; thus the set

$$W = \{\ell \in G_n \mid L \cap S(Q) = \varnothing\}$$

of lines L which do not meet $S(Q)$ is non-empty and Zariski-open.

For every $\ell \in W$ the restriction $Q|L$ is locally free, globally generated and has $c_1(Q|L) = 0$. By the Corollary to Theorem 3.2.1 in Ch. I it follows that $Q|L$ is trivial. Thus q_*p^*Q is a coherent sheaf of rank $r = \operatorname{rk} Q$ over G_n.

Since q is flat q_*p^*Q and thus also $q^*q_*p^*Q$ is torsion-free. The epimorphism of sheaves

$$h\colon q^*q_*p^*Q \to p^*Q$$

is even an isomorphism since its kernel, being a torsion-free sheaf of rank

$$\operatorname{rk}(q^*q_*p^*Q) - \operatorname{rk}(p^*Q) = 0,$$

is zero.

From the diagram (*) one now gets

$$(**) \qquad Q \cong \sigma_*\sigma^*Q \cong \sigma_*(i^*p^*)Q \cong \sigma_*i^*q^*q_*p^*Q \cong \sigma_*f^*(j^*q_*p^*Q).$$

We chose x so that Q is locally free in a neighborhood of x; thus

$$k^*i^*p^*Q$$

is trivial of rank $r = \operatorname{rk} Q$ over $\mathbb{F}(x) = p^{-1}(x)$. The isomorphism h gives

$$\mathcal{O}_{\mathbb{F}(x)}^{\oplus r} \cong k^*i^*p^*Q \cong k^*i^*q^*q_*p^*Q \cong k^*f^*j^*q_*p^*Q \cong (f \circ k)^*(j^*q_*p^*Q).$$

The map $f \circ k \colon \mathbb{F}(x) \to G(x)$ is however an isomorphism and hence $j^*q_*p^*Q$ is trivial. Finally using (**) we get the result

$$Q \cong \mathcal{O}_{\mathbb{P}_n}^{\oplus r}.$$

Thus the lemma is proved and we have seen that the tangent bundle $T_{\mathbb{P}_n}$ of $\mathbb{P}_n$ is stable. □

As we already know, restriction of $T_{\mathbb{P}_n}$ to any hyperplane $H \subset \mathbb{P}_n$ gives the equation

$$T_{\mathbb{P}_n}|H \cong T_H \oplus \mathcal{O}_H(1).$$

Since $\mu(\mathcal{O}_H(1)) = 1 \neq \mu(T_H) = \frac{n}{n-1}$, it follows that $T_{\mathbb{P}_n}|H$ is not semistable.

In Chapter I, §5 we constructed a 2-bundle for every 2-codimensional locally complete intersection $Y \subset \mathbb{P}_n$ if certain conditions on the normal bundle were fulfilled. We now give a geometric criterion which permits one to prove the stability or instability of such bundles.

LEMMA 1.3.4. *Let E be a holomorphic 2-bundle over $\mathbb{P}_n$ which belongs to the 2-codimensional locally complete intersection $Y \subset \mathbb{P}_n$ with* $\det N_{Y/\mathbb{P}_n} = \mathcal{O}_Y(k)$. *$E$ is stable if and only if $k > 0$ and Y lies in no hypersurface of degree $\leq \frac{k}{2}$. If k is even, then E is semistable if and only if $k \geq 0$ and Y lies in no hypersurface of degree $\leq \frac{k}{2} - 1$.*

PROOF. We begin with the exact sequence

$$0 \to \mathcal{O}_{\mathbb{P}_n} \to E \to J_Y(k) \to 0.$$

Let $k \equiv 0 \bmod 2$ (the case $k \equiv 1 \bmod 2$ can be treated completely analogously). Then

$$E_{\text{norm}} = E(-\tfrac{k}{2}).$$

We get the sequence

$$0 \to \mathcal{O}_{\mathbb{P}_n}(-\tfrac{k}{2}) \to E_{\text{norm}} \to J_Y(\tfrac{k}{2}) \to 0$$

and hence the exact sequence

$$0 \to H^0(\mathbb{P}_n, \mathcal{O}_{\mathbb{P}_n}(-\tfrac{k}{2})) \to H^0(\mathbb{P}_n, E_{\text{norm}}) \to H^0(\mathbb{P}_n, J_Y(\tfrac{k}{2})) \to 0,$$

i.e., $H^0(\mathbb{P}_n, E_{\text{norm}})$ vanishes if and only if k is positive and

$$H^0(\mathbb{P}_n, J_Y(\tfrac{k}{2})) = 0.$$

This means that E is stable if and only if $k > 0$ and Y lies in no hypersurface of degree $\leq \frac{k}{2}$. The criterion for semistability follows in the same way. □

We now want to investigate the examples of Chapter I, §5.

Example 1. We had extensions

$$0 \to \mathcal{O}_{\mathbb{P}_2} \to E \to J_Y(k) \to 0$$

with $k = 1, 2$, where Y consisted of $m > 1$ points

$$Y = \{x_1, \ldots, x_m\}.$$

For $k = 1$ the bundle E is always stable and for $k = 2$ it is always semistable and it is stable if and only if not all the points x_i lie on one line.

Example 2. In this example Y consisted of $d > 1$ disjoint lines L_i in $\mathbb{P}_3$. We had

$$\det N_{Y/\mathbb{P}_3} = \mathcal{O}_Y(2), \qquad \textit{i.e.}, \quad k = 2.$$

Since Y cannot lie in any plane, the corresponding bundles are always stable.

Example 3. If Y is a disjoint union of $r > 1$ elliptic curves of degree d_i in $\mathbb{P}_3$, then

$$\det N_{Y/\mathbb{P}_3} = \mathcal{O}_Y(4).$$

The associated bundle is stable if and only if Y does not lie in a hypersurface of degree 2. If for example Y consists of two plane curves

$$Y = C_1 \cup C_2, \qquad C_i \subset H_i \cong \mathbb{P}_2, \qquad i = 1, 2,$$

then Y lies in $D = H_1 \cup H_2$ and is thus not stable but is still semistable.

Example 4. If Y consists of $r > 1$ disjoint conics in $\mathbb{P}_3$, then

$$\det N_{Y/\mathbb{P}_3} = \mathcal{O}_Y(3).$$

The associated bundles are all stable.

1.4. Further results and open questions. The concept of stability due to Mumford and Takemoto, which was introduced in this paragraph, also makes sense for holomorphic vector bundles over projective algebraic manifolds $X \hookrightarrow \mathbb{P}_N$. For a holomorphic vector bundle E over X the degree of E is defined by

$$\deg(E) = (c_1(E) \cup c_1(\mathcal{O}_X(1))^{n-1})[X] \qquad n = \dim X.$$

For $X = \mathbb{P}_N$ we have $\deg(E) = c_1(E)$. With

$$\mu(E) = \frac{\deg(E)}{\operatorname{rk}(E)}$$

we can define stability and semistability as before.

One should remark that this concept of stability depends on the embedding (cf. [122]). From the characterization of stable 2-bundles on $\mathbb{P}_n$ in Lemma 1.2.5 and with the help of the semicontinuity theorem one sees immediately that stability and semistability are Zariski-open properties. This is still true for higher rank and arbitrary projective algebraic base space (cf. Maruyama [83]).

In [89] Mumford developed a general theory of stability. For vector bundles on curves this theory led to the definition above.

Bogomolov has proposed a further concept of stability (cf. [21], [26]). He calls a holomorphic r-bundle E over X *unstable* if there is a representation ρ of $GL(r, \mathbb{C})$ with determinant 1 such that the induced bundle $E^{(\rho)}$ has a nontrivial holomorphic section with zeros. For $X = \mathbb{P}_n$ this is the same as requiring the existence of a coherent subsheaf $F \subset E$, $0 < \operatorname{rk} F < \operatorname{rk} E$, with torsion-free quotient $Q = E/F$ such that

$$\mu(F) > \mu(E)$$

or $$\mu(F) = \mu(E) \qquad \text{and} \qquad F|\mathbb{P}_n \setminus S(Q)$$

cannot be extended to a subbundle of E (cf. [99]).

In particular one sees that a non-semistable bundle in the sense of Mumford is unstable in the sense of Bogomolov. Bogomolov shows [21] that a 2-bundle over a surface with $c_1^2 - 4c_2 > 0$ must be unstable. As a consequence of this the Chern numbers of a surface of general type must satisfy the inequality

$$c_1^2 \leq 4c_2$$

(cf. [21], [135]).

For vector bundles on curves all the concepts mentioned here coincide. The question which stability (resp. semistability) concept in the higher dimensional case is the right one depends largely on the question being asked. For example the semistability concept of Gieseker seems to be the best one if one is seeking "good" compactifications of the moduli spaces of stable bundles by means of semistable sheaves.

For r-bundles over $\mathbb{P}_n$, $r \geq 4$, stability is difficult to verify in concrete cases. For example it seems to be unknown whether the symmetric powers $S^m T_{\mathbb{P}_n}$ are stable (for $m = 2$, $n = 2$ this is true).

In closing we formulate the following problem.

PROBLEM 1.4.1. What topological conditions hold for stable (resp. semistable) r-bundles over $\mathbb{P}_n$?

For $r = 2$, $n = 2$ one has the answer of Schwarzenberger (Lemma 1.2.7)

$$c_1^2 - 4c_2 < 0$$

resp.

$$c_1^2 - 4c_2 \leq 0.$$

We shall have more to say about this in the next paragraph.

§2. The splitting behavior of stable bundles

We begin this paragraph with the construction of subsheaves in holomorphic vector bundles. We begin by showing: if E is a holomorphic vector bundle over $\mathbb{P}_n$ with generic splitting type $\underline{a}_E = (a_1, \ldots, a_r)$, $a_1 \geq \cdots \geq a_r$, and if one of the differences $a_s - a_{s+1}$ is greater than 1, then E contains a normal subsheaf F which on generic lines splits in the form

$$F|L = \bigoplus_{i=1}^{s} \mathcal{O}_L(a_i).$$

As an immediate application of this theorem we get a characterization of the generic splitting type of semistable bundles: if a semistable bundle E over $\mathbb{P}_n$ has the splitting type $\underline{a}_E = (a_1, \ldots, a_r)$ then $a_i - a_{i+1} \leq 1$ for all $i = 1, \ldots, r-1$.

For $r = 2$ this is the "classical" theorem of Grauert–Mülich. Just as immediate is the deduction that for a uniform n-bundle E over $\mathbb{P}_n$ of type $\underline{a}_E = (a_1, \ldots, a_n)$ we have $a_i - a_{i+1} \leq 1$ unless E splits. We use this latter result to derive a theorem of Van de Ven, which says that the tangent bundle of $\mathbb{P}_2$ is up to tensoring with line bundles the only non-splitting uniform 2-bundle over $\mathbb{P}_2$.

We then go on to show how for a 2-bundle E over $\mathbb{P}_n$ of type $\underline{a}_E = (0, 0)$ one can regard the set $S_E \subset G_n$ of jump lines as the support of a divisor of degree $c_2(E)$ in the Grassmann manifold of lines in $\mathbb{P}_n$. As a last application of the theorem of Grauert and Mülich we investigate the behavior of semistable 2-bundles under restriction to hyperplanes.

2.1. Construction of subsheaves. To examine properties of stable bundles it is important to be able to construct subsheaves of such bundles. We want to give a condition which guarantees that for an r-bundle E with generic splitting type $\underline{a}_E = (a_1, \dots, a_r)$, $a_1 \geq \cdots \geq a_r$, one can find a subsheaf of generic splitting type $(a_1, \dots, a_s)$. To this end we must do some preparatory work.

Let X be an n-dimensional complex manifold, E a holomorphic vector bundle of rank r over X. For s with $0 \leq s \leq r$ let

$$\pi\colon \mathrm{Gr}_s(E) \to X$$

be the holomorphic Grassmann bundle of s-dimensional linear subspaces of E; the fibre over x,

$$\pi^{-1}(x) = \mathrm{Gr}_s(E(x)),$$

is the Grassmann manifold of s-dimensional subspaces in $E(x)$.

Over $\mathrm{Gr}_s(E)$ we have the universal bundle sequence

$$0 \to F_E \to \pi^* E \to Q_E \to 0 \tag{1}$$

with the tautological bundle $F_E \subset \pi^* E$ whose total space is

$$F_E = \{(V, v) \in \mathrm{Gr}_s(E) \times E \mid v \in V\}.$$

For $s = 1$ the sequence (1) is just the relative Euler sequence over $\mathbb{P}(E)$. The differential

$$d\pi\colon T_{\mathrm{Gr}_s}(E) \to \pi^* T_X$$

is epimorphic; its kernel is the relative tangent bundle $T_{\mathrm{Gr}_s(E)/X}$. We have

LEMMA 2.1.1. *The relative tangent bundle is*

$$T_{\mathrm{Gr}_s(E)/X} \cong \mathcal{H}om(F_E, Q_E).$$

In general if $f\colon X \to Y$ is a surjective holomorphic submersion, then we call

$$T_{X/Y} = \ker(df\colon T_X \to f^* T_Y)$$

the relative tangent bundle.

The following Descente-Lemma will play an important rôle in the construction of subsheaves:

LEMMA 2.1.2. *Let X, Y be complex manifolds, $f\colon X \to Y$ a surjective holomorphic submersion with connected fibres, E a holomorphic r-bundle over Y. Let $\tilde{F} \subset f^* E$ be a subbundle of rank s in $f^* E$, $\tilde{Q} = f^* E / \tilde{F}$ its quotient. If*

$$\mathrm{Hom}(T_{X/Y}, \mathcal{H}om(\tilde{F}, \tilde{Q})) = 0,$$

then $\tilde{F}$ is of the form $\tilde{F} = f^*F$ for some holomorphic subbundle $F \subset E$ of rank s.

PROOF. The subbundle $\tilde{F} \subset f^*E$ defines a holomorphic section

$$g\colon X \to \mathrm{Gr}_s(f^*E) = X \times_Y \mathrm{Gr}_s(E)$$

in $\mathrm{Gr}_s(f^*E)$. The composition ϕ of g with

$$X \times_Y \mathrm{Gr}_s(E) \to \mathrm{Gr}_s(E)$$

gives a commutative diagram

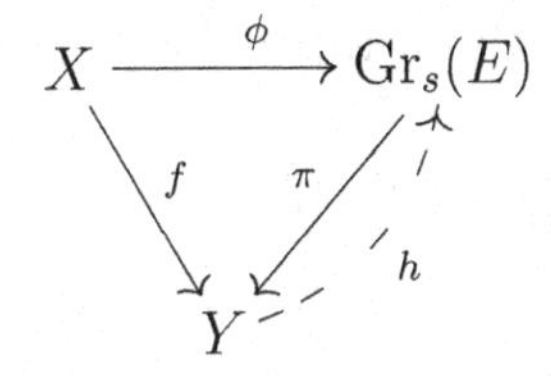

We wish to find a section $h\colon Y \to \mathrm{Gr}_s(E)$ with $\phi = h \circ f$. Then for the subbundle $F \subset E$ defined by h we have

$$f^*F = \phi^*F_E = \tilde{F},$$

and we have found the bundle we seek.

In order to construct $h\colon Y \to \mathrm{Gr}_s(E)$ we consider

$$d\phi\colon T_{X/Y} \to \phi^*T_{\mathrm{Gr}_s(E)/Y}.$$

By Lemma 2.1.1 we have

$$\begin{aligned}\phi^*T_{\mathrm{Gr}_s(E)/Y} &= \phi^*\mathcal{H}om(F_E, Q_E) = \mathcal{H}om(\phi^*F_E, \phi^*Q_E)\\ &= \mathcal{H}om(\tilde{F}, \tilde{Q}).\end{aligned}$$

Hence by assumption

$$d\phi \in \mathrm{Hom}(T_{X/Y}, \mathcal{H}om(\tilde{F}, \tilde{Q})) = 0$$

is zero, i.e., ϕ is constant on the fibres of f. Thus ϕ factors through f — that is there is a map

$$h\colon Y \to \mathrm{Gr}_s(E)$$

with $h \circ f = \phi$. h is holomorphic, because as a surjective holomorphic submersion f has local holomorphic sections. □

At this point we prove the following for later reference.

LEMMA 2.1.3. *Let $\mathbb{F}_n = \{(x, \ell) \in \mathbb{P}_n \times G_n \mid x \in L\}$ be the flag manifold, $\tilde{L} = q^{-1}(\ell) \subset \mathbb{F}_n$ for $\ell \in G_n$. For the relative tangent bundle $T_{\mathbb{F}_n/\mathbb{P}_n}$ we have*

$$T_{\mathbb{F}_n/\mathbb{P}_n}|\tilde{L} = \mathcal{O}_{\tilde{L}}(-1)^{\oplus(n-1)}.$$

PROOF. Over $\mathbb{F}_n = \mathbb{P}(T_{\mathbb{P}_n})$ we have the universal sequence

$$(*) \qquad 0 \to F_{T_{\mathbb{P}_n}} \to p^* T_{\mathbb{P}_n} \to Q_{T_{\mathbb{P}_n}} \to 0.$$

From Lemma 2.1.1 it follows that

$$T_{\mathbb{F}_n/\mathbb{P}_n} = \mathcal{H}om(F_{T_{\mathbb{P}_n}}, Q_{T_{\mathbb{P}_n}}),$$

and thus

$$T_{\mathbb{F}_n/\mathbb{P}_n}|\tilde{L} = \mathcal{H}om(F_{T_{\mathbb{P}_n}}|\tilde{L}, Q_{T_{\mathbb{P}_n}}|\tilde{L}).$$

If we identify $\tilde{L}$ by means of p with the line $L \subset \mathbb{P}_n$ determined by ℓ, then we see that over the q-fibre the sequence (*) restricts to the sequence

$$0 \to T_L \to T_{\mathbb{P}_n}|L \to N_{L/\mathbb{P}_n} \to 0.$$

Thus we have

$$T_{\mathbb{F}_n/\mathbb{P}_n}|\tilde{L} \cong \mathcal{H}om(\mathcal{O}_L(2), \mathcal{O}_L(1)^{\oplus(n-1)}) = \mathcal{O}_L(-1)^{\oplus(n-1)}. \qquad \square$$

We now come to the construction of subsheaves. Let E be a holomorphic r-bundle over $\mathbb{P}_n$ with generic splitting type

$$\underline{a}_E = (a_1, \dots, a_r), \qquad a_1 \geq \dots \geq a_r.$$

Further suppose that

$$a_s = 0, \quad a_{s+1} < 0 \qquad \text{for some} \quad s < r.$$

Since we want to examine the restriction of E to lines, we consider the standard diagram (Chap. I, §3.1):

$$\begin{array}{ccc} \mathbb{F}_n & \xrightarrow{q} & G_n \\ {\scriptstyle p}\downarrow & & \\ \mathbb{P}_n & & \end{array}$$

Here $\mathbb{F}_n = \mathbb{P}(T_{\mathbb{P}_n})$; G_n is the Grassmann manifold of lines in $\mathbb{P}_n$. Let $S_E \subset \mathbb{P}_n$ be the set of jump lines.

$$U_E = G_n \setminus S_E$$

is a non-empty Zariski-open set in G_n; for every point $\ell \in U_E$ we have

$$p^*E|q^{-1}(\ell) \cong E|L \cong \bigoplus_{i=1}^{r} \mathcal{O}_L(a_i).$$

q_*p^*E is a coherent sheaf over G_n which over U_E is locally free. The morphism

$$\Phi\colon q^*q_*p^*E \to p^*E$$

on the line $\tilde{L} = q^{-1}(\ell) \cong L$ for an $\ell \in U_E$ is given by the evaluation of the sections of $E|L$, i.e., the following diagram commutes:

$$\begin{array}{ccc} q^*q_*p^*E|\tilde{L} & \xrightarrow{\Phi|\tilde{L}} & p^*E|\tilde{L} \\ \downarrow \wr & & \downarrow \wr \\ H^0(L, E|L) \otimes \mathcal{O}_L & \xrightarrow{\text{ev}} & E|L \end{array}$$

Thus the image of $\Phi|\tilde{L}$ is the subbundle

$$\bigoplus_{i=1}^{s} \mathcal{O}_L(a_i) \subset E|L$$

of rank s. Hence over the open set $q^{-1}(U_E)$

$$\Phi\colon q^*q_*p^*E \to p^*E$$

is a morphism of constant rank s and thus its image $\operatorname{Im}\Phi \subset p^*E$ is over $q^{-1}(U_E)$ a subbundle of rank s.

Let

$$Q' = p^*E/\operatorname{Im}\Phi,$$

$T(Q')$ the torsion subsheaf of Q' and

$$\tilde{F} = \ker(p^*E \to Q'/T(Q')).$$

We thus obtain an exact sequence of sheaves

$$0 \to \tilde{F} \to p^*E \to \tilde{Q} \to 0$$

with a torsion-free sheaf $\tilde{Q} = Q'/T(Q')$. Outside the singularity set $S(\tilde{Q})$ of $\tilde{Q}$

$$\tilde{F} \subset p^*E$$

is even a subbundle.

Let $X = \mathbb{F}_n \setminus S(\tilde{Q})$. X is open in $\mathbb{F}_n$ and contains $q^{-1}(U_E)$. The projection $p\colon \mathbb{F}_n \to \mathbb{P}_n$ defines a commutative diagram

$$\begin{array}{ccc} X & \overset{i}{\hookrightarrow} & \mathbb{F}_n \\ \downarrow f & & \downarrow p \\ Y = p(X) & \overset{j}{\hookrightarrow} & \mathbb{P}_n \end{array}$$

with a surjective holomorphic submersion f with connected fibres.

To apply the Descente-Lemma to the subbundle

$$\tilde{F}|X \subset f^*(E|Y)$$

we must investigate

$$\operatorname{Hom}(T_{X/Y}, \mathcal{H}om(\tilde{F}|X, \tilde{Q}|X)).$$

Claim. If $a_{s+1} < -1$, then

$$\operatorname{Hom}(T_{X/Y}, \mathcal{H}om(\tilde{F}|X, \tilde{Q}|X))$$

vanishes.

PROOF. Let $\Omega^1_{X/Y} = T^*_{X/Y}$. Then $\Omega^1_{X/Y} = \Omega^1_{\mathbb{F}_n/\mathbb{P}_n}|X$ and

$$\operatorname{Hom}(T_{X/Y}, \mathcal{H}om(\tilde{F}|X, \tilde{Q}|X)) = H^0(X, \Omega^1_{\mathbb{F}_n/\mathbb{P}_n} \otimes \tilde{F}^* \otimes \tilde{Q}).$$

Since the restriction

$$H^0(X, \Omega^1_{\mathbb{F}_n/\mathbb{P}_n} \otimes \tilde{F}^* \otimes \tilde{Q}) \to H^0(q^{-1}(U_E), \Omega^1_{\mathbb{F}_n/\mathbb{P}_n} \otimes \tilde{F}^* \otimes \tilde{Q})$$

is injective, it suffices to show that

$$\Omega^1_{\mathbb{F}_n/\mathbb{P}_n} \otimes \tilde{F}^* \otimes \tilde{Q}$$

has no sections over $q^{-1}(U_E)$.

Let $\ell \in U_E$, $\tilde{L} = q^{-1}(\ell)$, $L \subset \mathbb{P}_n$ the line determined by ℓ. Then

$$\tilde{F}^*|\tilde{L} \cong \bigoplus_{i=1}^{s} \mathcal{O}_L(-a_i), \qquad \tilde{Q}|\tilde{L} \cong \bigoplus_{j=s+1}^{r} \mathcal{O}_L(a_j)$$

and by Lemma 2.1.3

$$\Omega^1_{\mathbb{F}_n/\mathbb{P}_n}|\tilde{L} \cong \mathcal{O}_L(1)^{\oplus(n-1)}$$

and thus

$$H^0(\tilde{L}, \Omega^1_{\mathbb{F}_n/\mathbb{P}_n} \otimes \tilde{F}^* \otimes \tilde{Q}|\tilde{L}) = 0 \qquad \text{if} \quad a_{s+1} < -1. \qquad \square$$

Let then E be a holomorphic r-bundle over $\mathbb{P}_n$ of type

$$\underline{a}_E = (a_1, \dots, a_r), \qquad a_1 \geq \dots \geq a_r$$
$$a_s = 0, \qquad a_{s+1} < -1;$$

then every section of $(\Omega^1_{\mathbb{F}_n/\mathbb{P}_n} \otimes \tilde{F}^* \otimes \tilde{Q})|X$ is zero over $q^{-1}(U_E)$ and hence over X. The hypothesis of the Descente-Lemma is satisfied and over the open set $Y \subset \mathbb{P}_n$ we get a subbundle

$$F' \subset E|Y$$

with $f^*F' = \tilde{F}|X \subset f^*(E|Y)$.

F' can be extended to a normal subsheaf F in E on all of $\mathbb{P}_n$: because $S(\tilde{Q})$ has codimension at least 2 and $\tilde{F}$ is a normal sheaf, we have for the inclusion $i\colon X \hookrightarrow \mathbb{F}_n$

$$i_* i^* \tilde{F} \cong \tilde{F}.$$

From the diagram

$$\begin{array}{ccc} X & \xhookrightarrow{i} & \mathbb{F}_n \\ {\scriptstyle f}\downarrow & & \downarrow{\scriptstyle p} \\ Y & \xhookrightarrow{j} & \mathbb{P}_n \end{array}$$

we deduce that

$$f_*\mathcal{O}_X = j^*j_*f_*\mathcal{O}_X = j^*p_*i_*\mathcal{O}_X = j^*p_*\mathcal{O}_{\mathbb{F}_n} = j^*\mathcal{O}_{\mathbb{P}_n} = \mathcal{O}_Y,$$

and thus

$$j_*F' = j_*(F' \otimes f_*\mathcal{O}_X) = j_*f_*f^*F' = p_*i_*(f^*F') = p_*i_*(\tilde{F}|X) = p_*\tilde{F}.$$

The extension

$$F = j_*F' = p_*\tilde{F}$$

is thus a normal coherent subsheaf of E with $F|Y = F'$. We summarize what we have just said in the following theorem.

THEOREM 2.1.4. *Let E be a holomorphic r-bundle over $\mathbb{P}_n$ of type $\underline{a}_E = (a_1, \ldots, a_r)$, $a_1 \geq \cdots \geq a_r$. If*

$$a_s - a_{s+1} \geq 2 \qquad \text{for some} \quad s < r,$$

then there is a normal subsheaf $F \subset E$ in E of rank s with the following properties: over the open set $V_E = p(q^{-1}(U_E)) \subset \mathbb{P}_n$ the sheaf F is a subbundle of E, which on the line $L \subset \mathbb{P}_n$ given by $\ell \in U_E$ has the form

$$F|L \cong \bigoplus_{i=1}^{s} \mathcal{O}_L(a_i).$$

PROOF. We can apply the construction described above to the bundle $E(-a_s)$ and obtain from the corresponding normal subsheaf $F \subset E(-a_s)$ by tensoring with $\mathcal{O}_{\mathbb{P}_n}(a_s)$ the desired sheaf. □

This theorem has far reaching consequences. We give first a series of immediate deductions.

COROLLARY 1. *For a semistable r-bundle E over $\mathbb{P}_n$ of type $\underline{a}_E = (a_1, \ldots, a_r)$, $a_1 \geq \cdots \geq a_r$, we have*

$$a_i - a_{i+1} \leq 1 \qquad \text{for} \quad i = 1, \ldots, r-1.$$

PROOF. If for some $s < r$ we had $a_s - a_{s+1} \geq 2$, then we could find a normal subsheaf $F \subset E$ which is of the form

$$F|L \cong \bigoplus_{i=1}^{s} \mathcal{O}_L(a_i)$$

over the general line $L \subset \mathbb{P}_n$. Then we would have

$$\mu(E) < \mu(F).$$

□

In particular we get the theorem of Grauert and Mülich:

COROLLARY 2 (Grauert–Mülich). *The splitting type of a semistable normalized 2-bundle E over $\mathbb{P}_n$ is*

$$\underline{a}_E = \begin{cases} (0,0) & \text{if } c_1(E) = 0 \\ (0,-1) & \text{if } c_1(E) = -1. \end{cases}$$

Just as simply one has

COROLLARY 3. *For a uniform n-bundle E over $\mathbb{P}_n$ of type*

$$\underline{a}_E = (a_1, \dots, a_n), \qquad a_1 \geq \cdots \geq a_n,$$

which does not split we have

$$a_i - a_{i+1} \leq 1 \qquad \text{for} \quad i = 1, \dots, n-1.$$

PROOF. If for some $s < n$ we had $a_s - a_{s+1} \geq 2$, then (because $V_E = \mathbb{P}_n$) there would be a uniform subbundle $F \subset E$ of type $\underline{a}_F = (a_1, \dots, a_s)$. The quotient bundle $Q = E/F$ would then be uniform of type $(a_{s+1}, \dots, a_n)$. According to Ch. I, §3.2 the bundles F and Q must be direct sums of line bundles. The exact sequence

$$0 \to F \to E \to Q \to 0$$

would therefore split and hence E too would be a direct sum of line bundles contrary to hypothesis. □

REMARK 2.1.5. According to Corollary 1 there is for a fixed rank r a finite number of possibilities for the generic splitting type of a normalized semistable r-bundle. For semistable normalized 2-bundles the generic splitting type is either $(0,0)$ or $(0,-1)$ according to whether $c_1 = 0$ or $c_1 = -1$.

A semistable 3-bundle over $\mathbb{P}_n$ which is so normalized that $c_1 = 0$, -1 or -2 can only have a generic splitting type from one of the following four possibilities:

$$\begin{array}{lll} (0,0,0), \quad (1,0,-1) & \text{if} & c_1 = 0 \\ \qquad (0,0,-1) & \text{if} & c_1 = -1 \\ \qquad (0,-1,-1) & \text{if} & c_1 = -2 \end{array}$$

We have already encountered semistable and stable 2-bundles over $\mathbb{P}_2$ of the types $(0,0)$ and $(0,-1)$. If E is a (semi)stable 2-bundle over $\mathbb{P}_n$, then with the help of the sequences

$$0 \to \det E \to E \otimes E \to S^2 E \to 0$$

and

$$0 \to \det E^* \to E^* \otimes E^* \to S^2 E^* \to 0$$

one sees that S^2E is a (semi)stable 3-bundle over $\mathbb{P}_n$ of type $(0,0,0)$ in case $c_1(E)=0$ and of type $(0,-1,-2)$ in case $c_1(E)=-1$. After normalizing one thus has the splitting type

$$(0,0,0), \quad (1,0,-1)$$

realized by stable bundles over $\mathbb{P}_2$.

With the method used in Ch. I, §6.4 one can construct stable 3-bundles E over $\mathbb{P}_2$ with $c_1=-1$. These bundles are constructed as extensions

$$0 \to \mathcal{O}_{\mathbb{P}_2} \oplus \mathcal{O}_{\mathbb{P}_2} \to E(1) \to \mathcal{J}_Y(2) \to 0$$

and have the type $(0,0,-1)$. If E is such a bundle, then $E^*(-1)$ is of type $(0,-1,-1)$. Thus over $\mathbb{P}_2$ all possible generic splitting types of semistable 3-bundles actually occur.

2.2. Applications of the theorem of Grauert and Mülich. The theorem of Grauert and Mülich asserts that normalized semistable 2-bundles are of type

$$(0,0) \quad \text{or} \quad (0,-1)$$

depending on their first Chern class. For $c_1=0$ we also have the converse of this statement.

LEMMA 2.2.1. *A 2-bundle E over $\mathbb{P}_n$ which splits generically in the form*

$$E|L \cong \mathcal{O}_L \oplus \mathcal{O}_L$$

is semistable.

PROOF. On generic lines L we have

$$h^0(L, E(-k)|L) = 0 \qquad \text{for} \quad k \geq 1.$$

If one now chooses a flag

$$L = L_1 \subset L_2 \subset \cdots \subset L_n = \mathbb{P}_n$$

of linear subspaces L_i of dimension i between L and $\mathbb{P}_n$, then from the cohomology sequence of

$$0 \to E(-k-1)|L_i \to E(-k)|L_i \to E(-k)|L_{i-1} \to 0$$

it follows by induction over i that

$$h^0(L_i, E(-k)|L_i) = 0 \qquad \text{for} \quad k \geq 1, \quad 1 \leq i \leq n.$$

In particular then $h^0(\mathbb{P}_n, E(-1)) = 0$, i.e., E is semistable. □

REMARK. More generally we have: an r-bundle E over $\mathbb{P}_n$ which is trivial on generic lines is semistable.

PROOF. If $F \subset E$ is a coherent subsheaf of rank s with torsion-free quotient E/F, then on a generic line $L \subset \mathbb{P}_n$

$$F|L = \mathcal{O}_L(a_i) \oplus \cdots \oplus \mathcal{O}_L(a_s) \subset E|L = \mathcal{O}_L^{\oplus r}$$

and

$$\mu(F) = \frac{a_1 + \cdots + a_s}{s}.$$

Thus $\mu(F) \leq 0 = \mu(E)$, because we must have $a_i \leq 0$. $\square$

This proof is simpler than the one given for the special case $r = 2$ in Lemma 2.2.1 but the argument used in the proof of 2.2.1 will be needed later (2.2.5).

For $c_1 = -1$ there are 2-bundles over $\mathbb{P}_2$ of generic type $(0, -1)$ which are not semistable. A trivial example is $E = \mathcal{O}_{\mathbb{P}_2} \oplus \mathcal{O}_{\mathbb{P}_2}(-1)$. Examples of such bundles which are in addition indecomposable can be obtained with the methods of Chapter I, §5 by choosing m simple points

$$Y = \{x_1, \dots, x_m\}$$

in $\mathbb{P}_2$ and considering extensions

$$0 \to \mathcal{O}_{\mathbb{P}_2} \to E \to J_Y(-1) \to 0.$$

Such extensions have the type $(0, -1)$; but since they have non-trivial sections, they are not semistable.

As a first application of the theorem of Grauert and Mülich we prove a theorem of Van de Ven.

THEOREM 2.2.2 (Van de Ven). *A uniform 2-bundle over $\mathbb{P}_n$ which does not split has the form $T_{\mathbb{P}_2}(a)$.*

PROOF. We already know (Ch. I., §3.2) that all uniform 2-bundles over $\mathbb{P}_n$ split if $n > 2$. Therefore let E be an indecomposable normalized uniform 2-bundle over $\mathbb{P}_2$. By Corollary 3 of Theorem 2.1.4 the bundle E restricted to any line L has the form

$$E|L = \begin{cases} \mathcal{O}_L \oplus \mathcal{O}_L & \text{if } c_1(E) = 0 \\ \mathcal{O}_L \oplus \mathcal{O}_L(-1) & \text{if } c_1(E) = -1. \end{cases}$$

In the former case $(c_1(E) = 0)$ it follows from the Theorem 3.2.1 in Chapter I that E is trivial. We are thus finished, if we can show the following.

Claim. The only indecomposable uniform 2-bundle over $\mathbb{P}_2$ of type $(0, -1)$ is $\Omega^1_{\mathbb{P}_2}(1)$.

In order to prove this claim we consider again the standard diagram:

$$\begin{array}{ccc} \mathbb{F}_2 & \xrightarrow{q} & \mathbb{P}_2^* = G_2 \\ {\scriptstyle p}\downarrow & & \\ \mathbb{P}_2 & & \end{array}$$

q_*p^*E is a line bundle over $\mathbb{P}_2$ — say

$$q_*p^*E \cong \mathcal{O}_{\mathbb{P}_2^*}(-m).$$

By pulling bundles back to the flag manifold we get an exact sequence of bundles

$$(*) \qquad 0 \to q^*\mathcal{O}_{\mathbb{P}_2^*}(-m) \to p^*E \to L \to 0.$$

By restricting to the p- and q-fibres one sees that

$$L \cong p^*\mathcal{O}_{\mathbb{P}_2}(-1) \otimes q^*\mathcal{O}_{\mathbb{P}_2^*}(m).$$

We calculate m.

Let
$$X = p^*(c_1(\mathcal{O}_{\mathbb{P}_2}(1))) \in H^2(\mathbb{F}_2, \mathbb{Z})$$
$$Y = q^*(c_1(\mathcal{O}_{\mathbb{P}_2^*}(1))) \in H^2(\mathbb{F}_2, \mathbb{Z}).$$

Using the Theorem of Leray and Hirsch one can compute the cohomology ring of $\mathbb{F}_2$ (cf. [31]); we have

$$H^*(\mathbb{F}_2, \mathbb{Z}) = \mathbb{Z}[X, Y](X^3, Y^3, X^2Y - Y^2X, XY - X^2 - Y^2).$$

From the sequence (*) it therefore follows that

$$\begin{aligned} c_2(E)X^2 = p^*c_2(E) &= c_2(p^*E) \\ &= -mY(mY - X) \\ &= (m - m^2)Y^2 + mX^2, \end{aligned}$$

i.e.,
$$(c_2(E) - m)X^2 = -m(m-1)Y^2.$$

This can only be the case if

$$c_2(E) = m \in \{0, 1\}.$$

The case $m = 0$ is not possible, because then under projection onto $\mathbb{P}_2$ the sequence (*) would become

$$0 \to \mathcal{O}_{\mathbb{P}_2} \to E \to \mathcal{O}_{\mathbb{P}_2}(-1) \to 0$$

and thus E would be decomposable. Hence $m = 1$.

Because the image sheaves $p_*q^*\mathcal{O}_{\mathbb{P}_2^*}(-1)$, $R^1p_*q^*\mathcal{O}_{\mathbb{P}_2^*}(-1)$ vanish, projecting (*) onto $\mathbb{P}_2$ gives an isomorphism

$$E \cong p_*(p^*\mathcal{O}_{\mathbb{P}_2}(-1) \otimes q^*\mathcal{O}_{\mathbb{P}_2^*}(1)).$$

Thus there is at most one uniform indecomposable 2-bundle of type $(0,-1)$ over $\mathbb{P}_2$. Because $\Omega^1_{\mathbb{P}_2}(1)$ is such a bundle the claim — and hence also the theorem of Van de Ven — is proved. □

As a second application of the theorem of Grauert and Mülich we show that the set $S_E \subset G_n$ of jump lines of a 2-bundle E of type $\underline{a}_E = (0,0)$ can be regarded in a canonical fashion as the support of a divisor D_E of degree $c_2(E)$. Let E be a 2-bundle over $\mathbb{P}_n$ of type $(0,0)$. E is then necessarily semistable; a line $L \subset \mathbb{P}_n$ is a jump line of E if and only if

$$h^0(L, E(-1)|L) > 0.$$

$H^0(L, E(-1)|L)$ is dual to $H^1(L, E(-1)|L)$; thus jump lines are also characterized by

$$h^1(L, E(-1)|L) > 0.$$

We consider the standard diagram

$$\begin{array}{ccc} \mathbb{F}_n & \xrightarrow{q} & G_n \\ {\scriptstyle p}\downarrow & & \\ \mathbb{P}_n & & \end{array}$$

and form the sheaf

$$R^1q_*p^*(E(-1)).$$

Over $U_E = G_n \setminus S_E$ the sheaf $R^1q_*p^*(E(-1))$ vanishes.

Claim. S_E is the support of the sheaf $R^1q_*p^*(E(-1))$.

Proof. Let $\ell \in G_n$, $U \subset G_n$ a Stein open neighborhood of ℓ in G_n with

$$q^{-1}(U) \cong U \times \tilde{L};$$

let $J_{\tilde{L}} \subset \mathcal{O}_{q^{-1}(U)}$ be the sheaf of ideals of $\tilde{L} \subset q^{-1}(U)$. Because $U \times \tilde{L}$ can be covered by two open Stein sets

$$H^2(q^{-1}(U), p^*E(-1) \otimes J_{\tilde{L}})$$

vanishes; thus the restriction map

$$H^1(q^{-1}(U), p^*E(-1)) \to H^1(\tilde{L}, p^*E(-1)|\tilde{L})$$

is surjective and hence also the canonical base-change homomorphism

$$(R^1q_*p^*E(-1))(\ell) \to H^1(L, E(-1)|L).$$

Therefore we have

$$\operatorname{supp} R^1q_*p^*E(-1) = S_E. \qquad \square$$

With the help of the sheaf

$$F = R^1 q_* p^* E(-1)$$

we provide S_E with a complex structure. To this end we choose a resolution

$$0 \to K \to \bigoplus_{i=1}^{N} \mathcal{O}_{\mathbb{P}_n}(k_i) \to E(-1) \to 0$$

with $k_i < 0$ (after twisting with $\mathcal{O}_{\mathbb{P}_n}(k)$ for suitable $k > 0$ the bundle $E(-1)$ becomes globally generated!) and we project the sequence

$$0 \to p^* K \to \bigoplus_i p^* \mathcal{O}_{\mathbb{P}_n}(k_i) \to p^* E(-1) \to 0$$

onto G_n. The result is an exact sequence of sheaves

$$0 \to R^1 q_* p^* K \xrightarrow{h} \bigoplus_i R^1 q_* p^* \mathcal{O}_{\mathbb{P}_n}(k_i) \to F \to 0,$$

for $q_* p^* E(-1)$ is torsion-free and zero outside of S_E while $R^2 q_* p^* K$ vanishes, since the q-fibres are one-dimensional.

The sheaves $R^1 q_* p^* \mathcal{O}_{\mathbb{P}_n}(k_i)$ and $R^1 q_* p^* K$ are locally free because of the base-change theorem, for $h^1(L, \mathcal{O}_L(k_i))$ and

$$h^1(L, K|L) = -\chi(L, K|L)$$

do not depend on L. We let

$$E_1 = R^1 q_* p^* K, \qquad E_2 = \bigoplus_i R^1 q_* p^* \mathcal{O}_{\mathbb{P}_n}(k_i).$$

$h\colon E_1 \hookrightarrow E_2$ is then a sheaf homomorphism between locally free sheaves of the same rank with cokernel F.

Let $\operatorname{Im}(\det h) \subset \det E_2$ be the image of the determinant

$$\det h\colon \det E_1 \hookrightarrow \det E_2.$$

Then

$$J_h = \operatorname{Im}(\det h) \otimes \det E_2^* \subset \mathcal{O}_{G_n}$$

is an invertible sheaf of ideals with

$$\operatorname{supp} \mathcal{O}_{G_n}/J_h = S_E.$$

We now define the divisor D_E by

$$D_E = (S_E, \mathcal{O}_{G_n}/J_h).$$

The line bundle defined by D_E is

$$[D_E] = J_h^* = \det E_2 \otimes \det E_1^*;$$

thus

$$\deg D_E = c_1(E_2) - c_1(E_1).$$

We must show that J_h is independent of the choice of resolution: suppose $J_h \subset \mathcal{O}_{G_n}$ is given in the point $\ell \in G_n$ by the equation

$$\det h = p_1^{k_1} \dots p_m^{k_m}$$

with distinct irreducible factors $p_i \in \mathcal{O}_{G_n,\ell}$. The elements p_i determine the prime factorization of the germ of S_E in $\ell \in G_n$ and are thus independent of the resolution.

In order to see that the multiplicity k_i with which p_i occurs is also independent of the choice of resolution we choose a point ℓ' near ℓ in which the analytic set defined by p_i is smooth and D_E is given by $p_i^{k_i}$.

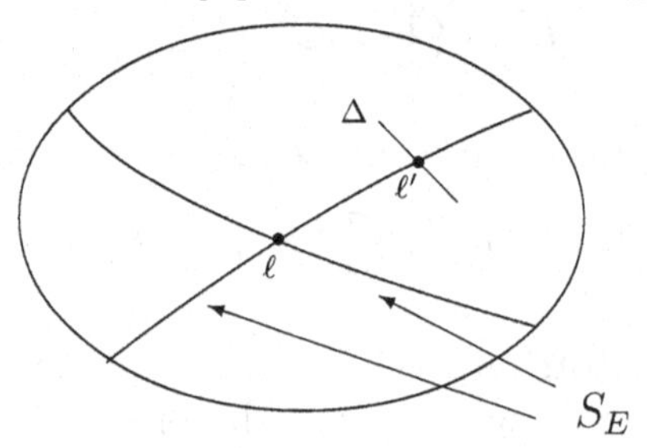

p_i can be extended to a coordinate system $\{p_i, z_2, \dots, z_d\}$ around ℓ'. Let Δ be the disk

$$\{(p_i, z_2, \dots, z_d) \in G_n \mid |p_i| < \varepsilon, \quad z_2 = \dots = z_d = 0\} \quad (\varepsilon \text{ small})$$

through the point ℓ' (picture above). Then k_i is the order of vanishing of $\det h|\Delta$ in the point ℓ'. Because $E_1|\Delta$ and $E_2|\Delta$ are locally free of equal rank, we have in the point ℓ' an exact sequence

$$0 \to \mathcal{O}_{\Delta,\ell'}^{\oplus r} \xrightarrow{h|\Delta} \mathcal{O}_{\Delta,\ell'}^{\oplus r} \to (F|\Delta)_{\ell'} \to 0.$$

Since $\mathcal{O}_{\Delta,\ell'}$ is a principal ideal ring, we can assume that $h|\Delta$ is given by a diagonal matrix of the form

$$\begin{pmatrix} p_i^{a_1} \cdot h_1 & & 0 \\ & \ddots & \\ 0 & & p_i^{a_r} \cdot h_r \end{pmatrix}$$

with units $h_j \in \mathcal{O}_{\Delta,\ell'}$. Thus

$$(F|\Delta)_{\ell'} \cong \bigoplus_{j=1}^{r} \mathcal{O}_{\Delta,\ell'}/(p_i^{a_j})$$

and hence

$$k_i = a_1 + \dots + a_r = \dim_{\mathbb{C}}(F|\Delta)_{\ell'}.$$

Thus the multiplicities of the prime factors of the divisor D_E are completely determined by the sheaf $F = R^1 q_* p^* E(-1)$.

To compute the degree of D_E it suffices to consider the case $n = 2$. For if $L_0 \subset \mathbb{P}_n$ is a generic line and $\mathbb{P}_2 \supset L_0$ a plane in $\mathbb{P}_n$ which contains L_0, then $E|\mathbb{P}_2$ is also semistable, since the splitting type of $E|\mathbb{P}_2$ is $\underline{a}_{E|\mathbb{P}_2} = (0,0)$. By restricting the resolution

$$0 \to K \to \bigoplus_i \mathcal{O}_{\mathbb{P}_n}(k_i) \to E(-1) \to 0$$

to $\mathbb{P}_2$ we get the resolution

$$(*) \qquad 0 \to K' \to \bigoplus_i \mathcal{O}_{\mathbb{P}_2}(k_i) \to E(-1)|\mathbb{P}_2 \to 0$$

of $E(-1)|\mathbb{P}_2$.

We consider the standard diagram associated with $\mathbb{P}_2 \subset \mathbb{P}_n$:

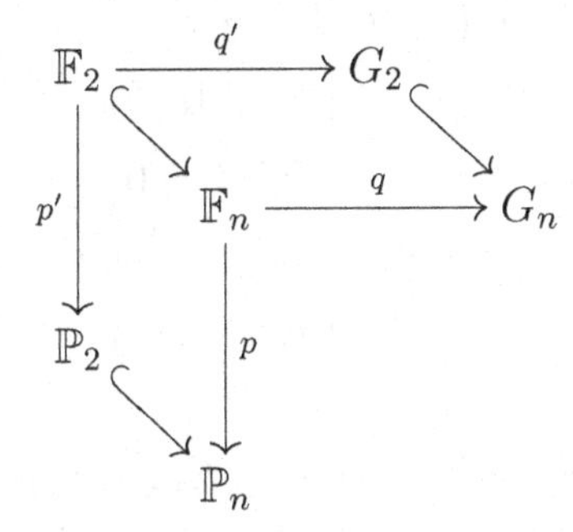

If we project the sequence (*) onto the Grassmann manifold G_2 of lines in $\mathbb{P}_2$, then with the help of the base-change theorem we get

$$\begin{aligned} E_1' &= R^1 q_*' p'^* K' \cong (R^1 q_* p^* K)|G_2 \\ E_2' &= R^1 q_*' p'^* (\bigoplus_i \mathcal{O}_{\mathbb{P}_2}(k_i)) \cong (R^1 q_* p^* (\bigoplus_i \mathcal{O}_{\mathbb{P}_n}(k_i)))|G_2 \\ F' &= R^1 q_*' p'^* (E(-1)|\mathbb{P}_2) \cong (R^1 q_* p^* E(-1))|G_2 = F|G_2. \end{aligned}$$

This shows that the sequence

$$0 \to E_1 \xrightarrow{h} E_2 \to F \to 0$$

restricts on G_2 to the sequence

$$0 \to E_1' \xrightarrow{h'} E_2' \to F' \to 0.$$

Thus for the line bundle belonging to the divisor $D_{E|\mathbb{P}_2}$ we have

$$[D_{E|\mathbb{P}_2}] = [D_E]|G_2$$

and thus

$$\deg D_{E|\mathbb{P}_2} = \deg D_E.$$

Hence we only need prove the equation

$$\deg D_E = c_2(E)$$

for bundles of type $(0,0)$ over $\mathbb{P}_2$. To this end we provide ourselves with a suitable resolution of $E(-1)$.

LEMMA 2.2.3. *Let E be a semistable 2-bundle over $\mathbb{P}_2$ with $c_1(E) = 0$. Then there is a resolution*

$$0 \to \bigoplus_{i=1}^{r} \mathcal{O}_{\mathbb{P}_2}(k_i) \to \bigoplus_{j=1}^{s} \mathcal{O}_{\mathbb{P}_2}(m_j) \to E(-1) \to 0$$

with $k_i, m_j < 0$.

PROOF. Let $n_1, \ldots, n_t \in \mathbb{Z}$ be integers with

$$H^1(\mathbb{P}_2, E(1+n_i)) \neq 0.$$

(Since $H^1(\mathbb{P}_2, E(k)) = 0$ for $k \gg 0$ and by Serre-duality also

$$H^1(\mathbb{P}_2, E(k)) = 0$$

for $k \ll 0$, there are only finitely many integers k with

$$H^1(\mathbb{P}_2, E(k)) \neq 0.)$$

Let

$$N_i = \dim_{\mathbb{C}} H^1(\mathbb{P}_2, E(1+n_i))$$

and $g_{i,1}, \ldots, g_{i,N_i} \in H^1(\mathbb{P}_2, E(1+n_i))$ be a basis. The tuple

$$g = (g_{ij}) \in H^1(\mathbb{P}_2, E(1) \otimes \bigoplus_{i=1}^{t} \mathcal{O}_{\mathbb{P}_2}(n_i)^{\oplus N_i})$$

defines an extension

$$(**) \qquad 0 \to E(1) \to E' \to \bigoplus_{i=1}^{t} \mathcal{O}_{\mathbb{P}_2}(-n_i)^{\oplus N_i} \to 0$$

with $H^1(\mathbb{P}_2, E'(n)) = 0$ for $n \neq n_i$ $(i = 1, \ldots, t)$. We now show that

$$H^1(\mathbb{P}_2, E'(n)) = 0$$

holds even for $n \in \{n_1, \ldots, n_t\}$. From the splitting criterion of Horrocks (Ch. I, 2.3.1) it then follows that E' splits.

To see that $H^1(\mathbb{P}_2, E'(n_i)) = 0$ we consider the diagram $(j = 1, \ldots, t,\ k = 1, \ldots, N_j)$

$$\begin{array}{ccc} H^0(\mathbb{P}_2, (\bigoplus_i \mathcal{O}_{\mathbb{P}_2}(-n_i)^{\oplus N_i}) \otimes (\bigoplus_i \mathcal{O}_{\mathbb{P}_2}(n_i)^{\oplus N_i})) & \xrightarrow{\delta} & H^1(\mathbb{P}_2, E(1) \otimes (\bigoplus_i \mathcal{O}_{\mathbb{P}_2}(n_i)^{\oplus N_i})) \\ \downarrow p_{jk} & & \downarrow p'_{jk} \\ H^0(\mathbb{P}_2, \bigoplus_i \mathcal{O}_{\mathbb{P}_2}(-n_i)^{\oplus N_i} \otimes \mathcal{O}_{\mathbb{P}_2}(n_j)) & \xrightarrow{\delta'} & H^1(\mathbb{P}_2, E(1+n_j)). \end{array}$$

p_{jk}, p'_{jk} are induced by the projections

$$\bigoplus_i \mathcal{O}_{\mathbb{P}_2}(n_i)^{\oplus N_i} \to \mathcal{O}_{\mathbb{P}_2}(n_j)$$

onto the kth component in $\mathcal{O}_{\mathbb{P}_2}(n_j)^{\oplus N_j}$. By construction

$$\delta(\mathrm{id}) = g$$

and thus

$$\delta'(p_{jk}(\mathrm{id})) = p'_{jk}(g) = g_{jk}.$$

Since the g_{jk}, $k = 1, \ldots, N_j$, generate the vector space

$$H^1(\mathbb{P}_2, E(1+n_j)),$$

we see that δ' is surjective.

From the cohomology sequence of the sequence (**) tensored with $\mathcal{O}(n_j)$ one deduces that

$$H^1(E'(n_j)) = 0.$$

Thus we have a resolution

$$0 \to \bigoplus_{i=1}^{r} \mathcal{O}_{\mathbb{P}_2}(k_i) \to \bigoplus_{j=1}^{s} \mathcal{O}_{\mathbb{P}_2}(m_j) \to E(-1) \to 0$$

$(\bigoplus_{i=1}^{r} \mathcal{O}_{\mathbb{P}_2}(k_i) = \bigoplus_{i=1}^{t} \mathcal{O}_{\mathbb{P}_2}(n_i)^{\oplus N_i})$. Since $H^0(\mathbb{P}_2, E(-1)) = 0$ one can assume that the integers k_i, m_j are negative. □

Now the degree of the divisor D_E can be easily computed. Proceeding from the resolution

$$0 \to \bigoplus_i \mathcal{O}_{\mathbb{P}_2}(k_i) \to \bigoplus_j \mathcal{O}_{\mathbb{P}_2}(m_j) \to E(-1) \to 0 \qquad (k_i, m_j < 0)$$

we get the equations

i) $\deg D_E = c_1(\bigoplus_j R^1 q_* p^* \mathcal{O}_{\mathbb{P}_2}(m_j)) - c_1(\bigoplus_i R^1 q_* p^* \mathcal{O}_{\mathbb{P}_2}(k_i))$

ii) $c_2(E(-1)) = \sum_{i<j} m_i m_j - \sum_{i<j} k_i k_j + 2\sum_i k_i$

iii) $-2 = c_1(E(-1)) = \sum_j m_j - \sum_i k_i$.

We must determine

$$c_1(R^1 q_* p^* \mathcal{O}_{\mathbb{P}_2}(k))$$

for $k < 0$.

Claim. For $k < 0$ we have $c_1(R^1 q_* p^* \mathcal{O}_{\mathbb{P}_2}(k)) = -\frac{k(k+1)}{2}$.

If we have shown this, then with equations i)–iii) we get

$$\begin{aligned}\deg D_E &= \tfrac{1}{2}\Big(\sum_i (k_i^2 + k_i) - \sum_j (m_j^2 + m_j)\Big)\\ &= \tfrac{1}{2}\Big(\sum_i k_i^2 - \sum_j m_j^2\Big) + 1\\ &= \tfrac{1}{2}\Big(\big(\sum_i k_i\big)^2 - \big(\sum_j m_j\big)^2 + 2c_2(E(-1)) - 4\big(\sum_i k_i\big)\Big) + 1\\ &= \tfrac{1}{2}\Big(\big(\sum_i k_i\big)^2 - \big(\big(\sum_i k_i\big) - 2\big)^2 - 4\sum_i k_i + 2c_2(E(-1))\Big) + 1\\ &= c_2(E(-1)) - 1 = c_2(E).\end{aligned}$$

We thus have to prove the above claim. We consider $\mathbb{F}_2$ as divisor in $\mathbb{P}_2 \times \mathbb{P}_2^*$ ($G_2 = \mathbb{P}_2^*$). Let $\mathcal{O}(a,b)$ be the tensor product

$$\mathcal{O}(a,b) = \bar{p}^*\mathcal{O}_{\mathbb{P}_2}(a) \otimes \bar{q}^*\mathcal{O}_{\mathbb{P}_2^*}(b),$$

where $\bar{p}\colon \mathbb{P}_2 \times \mathbb{P}_2^* \to \mathbb{P}_2$, $\bar{q}\colon \mathbb{P}_2 \times \mathbb{P}_2^* \to \mathbb{P}_2^*$ are the projections. Thus $p = \bar{p}|\mathbb{F}_2$, $q = \bar{q}|\mathbb{F}_2$. The line bundle of $\mathbb{F}_2$ over $\mathbb{P}_2 \times \mathbb{P}_2^*$ is

$$[\mathbb{F}_2] = \mathcal{O}(1,1)$$

and hence we get the exact sequence

$$0 \to \mathcal{O}(-1,-1) \to \mathcal{O}_{\mathbb{P}_2 \times \mathbb{P}_2^*} \to \mathcal{O}_{\mathbb{F}_2} \to 0.$$

By tensoring this sequence with $\mathcal{O}(k,0) = \bar{p}^*\mathcal{O}_{\mathbb{P}_2}(k)$, projecting onto $\mathbb{P}_2^*$ and using

$$R^i\bar{q}_*(\bar{p}^*\mathcal{O}_{\mathbb{P}_2}(a) \otimes \bar{q}^*\mathcal{O}_{\mathbb{P}_2^*}(b)) = H^i(\mathbb{P}_2, \mathcal{O}_{\mathbb{P}_2}(a)) \otimes \mathcal{O}_{\mathbb{P}_2^*}(b)$$

we get the exact sequence ($k < 0$)

$$\begin{aligned}0 \to R^1\bar{q}_*(\bar{p}^*\mathcal{O}_{\mathbb{P}_2}(k) \otimes \mathcal{O}_{\mathbb{F}_2}) \to H^2(\mathbb{P}_2, \mathcal{O}_{\mathbb{P}_2}(k-1)) \otimes \mathcal{O}_{\mathbb{P}_2^*}(-1)\\ \to H^2(\mathbb{P}_2, \mathcal{O}_{\mathbb{P}_2}(k)) \otimes \mathcal{O}_{\mathbb{P}_2^*} \to 0.\end{aligned}$$

For the first Chern class in this sequence one gets the equation

$$-h^2(\mathbb{P}_2, \mathcal{O}_{\mathbb{P}_2}(k-1)) = c_1(R^1\bar{q}_*(\bar{p}^*\mathcal{O}_{\mathbb{P}_2}(k) \otimes \mathcal{O}_{\mathbb{F}_2}))$$

and thus the desired result

$$c_1(R^1\bar{q}_*p^*\mathcal{O}_{\mathbb{P}_2}(k)) = -\frac{k(k+1)}{2}.$$

Thus we have proved:

THEOREM 2.2.4 (Barth). *The set S_E of jump lines of a semistable 2-bundle E over $\mathbb{P}_n$ with $c_1(E) = 0$ is purely 1-codimensional in G_n.*

The sheaf $F = R^1 q_* p^*(E(-1))$ defines a divisor D_E of degree $c_2(E)$ over G_n with

$$S_E = \operatorname{supp} F = \operatorname{supp} D_E.$$

As a last application of the theorem of Grauert and Mülich we investigate the behavior of semistable 2-bundles under restriction to hyperplanes.

THEOREM 2.2.5 (Maruyama, Barth). *Let E be a semistable 2-bundle over $\mathbb{P}_n$, $n \geq 3$. There is a non-empty Zariski-open subset $U \subset \mathbb{P}_n^*$ of the Grassmann manifold of hyperplanes $H \subset \mathbb{P}_n$ such that the restriction $E|H$ remains semistable for all hyperplanes H whose associated point $h \in \mathbb{P}_n^*$ lies in U.*

PROOF. We can assume that E is normalized.

i) If $c_1(E) = 0$, then the generic splitting type is $\underline{a}_E = (0,0)$. If $H \subset \mathbb{P}_n$ is a hyperplane which contains a generic line, then $E|H$ has type $\underline{a}_{E|H} = (0,0)$ and by Lemma 2.2.1 is thus semistable.

ii) Let $c_1(E) = -1$ (the concepts "stable" and "semistable" then coincide). We consider the standard diagram

$$\begin{array}{ccc} \mathbb{F}_n^* & \xrightarrow{q} & \mathbb{P}_n^* \\ {\scriptstyle p}\downarrow & & \\ \mathbb{P}_n & & \end{array}$$

with the flag manifold

$$\mathbb{F}_n^* = \{(x,h) \in \mathbb{P}_n \times \mathbb{P}_n^* \mid x \in H\}.$$

The point in $\mathbb{P}_n^*$ belonging to a hyperplane $H \subset \mathbb{P}_n$ will be denoted by h.

Let $L \subset \mathbb{P}_n$ be a generic line for E $(E|L = \mathcal{O}_L \oplus \mathcal{O}_L(-1))$,

$$\mathbb{P}_n^*(L) = \{h \in \mathbb{P}_n^* \mid L \subset H\}$$

the set of hyperplanes in $\mathbb{P}_n$ which contains L. We set

$$\mathbb{B}(L) = q^{-1}(\mathbb{P}_n^*(L)) = \{(x,h) \in \mathbb{P}_n \times \mathbb{P}_n^* \mid x \in H \supset L\} \subset \mathbb{F}_n^*.$$

The restriction σ of p to $\mathbb{B}(L)$ is a proper holomorphic mapping with connected fibres (for $n = 3$ it is the σ-process of $\mathbb{P}_3$ along L).

We then get the following diagram:

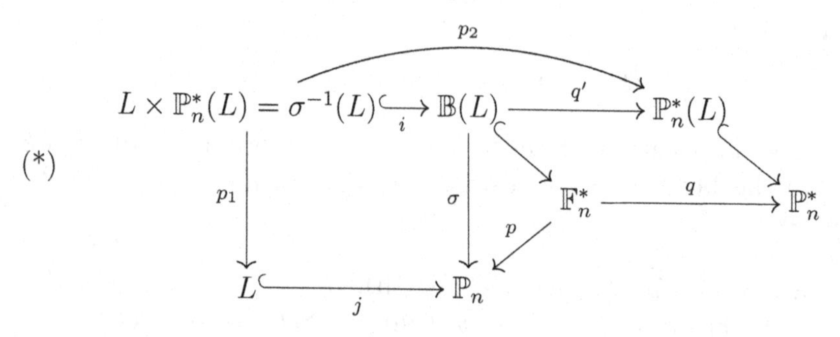

Because $E|L \cong \mathcal{O}_L \oplus \mathcal{O}_L(-1)$, we have for every linear subspace $H' \subset \mathbb{P}_n$ containing L

$$h^0(H', E(-1)|H') = 0.$$

One sees this as in the proof of Lemma 2.2.1.

We now assume that $E|H$ is not semistable for any hyperplane $H \subset \mathbb{P}_n$. Thus we assume

$$h^0(H, E|H) > 0$$

for every $h \in \mathbb{P}_n^*$.

Claim. For all hyperplanes $H \subset \mathbb{P}_n$ which contain L the restriction map

$$H^0(H, E|H) \xrightarrow{\sim} H^0(L, E|L)$$

is an isomorphism and in particular $h^0(H, E|H) = 1$.

To prove this we choose a flag

$$L = H_1 \subset H_2 \subset \cdots \subset H_{n-1} = H$$

from L to H. The cohomology sequences of

$$0 \to E(-1)|H_i \to E|H_i \to E|H_{i-1} \to 0$$

together with the fact that $H^0(H_i, E(-1)|H_i) = 0$ yield that all restriction maps

$$H^0(H_i, E|H_i) \to H^0(H_{i-1}, E|H_{i-1})$$

are injective $(i = 2, \dots, n-1)$. Thus the composition

$$H^0(H, E|H) \to H^0(L, E|L)$$

is also injective, therefore bijective for dimension reasons, and in particular $h^0(H, E|H) = h^0(L, E|L) = 1$.

From the base-change theorem it now follows that

$$q'_*\sigma^*E$$

is a line bundle over $\mathbb{P}_n^*(L)$. With the notation in diagram (*) we have

$$\begin{aligned} q'_* i_* i^* \sigma^* E &\cong p_{2*} i^* \sigma^* E \cong p_{2*} p_1^* (E|L) \\ &\cong p_{2*}(\mathcal{O}_{L \times \mathbb{P}_n^*(L)} \oplus p_1^* \mathcal{O}_L(-1)) \\ &\cong \mathcal{O}_{\mathbb{P}_n^*(L)} \oplus p_{2*} p_1^* \mathcal{O}_L(-1) = \mathcal{O}_{\mathbb{P}_n^*(L)}, \end{aligned}$$

because the image sheaf $p_{2*} p_1^* \mathcal{O}_L(-1)$ is zero.

The canonical homomorphism

$$q'_* \sigma^* E \to q'_*(i_* i^* \sigma^* E) \cong \mathcal{O}_{\mathbb{P}_n^*(L)}$$

is determined in the point $h \in \mathbb{P}_n^*(L)$ by the following diagram:

$$\begin{array}{ccc} (q'_* \sigma^* E)(h) & \longrightarrow & (q'_* i_* i^* \sigma^* E)(h) \\ \Big\| \wr & & \Big\| \wr \\ H^0(H, E|H) & & (p_{2*} p_1^*(E|L))(h) \\ & \underset{\text{res}}{\overset{\sim}{\searrow}} & \Big\| \wr \\ & & H^0(L, E|L) \end{array}$$

We had already seen that the restriction map

$$H^0(H, E|H) \to H^0(L, E|L)$$

is bijective; the morphism

$$q'_* \sigma^* E \to q'_* i_* i^* \sigma^* E \cong \mathcal{O}_{\mathbb{P}_n^*(L)}$$

is therefore an isomorphism, i.e., we have

$$q'_* \sigma^* E \cong \mathcal{O}_{\mathbb{P}_n^*(L)}.$$

Over $\mathbb{B}(L)$ with the canonical morphism

$$\mathcal{O}_{\mathbb{B}(L)} \cong q'^* q'_* \sigma^* E \to \sigma^* E$$

we thus have a monomorphism, which after projection onto $\mathbb{P}_n$ gives the monomorphism

$$\mathcal{O}_{\mathbb{P}_n} = \sigma_* \mathcal{O}_{\mathbb{B}(L)} \hookrightarrow \sigma_* \sigma^* E \cong E \otimes \sigma_* \mathcal{O}_{\mathbb{B}(L)} \cong E,$$

i.e., a nontrivial section in $H^0(E)$. This contradicts the semistability of E.

Thus the assumption that $E|H$ has sections for all hyperplanes H was false; hence there is a hyperplane H_0 over which $E|H_0$ is semistable. With the help of the semicontinuity theorem it then follows that $E|H$ must be semistable for all hyperplanes $h \in \mathbb{P}_n^*$ in a Zariski-open neighborhood of H_0. □

From this theorem and the inequality of Schwarzenberger for semistable 2-bundles over $\mathbb{P}_2$ (1.2.7) we get the following

COROLLARY. *For the discriminant* $\Delta_E = c_1^2(E) - 4c_2(E)$ *of a semistable 2-bundle* E *over* $\mathbb{P}_n$ *we have*

$$\Delta_E \leq 0.$$

Following 1.3.1 we mentioned the theorem of Barth that the restriction of a stable 2-bundle to a hyperplane is again stable (with the exception of the null-correlation bundles over $\mathbb{P}_3$). By Lemma 1.2.7 it then follows that for stable 2-bundles we always have

$$c_1^2 - 4c_2 < 0.$$

2.3. Historical remarks, further results, and open questions. Important ideas in this paragraph are due to Van de Ven. He showed [134] that a uniform 2-bundle over $\mathbb{P}_n$ of type (a, b) with $a - b \geq 2$ always has $\mathcal{O}_{\mathbb{P}_n}(a)$ as subbundle and consequently splits. Barth [12] extended these ideas to give a proof of the theorem of Grauert and Mülich, which was fundamental for the further generalizations. The generalized theorem of Grauert and Mülich (Theorem 2.1.4 and Corollary 1 of it) was proved by Spindler [119]. It is even true for bundles over projective algebraic manifolds [40]. One also finds the Descente-Lemma 2.1.2 there. From the theorem of Spindler one can deduce (cf. Maruyama [84], Elencwajg and Forster [34]) that the set of semistable vector bundles over $\mathbb{P}_n$ with fixed rank and fixed Chern classes forms a bounded family (cf. Grothendieck [52]). This implies finiteness properties of the moduli spaces.

We have seen that for (semi-)stable 3-bundles over $\mathbb{P}_2$ all splitting types which are possible by the theorem of Spindler actually can be realized.

PROBLEM 2.3.1. Investigate this question for other values of rank and base dimension.

The theorem of Van de Ven (Thm. 2.2.2) about uniform 2-bundles was proved here by the method of Elencwajg [31]. He used it first for uniform 3-bundles over $\mathbb{P}_2$. The theorem of Barth (Thm. 2.2.4) about the divisor of jump lines of a semistable 2-bundle over $\mathbb{P}_n$ can be proved much more simply when $n = 2$: for a semistable 2-bundle E over $\mathbb{P}_2$ with $c_1(E) = 0$ one considers the exact sequence

$$(*) \qquad 0 \to H^0(E(-1)|L) \to H^1(E(-2)) \xrightarrow{\alpha(L)} H^1(E(-1)).$$

Since $h^1(E(-1)) = h^1(E(-2)) = c_2$, it follows that

$$S_E = \{\ell \in \mathbb{P}_2^* | \det \alpha(L) = 0\}.$$

S_E is thus a curve of degree c_2 with the equation $\det \alpha = 0$.

For semistable 2-bundles E over $\mathbb{P}_n$ with $c_1(E) = -1$ the set S_E is not necessarily a hypersurface. In order to get over this difficulty at least on $\mathbb{P}_2$ Hulek [72] introduces the concept of jump lines of the 2 nd kind. A line $L \subset \mathbb{P}_2$ is a jump line of the 2 nd kind for E if

$$H^0(E|L^2) \neq 0.$$

Here $L^2 \subset \mathbb{P}_2$ is the first infinitesimal neighborhood of L in $\mathbb{P}_2$. Let C_E denote the set of jump lines of E of the second kind. Hulek shows that $C_E \neq \mathbb{P}_2^*$ and then with a sequence analogous to (*) that C_E has degree $2(c_2 - 1)$. One always has

$$S_E \subset \mathrm{Sing}(C_E)$$

and in general even equality.

PROBLEM 2.3.2. Investigate jump lines of the second kind for higher base dimension.

We conclude this section with some remarks about the restriction of (semi-)stable bundles to hyperplanes. Maruyama [86] has shown that quite generally the restriction of a semistable r-bundle over an n-dimensional projective algebraic manifold X with $r < n$ to a generic hyperplane section is again semistable. The theorem of Barth mentioned above takes care of the restriction question for stable bundles of rank 2 over $\mathbb{P}_n$. If E is a (semi-)stable bundle, $H \subset \mathbb{P}_n$ a hyperplane and $E|H$ again (semi-)stable, then E satisfies certain topological conditions. For example for a stable r-bundle over $\mathbb{P}_2$ with $c_1 = 0$ one necessarily has $c_2 \geq r$, as one can read off from the Riemann–Roch formula. This necessary condition is however by no means sufficient. For the second symmetric power S^2N of a null-correlation bundle N over $\mathbb{P}_3$ is stable, its restriction to every plane only semistable. However one has $c_2(S^2N) = 4$, $\mathrm{rk}\,(S^2N) = 3$.

PROBLEM 2.3.3. Investigate the behavior of (semi-)stable holomorphic vector bundles under restriction to hyperplanes.

Positive results for 2.3.3 have applications to necessary topological conditions for (semi-)stable bundles over $\mathbb{P}_n$. For example the restriction of semistable 3-bundles with $c_1 = 0$ to generic hyperplanes is again semistable [106] and using this one can sharpen the results of Elencwajg and Forster [34] to show

$$|c_3| \leq c_2^2 + c_2.$$

This inequality cannot be improved upon.

Finally it would be desirable to investigate the restriction of (semi-) stable bundles to nonlinear submanifolds. In particular the restriction to curves would be interesting, since one knows a great deal about stable vector bundles over curves.

§3. Monads

In this paragraph we describe a construction which permits one to investigate holomorphic vector bundles with the methods of linear algebra. This leads to the concept of monads. In the next paragraph monads will be used to construct the moduli spaces of stable 2-bundles over $\mathbb{P}_2$.

We begin by proving a fairly general existence theorem, the theorem of Beilinson. Then we give some examples which demonstrate the application of this theorem to the description of bundles by monads. In the last section of this paragraph we describe one possibility to construct the essentially only known stable 2-bundle over $\mathbb{P}_4$.

3.1. The theorem of Beilinson. Let X be a compact complex manifold.

DEFINITION 3.1.1. A *monad* over X is a complex

$$0 \to A \xrightarrow{a} B \xrightarrow{b} C \to 0$$

of holomorphic vector bundles over X which is exact at A and at C, such that $\mathrm{Im}\,(a)$ is a subbundle of B. The holomorphic vector bundle

$$E = \mathrm{Ker}\, b/\mathrm{Im}\, a$$

is the *cohomology of the monad.*

A monad $0 \to A \xrightarrow{a} B \xrightarrow{b} C \to 0$ has a socalled display: this is a commutative diagram with exact rows and columns:

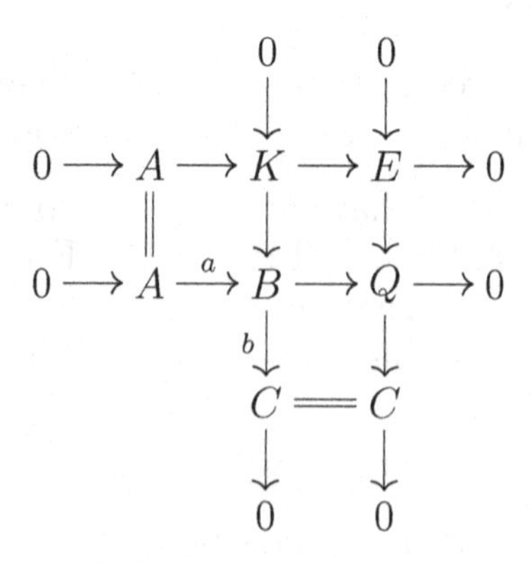

$$\begin{array}{ccccccccc} & & & & 0 & & 0 & & \\ & & & & \downarrow & & \downarrow & & \\ 0 & \longrightarrow & A & \longrightarrow & K & \longrightarrow & E & \longrightarrow & 0 \\ & & \| & & \downarrow & & \downarrow & & \\ 0 & \longrightarrow & A & \xrightarrow{a} & B & \longrightarrow & Q & \longrightarrow & 0 \\ & & & & {\scriptstyle b}\downarrow & & \downarrow & & \\ & & & & C & = & C & & \\ & & & & \downarrow & & \downarrow & & \\ & & & & 0 & & 0 & & \end{array}$$

Here $K = \mathrm{Ker}\, b$, $Q = \mathrm{Coker}\, a$. From the display one deduces the following

LEMMA 3.1.2. *If E is the cohomology of the monad*

$$0 \to A \xrightarrow{a} B \xrightarrow{b} C \to 0,$$

then the rank and total Chern class of E are given by

$$\operatorname{rk} E = \operatorname{rk} B - \operatorname{rk} A - \operatorname{rk} C$$

$$c(E) = c(B)c(A)^{-1}c(C)^{-1}.$$

Monads were first used by Horrocks. He showed for example that all holomorphic 2-bundles over $\mathbb{P}_3$ can be obtained as the cohomology of a monad of the special form

$$0 \to \bigoplus_i \mathcal{O}_{\mathbb{P}_3}(a_i) \to \bigoplus_j \mathcal{O}_{\mathbb{P}_3}(b_j) \to \bigoplus_k \mathcal{O}_{\mathbb{P}_3}(c_k) \to 0.$$

In Chapter I, §2, we saw that the cohomologically simplest bundles over $\mathbb{P}_n$ are the bundles $\Omega^p_{\mathbb{P}_n}(q)$ of twisted p-forms. These bundles are in a sense the building blocks for more complicated bundles. We have in fact

THEOREM 3.1.3 (Beilinson, Theorem I). *Let E be a holomorphic r-bundle over $\mathbb{P}_n$. There is a spectral sequence E_r^{pq} with E_1-term*

$$E_1^{pq} = H^q(\mathbb{P}_n, E(p)) \otimes \Omega^{-p}_{\mathbb{P}_n}(-p)$$

which converges to

$$E^i = \begin{cases} E & \text{for } i = 0 \\ 0 & \text{otherwise,} \end{cases}$$

i.e., $E_\infty^{pq} = 0$ *for* $p + q \neq 0$ *and*

$\bigoplus_{p=0}^{n} E_\infty^{-p,p}$ *is the associated graded sheaf of a filtration of E.*

PROOF. The proof has a geometric and an algebraic part. We begin with the geometry. We consider the projections

$$\begin{array}{ccc} \mathbb{P}_n \times \mathbb{P}_n & \xrightarrow{p_2} & \mathbb{P}_n \\ {\scriptstyle p_1}\downarrow & & \\ \mathbb{P}_n & & \end{array}$$

and abbreviate $A \boxtimes B = p_1^* A \otimes p_2^* B$ for any bundles A, B over $\mathbb{P}_n$. Taking the Euler sequence

$$0 \to \mathcal{O}_{\mathbb{P}_n}(-1) \to \mathcal{O}_{\mathbb{P}_n}^{\oplus(n+1)} \to Q \to 0$$

over $\mathbb{P}_n$ we consider the n-bundle

$$\mathcal{O}_{\mathbb{P}_n}(1) \boxtimes Q = \mathcal{H}om(p_1^* \mathcal{O}_{\mathbb{P}_n}(-1), p_2^* Q)$$

over $\mathbb{P}_n \times \mathbb{P}_n$. There is a canonical section s in this bundle defined as follows: let $v, w \in \mathbb{C}^{n+1} \setminus \{0\}$ and let $x \in \mathbb{P}(\mathbb{C}v)$, $y = \mathbb{P}(\mathbb{C}w)$ be the corresponding points in $\mathbb{P}_n$. Then one defines

$$s(x,y) \in \mathrm{Hom}_{\mathbb{C}}(\mathcal{O}_{\mathbb{P}_n}(-1)(x), Q(y))$$

by

$$s(x,y)(av) = av \mod \mathbb{C}w \in \mathbb{C}^{n+1}/\mathbb{C}w = Q(y) \quad \text{for} \quad a \in \mathbb{C}.$$

The section s corresponds to the identity under the canonical identification

$$\begin{aligned} H^0(\mathbb{P}_n \times \mathbb{P}_n, \mathcal{O}_{\mathbb{P}_n}(1) \boxtimes Q) &\cong H^0(\mathbb{P}_n, \mathcal{O}_{\mathbb{P}_n}(1)) \otimes H^0(\mathbb{P}_n, Q) \\ &\cong (\mathbb{C}^{n+1})^* \otimes \mathbb{C}^{n+1} \\ &\cong \mathrm{Hom}_{\mathbb{C}}(\mathbb{C}^{n+1}, \mathbb{C}^{n+1}). \end{aligned}$$

One sees directly that $s(x,y)$ vanishes if and only if v and w are linearly dependent, i.e., if and only if $x = y$. Thus the zero locus of s is the diagonal $\Delta \subset \mathbb{P}_n \times \mathbb{P}_n$. Hence s defines a locally free resolution (the Koszul complex) of the sheaf $\mathcal{O}_\Delta = \mathcal{O}_{\mathbb{P}_n \times \mathbb{P}_n}/J_\Delta$,

$$\begin{aligned} (1) \quad 0 \to \Lambda^n(\mathcal{O}_{\mathbb{P}_n}(-1) \boxtimes Q^*) &\to \Lambda^{n-1}(\mathcal{O}_{\mathbb{P}_n}(-1) \boxtimes Q^*) \to \\ &\cdots \to \mathcal{O}_{\mathbb{P}_n}(-1) \boxtimes Q^* \to \mathcal{O}_{\mathbb{P}_n \times \mathbb{P}_n} \to \mathcal{O}_\Delta \to 0. \end{aligned}$$

By tensoring this sequence with $p_1^* E$ we get the complex of sheaves

$$\begin{aligned} (2) \quad 0 \to E(-n) \boxtimes \Omega^n_{\mathbb{P}_n}(n) &\to E(-n+1) \boxtimes \Omega^{n-1}_{\mathbb{P}_n}(n-1) \to \\ &\cdots \to E(-1) \boxtimes \Omega^1_{\mathbb{P}_n}(1) \to E \boxtimes \mathcal{O}_{\mathbb{P}_n} \to 0. \end{aligned}$$

To save writing we set

$$C^{-k} = E(-k) \boxtimes \Omega^k_{\mathbb{P}_n}(k) \qquad (k = 0, \ldots, n)$$

and write $C^\bullet$ for the complex

$$(3) \qquad 0 \to C^{-n} \to C^{-n+1} \to \cdots \to C^0 \to 0.$$

Let $\mathbb{R}^i p_{2*}(C^\bullet)$ be the ith hyperdirect image (cf. Grothendieck [53], p. 406). $\mathbb{R}^i p_{2*}(C^\bullet)$ is defined as follows: one chooses a Cartan–Eilenberg resolution $L^{\bullet\bullet}$ of the complex $C^\bullet$, applies p_{2*} to the double complex $L^{\bullet\bullet}$ and defines $\mathbb{R}^i p_{2*}(C^\bullet)$ as the ith cohomology sheaf of the total chain complex associated to $p_{2*}(L^{\bullet\bullet})$.

To the double complex $p_{2*}(L^{\bullet\bullet})$ there belong two spectral sequences $'E$ and $''E$ with E_2 terms

$$\begin{aligned} 'E_2^{pq} &= H^p(R^q p_{2*}(C^\bullet)) \\ ''E_2^{pq} &= R^p p_{2*}(H^q(C^\bullet)), \end{aligned}$$

which both converge to $\mathbb{R}^{p+q} p_{2*}(C^\bullet)$.

Since $C^\bullet$ is a locally free resolution of $p_1^*E|\Delta$, it follows that

$$H^q(C^\bullet) \cong \begin{cases} p_1^*E|\Delta & \text{for } q = 0 \\ 0 & \text{otherwise} \end{cases}$$

and therefore

$$''E_2^{pq} = R^p p_{2*}(H^q(C^\bullet)) \cong \begin{cases} E & \text{for } p = q = 0 \\ 0 & \text{otherwise.} \end{cases}$$

$''E_2^{pq}$ converges to $\mathbb{R}^{p+q} p_{2*}(C^\bullet)$ and thus

$$\mathbb{R}^i p_{2*}(C^\bullet) \cong \begin{cases} E & \text{for } i = 0 \\ 0 & \text{otherwise.} \end{cases}$$

The E_1-term of the first spectral sequence is

$$\begin{aligned} 'E_1^{pq} = R^q p_{2*}(C^p) &= R^q p_{2*}(E(p) \boxtimes \Omega_{\mathbb{P}_n}^{-p}(-p)) \\ &= R^q p_{2*}(p_1^* E(p)) \otimes \Omega_{\mathbb{P}_n}^{-p}(-p) \\ &= H^q(\mathbb{P}_n, E(p)) \otimes \Omega_{\mathbb{P}_n}^{-p}(-p). \end{aligned}$$ □

If instead of the complex $C^\bullet$ with

$$C^{-k} = E(-k) \boxtimes \Omega_{\mathbb{P}_n}^k(k)$$

one considers the complex $D^\bullet$ with

$$D^{-k} = \mathcal{O}_{\mathbb{P}_n}(-k) \boxtimes (E \otimes \Omega_{\mathbb{P}_n}^k(k)),$$

then one obtains from the spectral sequences belonging to the hypercohomology $\mathbb{R}^i p_{1*}(D^\bullet)$ the following theorem.

THEOREM 3.1.4 (Beilinson, Theorem II). *Let E be a holomorphic r-bundle over $\mathbb{P}_n$. There is a spectral sequence with E_1-term*

$$E_1^{pq} = H^q(\mathbb{P}_n, E \otimes \Omega_{\mathbb{P}_n}^{-p}(-p)) \otimes \mathcal{O}_{\mathbb{P}_n}(p),$$

which converges to

$$E^i = \begin{cases} E & \text{if } i = 0 \\ 0 & \text{otherwise,} \end{cases}$$

i.e., $E_\infty^{p,q} = 0$ for $p + q \neq 0$ and $\bigoplus_{p=0}^{n} E_\infty^{-p,p}$ is the associated graded sheaf of a filtration of E.

For every bundle E we thus have two spectral sequences whose E_∞-terms belong to suitable filtrations of E. Furthermore $E_r^{pq} = 0$ if $q > n$ or $q < 0$ as well as for $p < -n$ or $p > 0$.

3.2. Examples. We want to show with several examples how one can apply the two spectral sequences to investigate a bundle E.

EXAMPLE 1. Let E be a stable 2-bundle over $\mathbb{P}_2$ with $c_1(E) = -1$, $c_2(E) = 1$. Then

$$E \cong \Omega^1_{\mathbb{P}_2}(1).$$

To see this we employ the spectral sequence of Theorem I with

$$E_1^{pq} = H^q(\mathbb{P}_2, E(p)) \otimes \Omega^{-p}_{\mathbb{P}_2}(-p).$$

Because E is stable and normalized we have

$$E_1^{p,0} = 0 \quad \text{and} \quad E_1^{p,2} = 0 \quad \text{for} \quad p = 0, -1, -2.$$

Thus the diagram of the E_1-term looks as follows:

0	0	0
$E_1^{-2,1}$	$E_1^{-1,1}$	$E_1^{0,1}$
0	0	0

(axes: q vertical, p horizontal)

The differentials $d_1^{pq}\colon E_1^{p,q} \to E_1^{p+1,q}$ give the complex

$$E_1^{-2,1} \xrightarrow{a} E_1^{-1,1} \xrightarrow{b} E_1^{0,1}.$$

Let $K = \operatorname{Ker} a$, $L = \operatorname{Ker} b / \operatorname{Im} a$, $M = \operatorname{Coker} b$. Then the diagram of the E_2-term looks as follows:

E_2^{pq}

0	0	0
K	L	M
0	0	0

(axes: q vertical, p horizontal)

All differentials $d_2^{pq}\colon E_2^{pq} \to E_2^{p+2,q-1}$ vanish and thus

$$K = E_\infty^{-2,1}, \quad L = E_\infty^{-1,1}, \quad M = E_\infty^{0,1}.$$

By Theorem I of Beilinson it follows that

$$K = M = 0 \quad \text{and} \quad E = L.$$

In other words

$$0 \to E_1^{-2,1} \xrightarrow{a} E_1^{-1,1} \xrightarrow{b} E_1^{0,1} \to 0$$

is a monad whose cohomology is E. But

$$E_1^{-2,1} = H^1(\mathbb{P}_2, E(-2)) \otimes \mathcal{O}_{\mathbb{P}_2}(-1)$$

and

$$E_1^{0,1} = H^1(\mathbb{P}_2, E) \otimes \mathcal{O}_{\mathbb{P}_2}.$$

From the Riemann–Roch formula for E one gets

$$h^1(\mathbb{P}_2, E) = h^1(\mathbb{P}_2, E(-2)) = 0 \quad \text{and} \quad h^1(\mathbb{P}_2, E(-1)) = 1;$$

thus

$$E \cong H^1(\mathbb{P}_2, E(-1)) \otimes \Omega^1_{\mathbb{P}_2}(1) \cong \Omega^1_{\mathbb{P}_2}(1).$$

EXAMPLE 2. Let E be a stable 2-bundle over $\mathbb{P}_2$ with $c_1(E) = 0$ and $c_2(E) = 2$.

In order to investigate E we consider the spectral sequence I with

$$E_1^{p,q} = H^q(\mathbb{P}_2, E(p)) \otimes \Omega^{-p}_{\mathbb{P}_2}(-p).$$

Again we have

$$E_1^{p,0} = E_1^{p,2} = 0 \quad \text{for} \quad p = 0, -1, -2$$

and

$$E_\infty^{p,q} = E_2^{p,q}.$$

As in example 1 this means that E is the cohomology of a monad

$$0 \to H^1(\mathbb{P}_2, E(-2)) \otimes \mathcal{O}_{\mathbb{P}_2}(-1) \to H^1(\mathbb{P}_2, E(-1)) \otimes \Omega^1_{\mathbb{P}_2}(1) \\ \to H^1(\mathbb{P}_2, E) \otimes \mathcal{O}_{\mathbb{P}_2} \to 0.$$

The Riemann–Roch formula gives

$$h^1(\mathbb{P}_2, E) = -\chi(\mathbb{P}_2, E) = c_2 - 2 = 0.$$

Similarly one computes

$$h^1(\mathbb{P}_2, E(-2)) = h^1(\mathbb{P}_2, E(-1)) = 2.$$

Thus one has an exact sequence

$$0 \to \mathcal{O}_{\mathbb{P}_2}(-1)^{\oplus 2} \to \Omega^1_{\mathbb{P}_2}(1)^{\oplus 2} \to E \to 0.$$

Every stable 2-bundle E over $\mathbb{P}_2$ with $c_1(E) = 0$, $c_2(E) = 2$ is thus the cokernel of a bundle monomorphism

$$\alpha\colon \mathcal{O}_{\mathbb{P}_2}(-1)^{\oplus 2} \hookrightarrow \Omega^1_{\mathbb{P}_2}(1)^{\oplus 2}.$$

EXAMPLE 3. Let E be a holomorphic r-bundle over $\mathbb{P}_2$ with

$$h^0(\mathbb{P}_2, E(-1)) = h^0(\mathbb{P}_2, E^*(-1)) = 0.$$

(A semistable normalized r-bundle over $\mathbb{P}_2$ for example satisfies these conditions.) We show that E is then the cohomology of a monad. For this we apply the spectral sequence of Theorem II to $E(-1)$. The E_1-term of this spectral sequence is

$$E_1^{p,q} = H^q(\mathbb{P}_2, E(-1) \otimes \Omega_{\mathbb{P}_2}^{-p}(-p)) \otimes \mathcal{O}_{\mathbb{P}_2}(p).$$

For $p = 0$ we have

$$E_1^{0,q} = \begin{cases} 0 & \text{for } q = 0 \\ H^1(\mathbb{P}_2, E(-1)) \otimes \mathcal{O}_{\mathbb{P}_2} & \text{for } q = 1 \\ 0 & \text{for } q = 2. \end{cases}$$

In order to calculate $E_1^{-1,q}$ we consider the dual Euler sequence

$$0 \to \Omega_{\mathbb{P}_2}^1(1) \to \mathcal{O}_{\mathbb{P}_2}^{\oplus 3} \to \mathcal{O}_{\mathbb{P}_2}(1) \to 0$$

over $\mathbb{P}_2$. By tensoring with $E(-1)$ we get

$$0 \to E \otimes \Omega_{\mathbb{P}_2}^1 \to E(-1)^{\oplus 3} \to E \to 0$$

and thus $h^0(E \otimes \Omega_{\mathbb{P}_2}^1) = 0$. Similarly it follows that

$$h^2(E \otimes \Omega_{\mathbb{P}_2}^1) = 0.$$

Thus

$$E_1^{-1,q} = \begin{cases} 0 & \text{for } q = 0 \\ H^1(\mathbb{P}_2, E \otimes \Omega_{\mathbb{P}_2}^1) \otimes \mathcal{O}_{\mathbb{P}_2}(-1) & \text{for } q = 1 \\ 0 & \text{for } q = 2. \end{cases}$$

The calculation of $E_1^{-2,q}$ gives

$$E_1^{-2,q} = \begin{cases} 0 & \text{for } q = 0 \\ H^1(\mathbb{P}_2, E(-2)) \otimes \mathcal{O}_{\mathbb{P}_2}(-2) & \text{for } q = 1 \\ 0 & \text{for } q = 2. \end{cases}$$

Thus we get the following E_1 diagram:

0	0	0
$E_1^{-2,1}$	$E_1^{-1,1}$	$E_1^{0,1}$
0	0	0

(axes: p horizontal, q vertical)

This shows that $E_2^{p,q} = E_\infty^{pq}$, i.e., $E(-1)$ is the cohomology of the monad

$$0 \to H^1(\mathbb{P}_2, E(-2)) \otimes \mathcal{O}_{\mathbb{P}_2}(-2) \xrightarrow{a} H^1(\mathbb{P}_2, E \otimes \Omega_{\mathbb{P}_2}^1) \otimes \mathcal{O}_{\mathbb{P}_2}(-1)$$
$$\xrightarrow{b} H^1(\mathbb{P}_2, E(-1)) \otimes \mathcal{O}_{\mathbb{P}_2} \to 0.$$

By tensoring with $\mathcal{O}_{\mathbb{P}_2}(1)$ we get the presentation of E as the cohomology of a monad of the form

$$0 \to H \otimes \mathcal{O}_{\mathbb{P}_2}(-1) \to K \otimes \mathcal{O}_{\mathbb{P}_2} \to L \otimes \mathcal{O}_{\mathbb{P}_2}(1) \to 0,$$

where H, K, L are finite-dimensional $\mathbb{C}$-vector spaces.

EXAMPLE 4. Let E be a holomorphic r-bundle over $\mathbb{P}_2$ with

$$h^0(\mathbb{P}_2, E) = h^0(\mathbb{P}_2, E^*(-1)) = 0$$

(a stable normalized r-bundle over $\mathbb{P}_2$ for example will satisfy these conditions). Then E is the cohomology of a monad

$$0 \to H^1(E(-2)) \otimes \mathcal{O}_{\mathbb{P}_2}(-1) \to H^1(E(-1)) \otimes \Omega^1_{\mathbb{P}_2}(1) \to H^1(E) \otimes \mathcal{O}_{\mathbb{P}_2} \to 0.$$

To see this one applies Theorem I of Beilinson to E as in Example 2 and gets

$$E_1^{p,0} = E_1^{p,2} = 0 \quad \text{for all } p$$

and

$$E_\infty^{p,q} = E_2^{p,q}.$$

As in Example 1 the claim follows from this.

EXAMPLE 5. Let $E \cong E^*$ be a self-dual holomorphic r-bundle over $\mathbb{P}_3$ with $H^0(\mathbb{P}_3, E) = 0$ and $H^1(\mathbb{P}_3, E(-2)) = 0$. Then E is the cohomology of a monad.

To show this we consider the spectral sequence II applied to $E(-2)$. With the help of the two exact sequences

$$0 \to \Omega^1_{\mathbb{P}_3} \to \mathcal{O}_{\mathbb{P}_3}(-1)^{\oplus 4} \to \mathcal{O}_{\mathbb{P}_3} \to 0$$
$$0 \to \Omega^2_{\mathbb{P}_3} \to \mathcal{O}_{\mathbb{P}_3}(-2)^{\oplus 6} \to \Omega^1_{\mathbb{P}_3} \to 0$$

over $\mathbb{P}_3$ the by now familiar argument gives

$$E_1^{p,0} = E_1^{p,3} = 0 \quad \text{for} \quad p = 0, -1, -2, -3.$$

In the $E_1^{p,1}$-row the condition $h^1(\mathbb{P}_3, E(-2)) = 0$ plays a rôle; we get

$$E_1^{p,1} = 0 \quad \text{for} \quad p = 0, -1, -2.$$

Finally

$$E_1^{0,2} = H^2(\mathbb{P}_3, E(-2)) \otimes \mathcal{O}_{\mathbb{P}_3}$$

also vanishes, since $H^2(\mathbb{P}_3, E(-2))$ is dual to $H^1(\mathbb{P}_3, E(-2))$. Thus we get the E_1-diagram:

0	0	0	0
$E_1^{-3,2}$	$E_1^{-2,2}$	$E_1^{-1,2}$	0
$E_1^{-3,1}$	0	0	0
0	0	0	0

Again E_2 is already the E_∞-term; from the theorem of Beilinson it thus follows that

$$E_1^{-3,1} = 0, \qquad \text{i.e.,} \quad H^1(\mathbb{P}_3, E(-3)) = 0.$$

$$0 \to E_1^{-3,2} \xrightarrow{d_1^{-3,2}} E_1^{-2,2} \xrightarrow{d_1^{-2,2}} E_1^{-1,2} \to 0$$

is a monad with cohomology $E(-2)$. By tensoring with $\mathcal{O}_{\mathbb{P}_3}(2)$ we get from this a description of E as cohomology of the monad

$$0 \to H^2(\mathbb{P}_3, E(-3)) \otimes \mathcal{O}_{\mathbb{P}_3}(-1) \xrightarrow{a} H^2(\mathbb{P}_3, E \otimes \Omega^2_{\mathbb{P}_3}) \otimes \mathcal{O}_{\mathbb{P}_3}$$
$$\xrightarrow{b} H^2(\mathbb{P}_3, E \otimes \Omega^1_{\mathbb{P}_3}(-1)) \otimes \mathcal{O}_{\mathbb{P}_3}(1) \to 0.$$

Conversely if E is the cohomology of a monad

$$0 \to H \otimes \mathcal{O}_{\mathbb{P}_3}(-1) \xrightarrow{a} K \otimes \mathcal{O}_{\mathbb{P}_3} \xrightarrow{b} L \otimes \mathcal{O}_{\mathbb{P}_3}(1) \to 0$$

over $\mathbb{P}_3$, then from the display of the monad one gets

$$h^1(\mathbb{P}_3, E(-k)) = 0 \qquad \text{for all} \quad k \geq 2.$$

EXAMPLE 6. We consider a stable 2-bundle E over $\mathbb{P}_3$ with $c_1(E) = 0$, $c_2(E) = 1$ and $H^1(\mathbb{P}_3, E(-2)) = 0$.

Remark. In §4.1 we shall show that the hypothesis $H^1(\mathbb{P}_3, E(-2)) = 0$ is automatically satisfied for stable 2-bundles over $\mathbb{P}_3$ with $c_1(E) = 0$, $c_2(E) = 1$.

The spectral sequence I with

$$E_1^{p,q} = H^q(\mathbb{P}_3, E(p)) \otimes \Omega^{-p}_{\mathbb{P}_3}(-p)$$

gives the following E_1-diagram:

E_1^{pq}

0	0	0	0
$E_1^{-3,2}$	0	$E_1^{-1,2}$	$E_1^{0,2}$
$E_1^{-3,1}$	0	$E_1^{-1,1}$	$E_1^{0,1}$
0	0	0	0

The differentials $d_1^{p,q}$ vanish with the possible exception of

$$\alpha = d_1^{-1,2} \colon E_1^{-1,2} \to E_1^{0,2}$$
$$\beta = d_1^{-1,1} \colon E_1^{-1,1} \to E_1^{0,1}.$$

One then gets the following E_2-diagram:

E_2^{pq}

0	0	0	0
$E_1^{-3,2}$	0	$\operatorname{Ker}\alpha$	$\operatorname{Coker}\alpha$
$E_1^{-3,1}$	0	$\operatorname{Ker}\beta$	$\operatorname{Coker}\beta$
0	0	0	0

The only possibly non-vanishing differential $d_1^{pq} \colon E_2^{p,q} \to E_2^{p+2,q-1}$ is

$$\gamma = d_2^{-3,2} \colon E_2^{-3,2} = E_1^{-3,2} \to \ker\beta = E_2^{-1,1}.$$

The $E_3 = E_\infty$-diagram is thus:

$E_\infty^{pq} = E_3^{pq}$

0	0	0	0
$\operatorname{Ker}\gamma$	0	$\operatorname{Ker}\alpha$	$\operatorname{Coker}\alpha$
$E_1^{-3,1}$	0	$\frac{\operatorname{Ker}\beta}{\operatorname{Im}\gamma}$	$\operatorname{Coker}\beta$
0	0	0	0

From the theorem of Beilinson it follows that

$$\operatorname{Ker}\alpha = 0, \quad \operatorname{Coker}\alpha = 0$$
$$\operatorname{Coker}\beta = 0,$$
$$\operatorname{Ker}\gamma = 0, \quad E_1^{-3,1} = 0,$$
$$E \cong \operatorname{Ker}\beta/\operatorname{Im}\gamma.$$

Thus we obtain the following diagram with exact rows and columns:

$$
\begin{array}{ccccccc}
 & & & & 0 & & \\
 & & & & \downarrow & & \\
0 \longrightarrow & H^2(E(-3)) \otimes \mathcal{O}_{\mathbb{P}_3}(-1) & \xrightarrow{\gamma} & & \operatorname{Ker}\beta & \longrightarrow & E \longrightarrow 0 \\
 & & & & \downarrow & & \\
(*) & & & & H^1(E(-1)) \otimes \Omega^1_{\mathbb{P}_3}(1) & & \\
 & & & & \downarrow & & \\
 & & & & H^1(E) \otimes \mathcal{O}_{\mathbb{P}_3} & & \\
 & & & & \downarrow & & \\
 & & & & 0 & &
\end{array}
$$

Now we have $h^0(E) = 0$, $h^3(E) = h^0(E(-4)) = 0$ and $h^2(E) = h^1(E(-4)) = 0$ (cf. Example 5). Thus it follows that $h^1(E) = -\chi(E)$. The Riemann–Roch formula gives $\chi(E) = 2 - 2c_2 = 0$, so $h^1(E) = 0$. Therefore

$$\ker \beta = H^1(E(-1)) \otimes \Omega^1(1).$$

Furthermore $h^2(E(-3)) = h^1(E(-1))$ and $h^q(E(-1)) = 0$ for $q \neq 1$. The Riemann–Roch formula for $E(-1)$ then gives

$$h^1(E(-1)) = -\chi(E(-1)) = c_2 = 1.$$

Thus the diagram (*) is equivalent to the exact sequence

$$0 \to \mathcal{O}_{\mathbb{P}_3}(-1) \to \Omega^1_{\mathbb{P}_3}(1) \to E \to 0;$$

i.e., the stable 2-bundles E over $\mathbb{P}_3$ with $c_1 = 0$, $c_2 = 1$ and $h^1(E(-2)) = 0$ are precisely the cokernels of bundle monomorphisms

$$a\colon \mathcal{O}_{\mathbb{P}_3}(-1) \hookrightarrow \Omega^1_{\mathbb{P}_3}(1).$$

3.3. A stable 2-bundle over $\mathbb{P}_4$. Horrocks and Mumford have shown that there is a holomorphic 2-bundle E over $\mathbb{P}_4$ with Chern classes

$$c_1(E) = 5, \qquad c_2(E) = 10.$$

A generic section of this bundle has as zero surface a 2-dimensional complex torus embedded in $\mathbb{P}_4$ without singularities. We shall now show how one can describe this bundle in a simple way with the help of monads.

In order to see which monad to take we assume we have a stable 2-bundle E over $\mathbb{P}_4$ with $c_1(E) = 5$ and with a section $s \in H^0(\mathbb{P}_4, E)$ whose zero locus $Y \subset \mathbb{P}_4$ is a smooth 2-dimensional complex torus. s then gives an exact sequence

$$(*) \qquad 0 \to \mathcal{O}_{\mathbb{P}_4} \to E \to J_Y(5) \to 0.$$

The stability of E means

$$(**) \qquad H^0(\mathbb{P}_4, J_Y(2)) = 0.$$

Remark. The 2-bundle E belonging to a smooth torus $Y \subset \mathbb{P}_4$ is automatically stable with $c_1(E) = 5$, $c_2(E) = 10$.

We now consider the spectral sequence I of Beilinson for the normalized bundle $E' = E(-3)$, $c_1(E') = -1$:

$$E_1^{pq} = H^q(\mathbb{P}_4, E'(p)) \otimes \Omega_{\mathbb{P}_4}^{-p}(-p).$$

Since E is stable we have

$$E_1^{p,0} = 0 \qquad \text{for} \quad -4 \le p \le 0$$
$$E_1^{p,4} = 0 \qquad \text{for} \quad -4 \le p \le 0.$$

We now calculate the first and third rows ($q = 1, 3$) of the E_1-diagram.

From the sequence (*) follows

$$H^1(E'(p)) \cong H^1(J_Y(p+2)) \qquad \text{for} \quad p < 3.$$

The sequence

$$0 \to J_Y(p+2) \to \mathcal{O}_{\mathbb{P}_4}(p+2) \to \mathcal{O}_Y(p+2) \to 0$$

gives us the cohomology sequence

$$0 \to H^0(J_Y(p+2)) \to H^0(\mathcal{O}_{\mathbb{P}_4}(p+2)) \\ \to H^0(\mathcal{O}_Y(p+2)) \to H^1(J_Y(p+2)) \to 0.$$

Because of (**) it then follows that

$$h^1(\mathbb{P}_4, J_Y(p+2)) = \begin{cases} h^0(Y, \mathcal{O}_Y(2)) - 15 & \text{for } p = 0 \\ h^0(Y, \mathcal{O}_Y(1)) - 5 & \text{for } p = -1 \\ 0 & \text{for } p = -2, -3, -4. \end{cases}$$

Since $\Omega_Y^2 = \mathcal{O}_Y$ and $\mathcal{O}_Y(k)$ is a positive line bundle over Y for $k \ge 1$, it follows from the Kodaira vanishing theorem (cf. Griffiths and Harris [49], p. 154) that

$$H^i(Y, \mathcal{O}_Y(k)) = 0 \qquad \text{for} \quad i > 0, \quad k \ge 1.$$

The Riemann–Roch formula for surfaces (cf. Griffiths and Harris [49], pp. 472, 600) then gives

$$\begin{aligned} h^0(Y, \mathcal{O}_Y(k)) &= \chi(\mathcal{O}_Y(k)) \\ &= \chi(\mathcal{O}_Y) + \tfrac{1}{2} c_1(\mathcal{O}_Y(k))^2 \\ &= \tfrac{1}{2} k^2 \cdot \deg Y = 5k^2 \quad (\deg Y = 10); \end{aligned}$$

thus

$$h^1(\mathbb{P}_4, J_Y(2)) = 5$$
$$h^1(\mathbb{P}_4, J_Y(1)) = 0.$$

Hence we have computed the terms $E_1^{p,1}$ and by Serre duality also the terms $E_1^{p,3}$:

$$E_1^{p,1} \cong \begin{cases} 0 & \text{for } p \neq 0 \\ \mathcal{O}_{\mathbb{P}_4}^{\oplus 5} & \text{for } p = 0 \end{cases}$$

$$E_1^{p,3} \cong \begin{cases} 0 & \text{for } p \neq -4 \\ \mathcal{O}_{\mathbb{P}_4}(-1)^{\oplus 5} & \text{for } p = -4. \end{cases}$$

We still have to determine the terms $E_1^{p,2}$. We have

$$H^2(\mathbb{P}_4, E'(p)) \cong H^2(\mathbb{P}_4, J_Y(p+2)) \cong H^1(Y, \mathcal{O}_Y(p+2)).$$

By the Kodaira vanishing theorem

$$H^1(Y, \mathcal{O}_Y(p+2)) = 0 \qquad \text{for} \quad p \geq -1,$$

and thus

$$H^2(\mathbb{P}_4, E') = H^2(\mathbb{P}_4, E'(-1)) = 0.$$

Since

$$H^2(\mathbb{P}_4, E'(-p-4)) \cong H^2(\mathbb{P}_4, E'(p)),$$

we also have

$$H^2(\mathbb{P}_4, E'(-4)) = H^2(\mathbb{P}_4, E'(-3)) = 0.$$

The Hodge decomposition of $H^1(Y, \mathbb{C})$ (cf. Griffiths, Harris [49], p. 116) then gives

$$h^1(Y, \mathbb{C}) = h^1(Y, \mathcal{O}_Y) + h^0(Y, \Omega_Y^1),$$

and hence

$$4 = h^1(Y, \mathcal{O}_Y) + h^0(Y, \mathcal{O}_Y^{\oplus 2}).$$

It follows that

$$h^2(E'(-2)) = h^1(Y, \mathcal{O}_Y) = 2,$$

$$E_1^{p,2} \cong \begin{cases} 0 & \text{for } p \neq -2 \\ \Omega_{\mathbb{P}_4}^2(2)^{\oplus 2} & \text{for } p = -2. \end{cases}$$

We get the following E_1-diagram:

q

0	0	0	0	0
$\mathcal{O}_{\mathbb{P}_4}(-1)^{\oplus 5}$	0	0	0	0
0	0	$\Omega^2_{\mathbb{P}_4}(2)^{\oplus 2}$	0	0
0	0	0	0	$\mathcal{O}_{\mathbb{P}_4}^{\oplus 5}$
0	0	0	0	0

p

By the theorem of Beilinson then E' is the cohomology of a monad of the form

$$0 \to \mathcal{O}_{\mathbb{P}_4}(-1)^{\oplus 5} \to \Omega^2_{\mathbb{P}_4}(2)^{\oplus 2} \to \mathcal{O}_{\mathbb{P}_4}^{\oplus 5} \to 0.$$

We now try to go in the opposite direction and construct a monad of the form

$$M(a,b): \qquad 0 \to \mathcal{O}_{\mathbb{P}_4}^{\oplus 5} \xrightarrow{a} \Omega^2_{\mathbb{P}_4}(3)^{\oplus 2} \xrightarrow{b} \mathcal{O}_{\mathbb{P}_4}(1)^{\oplus 5} \to 0.$$

Let V be a 5-dimensional vector space, $\mathbb{P} = \mathbb{P}(V)$ and

$$0 \to \mathcal{O}_{\mathbb{P}}(-1) \to V \otimes \mathcal{O}_{\mathbb{P}} \xrightarrow{\pi} Q \to 0$$

the Euler sequence. Then we have

$$\Omega^2_{\mathbb{P}}(3) \cong \Lambda^2 Q$$
$$\mathcal{O}_{\mathbb{P}}(1) \cong \Lambda^4 Q;$$

thus we seek a monad of the form

$$M(a,b): \qquad 0 \to V \otimes \mathcal{O}_{\mathbb{P}} \xrightarrow{a} (\Lambda^2 Q)^{\oplus 2} \xrightarrow{b} V^* \otimes \Lambda^4 Q \to 0.$$

The second exterior power of the Euler sequence is

$$(1) \qquad 0 \to Q \otimes \mathcal{O}_{\mathbb{P}}(-1) \to \Lambda^2 V \otimes \mathcal{O}_{\mathbb{P}} \xrightarrow{\Lambda^2 \pi} \Lambda^2 Q \to 0.$$

To define a it thus suffices to give two linear maps

$$a^+, a^- : V \to \Lambda^2 V$$

such that the composition

$$\begin{array}{ccc} V & \xrightarrow{(a^+,a^-)} & (\Lambda^2 V)^{\oplus 2} \\ & {\scriptstyle a(x)} \searrow & \downarrow {\scriptstyle (\Lambda^2\pi(x))^{\oplus 2}} \\ & & (\Lambda^2 Q(x))^{\oplus 2} \end{array}$$

is injective for every point $x = \mathbb{P}(\mathbb{C}u) \in \mathbb{P}$.

Let $\{v_0, \ldots, v_4\} \subset V$ be a basis. We define

$$a^+(v_i) = v_{i+2} \wedge v_{i+3} \qquad (i \bmod 5)$$
$$a^-(v_j) = v_{j+1} \wedge v_{j+4} \qquad (j \bmod 5).$$

Claim. The composition

$$a = (\Lambda^2\pi)^{\oplus 2} \circ ((a^+, a^-) \otimes \mathrm{id}_{\mathcal{O}_{\mathbb{P}}})$$

is a monomorphism of bundles.

PROOF. Let $x = \mathbb{P}(\mathbb{C}u) \in \mathbb{P}$. The sequence (1) shows: for any $v \in V$ we have

$$(\Lambda^2\pi(x))^{\oplus 2} \circ (a^+, a^-)(v) = 0$$

if and only if there are vectors $w^+, w^- \in V$ with

$$a^+(v) = w^+ \wedge u$$
$$a^-(v) = w^- \wedge u.$$

In particular it then follows that

$$a^+(v) \wedge a^-(v) = 0.$$

By definition of a^+, a^- however we have for $v = \sum x_i v_i$ that

$$a^+(v) \wedge a^-(v) = \sum (-1)^i x_i^2 v_0 \wedge \cdots \wedge \hat{v}_i \wedge \cdots \wedge v_4.$$

From $a^+(v) \wedge a^-(v) = 0$ it therefore follows that $v = 0$, i.e., $a(x)$ is injective for all $x \in \mathbb{P}$. □

We must still define an epimorphism of bundles

$$b\colon (\Lambda^2 Q)^{\oplus 2} \to V^* \otimes \Lambda^4 Q$$

with $b \circ a = 0$. The map b will be essentially the dual of a. We identify $(\Lambda^2 Q)^{\oplus 2}$ with

$$\mathcal{H}om((\Lambda^2 Q)^{\oplus 2}, \Lambda^4 Q)$$

by means of the following isomorphism

$$\Phi\colon (\Lambda^2 Q)^{\oplus 2} \to \mathcal{H}om((\Lambda^2 Q)^{\oplus 2}, \Lambda^4 Q)$$
$$\Phi(\xi, \eta)(\alpha, \beta) = -\eta \wedge \alpha + \xi \wedge \beta.$$

(Φ is $\phi \otimes \left(\begin{smallmatrix} 0 & 1 \\ -1 & 0 \end{smallmatrix}\right)$, where $\phi\colon \Lambda^2 Q \to \mathcal{H}om(\Lambda^2 Q, \Lambda^4 Q)$ is the canonical isomorphism $\phi(\xi)(\alpha) = \xi \wedge \alpha$.)

Now let b be the composition

$$b = a^* \circ \Phi,$$

$$b\colon (\Lambda^2 Q)^{\oplus 2} \xrightarrow{\Phi} \mathcal{H}om(\Lambda^2 Q)^{\oplus 2}, \Lambda^4 Q) \xrightarrow{a^*} \mathcal{H}om(V \otimes \mathcal{O}_{\mathbb{P}}, \Lambda^4 Q).$$

For $(\xi,\eta) \in (\Lambda^2 Q(x))^{\oplus 2}$, $v \in V$, we thus have

$$b(x)(\xi,\eta)(v) = -\eta \wedge (\Lambda^2\pi(x))(a^+(v)) + \xi \wedge (\Lambda^2\pi(x))(a^-(v)).$$

Since a is a bundle monomorphism, b is by construction a bundle epimorphism. To show that $b \circ a = 0$ we calculate

$$b(x)(a(x)(v_i))(v_j) \in \Lambda^4 Q(x) \qquad \text{for} \qquad i,j = 0,1,\dots,4;\ x \in \mathbb{P}.$$

We have

$$\begin{aligned} b(x)(a(x)(v_i))(v_j) &= -\Lambda^2\pi(x)(a^-(v_i)) \wedge \Lambda^2\pi(x)(a^+(v_j)) \\ &\qquad + \Lambda^2\pi(x)(a^+(v_i)) \wedge \Lambda^2\pi(x)(a^-(v_j)) \\ &= \Lambda^4\pi(x)(-a^-(v_i) \wedge a^+(v_j) + a^+(v_i) \wedge a^-(v_j)) \\ &= \Lambda^4\pi(x)(-v_{i+1} \wedge v_{i+4} \wedge v_{j+2} \wedge v_{j+3} \\ &\qquad + v_{i+2} \wedge v_{i+3} \wedge v_{j+1} \wedge v_{j+4}) \\ &\qquad\qquad (i,j \bmod 5). \end{aligned}$$

For $i \neq j$ the summands

$$v_{i+1} \wedge v_{i+4} \wedge v_{j+2} \wedge v_{j+3}$$
$$v_{i+2} \wedge v_{i+3} \wedge v_{j+1} \wedge v_{j+4}$$

both vanish; for $i = j$ they are equal and thus their difference is zero. Thus we have shown that

$$b(x) \circ a(x) = 0 \qquad \text{for all } x \in \mathbb{P}.$$

The desired monad

$$M: \qquad 0 \to V \otimes \mathcal{O}_{\mathbb{P}} \xrightarrow{a} (\Lambda^2 Q)^{\oplus 2} \xrightarrow{b} V^* \otimes \Lambda^4 Q \to 0$$

has thus been constructed.

Let E'' be the cohomology of M, $E = E''(2)$. From the display of the monad $M(2)$ got by tensoring M with $\mathcal{O}_{\mathbb{P}}(2)$ we immediately get the

THEOREM. *The cohomology bundle E of the monad*

$$M(2): \quad 0 \to V \otimes \mathcal{O}_{\mathbb{P}}(2) \xrightarrow{a(2)} \Lambda^2 Q^{\oplus 2} \otimes \mathcal{O}_{\mathbb{P}}(2) \xrightarrow{b(2)} V^* \otimes \mathcal{O}_{\mathbb{P}}(3) \to 0$$

is a stable 2-bundle over $\mathbb{P}_4$ with the Chern classes

$$c_1(E) = 5, \qquad c_2(E) = 10.$$

Since $h^2(\mathbb{P}_4, E) = h^3(\mathbb{P}_4, E) = h^4(\mathbb{P}_4, E) = 0$, it follows from the Riemann–Roch formula that

$$h^0(\mathbb{P}_4, E) \geq 2.$$

To see that a generic section has a smooth zero locus some further considerations are needed.

3.4. Historical remarks. Monads were first introduced by Horrocks. Their usefulness is not obvious at first glance. In the next paragraph we shall see how effective they are in the construction (and study of the properties) of moduli spaces for stable 2-bundles over $\mathbb{P}_2$. In this concrete situation monads were investigated and used by Drinfeld, Manin [28] and Barth, Hulek [15]. The theorem of Beilinson was proved in special cases by Drinfeld and Manin.

That $\Omega^1_{\mathbb{P}_2}(1)$ is the only stable 2-bundle over $\mathbb{P}_2$ with $c_1 = -1$, $c_2 = 1$ (cf. Example 1) was already known to Takemoto [122]. The Examples 3) and 4) will play an essential rôle in the next paragraph.

The bundles considered in Example 5) have to do with instantons. We shall have more to say about them in the concluding remarks of the next paragraph. Here we content ourselves with the following observation.

REMARK (FERRAND). Let E be a semistable 2-bundle over $\mathbb{P}_3$ with $c_1(E) = 0$ and $H^1(\mathbb{P}_3, E(-2)) = 0$. Then E is either stable or trivial.

PROOF. Suppose E is not stable. Then there is a section $s \in H^0(\mathbb{P}_3, E)$, $s \neq 0$. If s has no zeros, then E is trivial. Otherwise one has an exact sequence

$$0 \to \mathcal{O} \xrightarrow{s} E \to J_Y \to 0,$$

where the zero locus Y of s is a locally complete intersection curve. We have

$$\omega_Y = \mathcal{O}_Y(-4).$$

From $H^1(E(-2)) = 0$ it follows that $H^1(J_Y(-2)) = 0$ and thus that $H^0(\mathcal{O}_Y(-2)) = 0$. Without restriction we may assume that Y is irreducible and reduced. From Serre duality it follows that

$$p_a(Y) = h^1(Y, \mathcal{O}_Y) = h^0(\omega_Y).$$

Altogether this yields $p_a(Y) = 0$. Then Y is a $\mathbb{P}_1$ (cf. Hartshorne [56], p. 298). That is a contradiction to $\omega_Y = \mathcal{O}_Y(-4)$. □

It was Mumford who realized that there is a 2-dimensional smooth torus, i.e., an abelian surface, in $\mathbb{P}_4$. Then Horrocks [67], proceeding from this knowledge, computed the form of the monad which has as cohomology the bundle E belonging to the torus. The construction of the monad (i.e., of the maps) is carried out by Horrocks and Mumford using the tools of representation theory. The explicit description given here we learned from Barth (cf. also [16]).

It seems difficult to verify the deeper results about the Horrocks–Mumford bundle without representation theory, for example that the zero locus of a generic section is a smooth surface. By a theorem of

Kodaira this surface must then automatically be a torus. Horrocks and Mumford then further show that every smooth 2-dimensional torus in $\mathbb{P}_4$ is projectively equivalent to the zero locus of a section of E. Moreover they compute the complete cohomology of E.

§4. Moduli of stable 2-bundles

In this paragraph we construct with the help of monads the moduli spaces $M_{\mathbb{P}_2}(c_1, c_2)$ for stable 2-bundles over $\mathbb{P}_2$ with Chern classes c_1, c_2. We show that these spaces are smooth and that they are even fine moduli spaces for c_1 odd or c_1 even and $c_2 - c_1^2/4$ odd.

Next we translate the monads used for the construction into so-called Kronecker modules and with these show that the spaces

$$M_{\mathbb{P}_2}(c_1, c_2)$$

are irreducible for even c_1. Then we describe the simplest examples, the spaces $M_{\mathbb{P}_2}(-1, 2)$ and $M_{\mathbb{P}_2}(0, 2)$, as concrete subsets of projective spaces. As a last example we consider the moduli space of the null-correlation bundles; these are precisely the stable 2-bundles over $\mathbb{P}_3$ with Chern classes $c_1 = 0$, $c_2 = 1$.

4.1. Construction of the moduli spaces for stable 2-bundles over $\mathbb{P}_2$. In Chapter I §6 we described the topological classification of complex bundles over $\mathbb{P}_1$, $\mathbb{P}_2$ and $\mathbb{P}_3$. We saw that the topological type of a bundle is determined by finitely many discrete invariants — rank, Chern classes, α-invariant. This results from the fact that the topological classification is essentially a homotopy-theoretical problem.

In this last paragraph we wish to investigate the different structures of a fixed topological bundle. In general the holomorphic classification will no longer be given by discrete invariants; the holomorphic structures of a complex bundle depend on continuous parameters (moduli). The moduli problem consists of finding a complex (moduli) space whose points correspond exactly to the different holomorphic bundles of a fixed topological type and which has moreover certain functorial properties.

Maruyama has shown [84] that the set of isomorphism classes of Gieseker stable torsion-free sheaves with a fixed Hilbert polynomial can be parametrized by a so-called coarse moduli space. In particular the *existence* of a moduli space for *stable* bundles with given Chern classes is guaranteed. In this paragraph we want to describe explicitly some of these moduli spaces.

We begin with the formal definition of a moduli space: let S be a complex space, $c_1, \ldots, c_r \in \mathbb{Z}$. A family of stable r-bundles over $\mathbb{P}_n$

with Chern classes $c_1, \ldots, c_r$ and parametrized by S is a holomorphic r-bundle E over $S \times \mathbb{P}_n$ such that for all $s \in S$ the bundle

$$E(s) = E|_{\{s\}\times\mathbb{P}_n}$$

over $\mathbb{P}_n \cong \{s\} \times \mathbb{P}_n$ is stable and has Chern classes

$$c_i(E(s)) = c_i.$$

Let $p\colon S\times\mathbb{P}_n \to S$ be the projection onto the first factor. Two families E and E' parametrized by S are equivalent if there is a line bundle L over S such that

$$E' \cong E \otimes p^*L.$$

We denote by $\Sigma^{\mathbb{P}_n}_{c_1,\ldots,c_r}(S)$ the set of equivalence classes of families of stable r-bundles over $\mathbb{P}_n$ with Chern classes $c_1, \ldots, c_r$ and parametrized by S, i.e.,

$$\begin{aligned}\Sigma^{\mathbb{P}_n}_{c_1,\ldots,c_r}(S) = \{[E] \mid\ & E \text{ holomorphic } r\text{-bundle over } S \times \mathbb{P}_n,\\ & E(s) \text{ stable, } c_i(E(s)) = c_i,\ 1 \le i \le r\}.\end{aligned}$$

This defines a contravariant functor

$$\Sigma^{\mathbb{P}_n}_{c_1,\ldots,c_r} : \underline{\mathrm{An}} \to \underline{\mathrm{Ens}}$$

from the category of complex spaces to the category of sets.

Definition 4.1.1. A *fine moduli space* for stable r-bundles over $\mathbb{P}_n$ with given Chern classes $c_1, \ldots, c_r \in \mathbb{Z}$ is a complex space $M = M_{\mathbb{P}_n}(c_1, \ldots, c_r)$ together with a bundle U over $M \times \mathbb{P}_n$ such that the contravariant functor $\Sigma^{\mathbb{P}_n}_{c_1,\ldots,c_r}$ is represented by (M, U).

Le Potier has shown [97] that there is no fine moduli space for stable 2-bundles over $\mathbb{P}_2$ with $c_1 = 0$ and c_2 even.

Definition 4.1.2. A complex space $M = M_{\mathbb{P}_n}(c_1, \ldots, c_r)$ is a *coarse moduli space* for $\Sigma^{\mathbb{P}_n}_{c_1,\ldots,c_r}$ if the following conditions are satisfied:

i) there is a natural transformation of contravariant functors

$$\Sigma^{\mathbb{P}_n}_{c_1,\ldots,c_r} \to \mathrm{Hom}(-, M_{\mathbb{P}_n}(c_1, \ldots, c_r)),$$

which is bijective for any (reduced) point x_0;

ii) for every complex space N and every natural transformation $\Sigma^{\mathbb{P}_n}_{c_1,\ldots,c_r} \to \mathrm{Hom}(-, N)$ there is a unique holomorphic mapping

$$f\colon M_{\mathbb{P}_n}(c_1, \ldots, c_r) \to N$$

for which the diagram

$$\begin{array}{ccc} \Sigma^{\mathbb{P}_n}_{c_1,\dots,c_r} & \longrightarrow & \mathrm{Hom}(-, M_{\mathbb{P}_n}(c_1,\dots,c_r)) \\ & \searrow \quad \swarrow f_* & \\ & \mathrm{Hom}(-,N) & \end{array}$$

commutes.

We now wish to construct the moduli space $M_{\mathbb{P}_2}(c_1, c_2)$ for stable 2-bundles over $\mathbb{P}_2$ with Chern classes c_1, c_2. We may of course restrict ourselves to the two cases $c_1 = 0$ and $c_1 = -1$. We shall describe the moduli spaces $M_{\mathbb{P}_2}(0, n)$ explicitly and then give the changes necessary for $M_{\mathbb{P}_2}(-1, n)$.

Construction of $M_{\mathbb{P}_2}(0, n)$: Let V be a 3-dimensional $\mathbb{C}$-vector space, $\mathbb{P} = \mathbb{P}(V)$, E a stable 2-bundle over $\mathbb{P}$ with

$$c_1(E) = 0, \quad c_2(E) = n.$$

From the theorem of Beilinson (applied to $E(-1)$) we obtain a spectral sequence with E_1-term

$$E_1^{p,q} = H^q(\mathbb{P}, E(-1) \otimes \Omega_{\mathbb{P}}^{-p}(-p)) \otimes \mathcal{O}_{\mathbb{P}}(p).$$

In Example 3 in §3 we saw that this spectral sequence leads to a monad

$$0 \to H \otimes \mathcal{O}_{\mathbb{P}}(-1) \xrightarrow{a} K \otimes \mathcal{O}_{\mathbb{P}} \xrightarrow{b} H' \otimes \mathcal{O}_{\mathbb{P}}(1) \to 0$$

whose cohomology is the bundle E. Explicitly we had

$$\begin{aligned} H &\cong H^1(\mathbb{P}, E(-2)) \\ K &\cong H^1(\mathbb{P}, E \otimes \Omega^1_{\mathbb{P}}) \\ H' &\cong H^1(\mathbb{P}, E(-1)) \cong H^*. \end{aligned}$$

E is thus the cohomology of a monad of the form

$$0 \to H \otimes \mathcal{O}_{\mathbb{P}}(-1) \xrightarrow{a} K \otimes \mathcal{O}_{\mathbb{P}} \xrightarrow{b} H^* \otimes \mathcal{O}_{\mathbb{P}}(1) \to 0.$$

To see which monads lead to isomorphic bundles we prove the following

LEMMA 4.1.3. *Let $E = H(M)$, $E' = H(M')$ be the cohomology bundles of the two monads*

$$\begin{aligned} M: &\qquad 0 \to A \xrightarrow{a} B \xrightarrow{b} C \to 0 \\ M': &\qquad 0 \to A' \xrightarrow{a'} B' \xrightarrow{b'} C' \to 0 \end{aligned}$$

over a complex manifold X. The mapping

$$h\colon \mathrm{Hom}(M, M') \to \mathrm{Hom}(E, E')$$

which associates to each homomorphism of monads the induced homomorphism of cohomology bundles is bijective if the following hypotheses are satisfied:

$$\operatorname{Hom}(B, A') = \operatorname{Hom}(C, B') = 0$$
$$H^1(X, C^* \otimes A') = H^1(X, B^* \otimes A') = H^1(X, C^* \otimes B')$$
$$= H^2(X, C^* \otimes A') = 0.$$

PROOF. The monad M gives exact sequences

$$0 \to \ker b \to B \to C \to 0$$
$$0 \to A \to \ker b \to E \to 0$$

and analogously for M'. A given homomorphism

$$\phi \colon E \to E'$$

induces a homomorphism

$$\phi' \colon \ker b \to \ker b' / \operatorname{im} a',$$

that is an element ϕ' in

$$H^0(X, (\ker b)^* \otimes E').$$

From the exact sequence

$$0 \to (\ker b)^* \otimes A' \to (\ker b)^* \otimes \ker b' \to (\ker b)^* \otimes E' \to 0$$

we get the exact cohomology sequence

$$0 \to H^0(X, (\ker b)^* \otimes A') \to H^0(X, (\ker b)^* \otimes \ker b')$$
$$\to H^0(X, (\ker b)^* \otimes E') \to H^1(X, (\ker b)^* \otimes A').$$

From the hypothesis of the lemma one sees that

$$H^0(X, (\ker b)^* \otimes A'), \qquad H^1(X, (\ker b)^* \otimes A')$$

vanish, i.e., ϕ' can be uniquely lifted to a homomorphism

$$\psi \colon \ker b \to \ker b'.$$

We consider now the composition

$$\Psi \colon \ker b \to \ker b' \to B'.$$

Ψ has a unique extension to a homomorphism

$$\Phi \colon B \to B'$$

for the appropriate obstruction groups

$$H^0(X, C^* \otimes B'), \qquad H^1(X, C^* \otimes B')$$

vanish by hypothesis. The mapping

$$h\colon \operatorname{Hom}(M, M') \to \operatorname{Hom}(E, E')$$

is thus bijective. □

The hypotheses are satisfied for example for pairs (M, M') of monads of the form

$$0 \to H \otimes \mathcal{O}_{\mathbb{P}}(-1) \to K \otimes \mathcal{O}_{\mathbb{P}} \to H^* \otimes \mathcal{O}_{\mathbb{P}}(1) \to 0$$

over $\mathbb{P} = \mathbb{P}(V)$.

We draw the following two conclusions from this lemma.

COROLLARY 1. *If the hypotheses of the lemma are satisfied for the pairs (M, M'), (M, M), (M', M') and (M', M), then the isomorphisms of the monads M, M' correspond bijectively (under h) to the isomorphisms of the cohomology bundles E, E'.*

COROLLARY 2. *Let*

$$M\colon \qquad 0 \to A \xrightarrow{a} B \xrightarrow{b} C \to 0$$

be a monad with cohomology $E = H(M)$,

$$M^*\colon \qquad 0 \to C^* \xrightarrow{b^*} B^* \xrightarrow{a^*} A^* \to 0$$

the dual monad. Further suppose the hypotheses of the lemma to be satisfied for the pairs (M, M^*), (M^*, M^*), (M, M), (M^*, M). If

$$f\colon E \to E^*$$

is an isomorphism of the cohomology bundles with

$$f^* = -f,$$

then there are isomorphisms

$$h\colon C \to A^*$$
$$q\colon B \to B^*$$

such that

$$q^* = -q, \qquad h \circ b = a^* \circ q.$$

To see this one applies Corollary 1 to the pair (M, M^*) to get a commutative diagram

$$\begin{array}{ccccccccc} 0 & \longrightarrow & A & \xrightarrow{a} & B & \xrightarrow{b} & C & \longrightarrow & 0 \\ & & \downarrow{\scriptstyle g} & & \downarrow{\scriptstyle q} & & \downarrow{\scriptstyle h} & & \\ 0 & \longrightarrow & C^* & \xrightarrow{b^*} & B^* & \xrightarrow{a^*} & A^* & \longrightarrow & 0, \end{array}$$

i.e., $h \circ b = a^* \circ q$.

The isomorphism of monads

$$(h^*, q^*, g^*) \colon M \to M^*$$

induces the isomorphism

$$f^* \colon E \to E^*.$$

Because $f^* = -f$ it follows that

$$(h^*, q^*, g^*) = -(g, q, h),$$

i.e., $q^* = -q$.

We wish to apply these considerations to stable 2-bundles E over $\mathbb{P} = \mathbb{P}(V)$ with $c_1(E) = 0$. If E is such a bundle, then

$$f \colon E \to E^*$$
$$e \mapsto e \wedge -$$

defines an isomorphism from E to E^* for which

$$f^*(e^{**})(e') = e^{**}(f(e')) = e' \wedge e = -e \wedge e' = -f(e)(e'),$$

i.e., $f^* = -f$.

E is the cohomology of a monad

$$(*) \qquad 0 \to H \otimes \mathcal{O}_{\mathbb{P}}(-1) \xrightarrow{a} K \otimes \mathcal{O}_{\mathbb{P}} \xrightarrow{b} H' \otimes \mathcal{O}_{\mathbb{P}}(1) \to 0.$$

By Corollary 2 there is a non-degenerate symplectic form

$$q \colon K \to K^*$$

on K (i.e., with $q^* = -q$) and an isomorphism

$$h \colon H' \otimes \mathcal{O}_{\mathbb{P}}(1) \to H^* \otimes \mathcal{O}_{\mathbb{P}}(1)$$

with

$$b = h^{-1} \circ a^* \circ q.$$

We now replace $H' \otimes \mathcal{O}_{\mathbb{P}}(1)$ by $H^* \otimes \mathcal{O}_{\mathbb{P}}(1)$ and have E as the cohomology bundle of a *self-dual* monad

$$0 \to H \otimes \mathcal{O}_{\mathbb{P}}(-1) \xrightarrow{a} K \otimes \mathcal{O}_{\mathbb{P}} \xrightarrow{a^* \circ (q \otimes \mathrm{id})} H^* \otimes \mathcal{O}_{\mathbb{P}}(1) \to 0.$$

The monomorphism of bundles

$$a \colon H \otimes \mathcal{O}_{\mathbb{P}}(-1) \to K \otimes \mathcal{O}_{\mathbb{P}}$$

corresponds to an element in

$$H^0(\mathbb{P}, H^* \otimes \mathcal{O}_{\mathbb{P}}(1) \otimes K) \cong H^* \otimes K \otimes V^*.$$

We regard this element as a linear mapping

$$\alpha \colon V \to L(H, K)$$

of V into the vector space of linear mappings from H to K, that is

$$\alpha \in L(V, L(H, K)).$$

Explicitly

$$\alpha(v)(h) = a(x)(h \otimes v)$$

for

$$v \in V, \qquad x = \mathbb{P}(\mathbb{C}v) \in \mathbb{P}, \qquad h \in H.$$

For $v \in V$ let

$$\alpha(v)^t \colon K \to H^*$$

be the transposed map of $\alpha(v)$ (with respect to q):

$$\alpha(v)^t(k)(h) = q(k)(\alpha(v)(h)).$$

The linear mappings α which we thus associate to stable 2-bundles E over $\mathbb{P}(V)$ with $c_1(E) = 0$ have the following properties:

(E1): $\alpha(v) \colon H \to K$ is injective for $v \neq 0$;
(E2): $\alpha(v)^t \circ \alpha(v) \colon H \to H^*$ is the zero mapping for all $v \in V$;
(E3): the map $\hat{\alpha} \colon V \otimes H \to K$ is surjective, where $\hat{\alpha}(v \otimes h) = \alpha(v)(h)$.

(E1) is clear, since a is injective on every fibre of the vector bundle. (E2) means that the composition

$$a^* \circ q \circ a$$

is the zero mapping on every fibre. (E3) is equivalent to the stability of the bundle E: from the sequences

$$0 \to H \otimes \mathcal{O}_{\mathbb{P}}(-1) \xrightarrow{a} \ker(a^* \circ q) \to E \to 0$$

$$0 \to \ker(a^* \circ q) \to K \otimes \mathcal{O}_{\mathbb{P}} \xrightarrow{a^* \circ q} H^* \otimes \mathcal{O}_{\mathbb{P}}(1) \to 0$$

one sees namely that

$$\begin{aligned} H^0(\mathbb{P}, E) &= H^0(\mathbb{P}, \ker(a^* \circ q)) \\ &= \ker(K \to H^* \otimes V^*). \end{aligned}$$

E is thus stable if and only if

$$H^0(a^* \circ q) \colon K \to H^* \otimes V^*$$

is injective, i.e., if and only if

$$\hat{\alpha} \colon V \otimes H \to K$$

is surjective.

This construction can now be easily reversed: let V, H, K be $\mathbb{C}$-vector spaces with $\dim_{\mathbb{C}} V = 3$, $\dim_{\mathbb{C}} H = n$, $\dim_{\mathbb{C}} K = 2n+2$. We set $\mathbb{P} = \mathbb{P}(V)$ and choose on K a *fixed* non-degenerate symplectic form

$$q \colon K \to K^*,$$

i.e., $q^* = -q$. Let $Sp(q)$ be the associated automorphism group, i.e.,

$$Sp(q) = \{\phi \in GL(K) \mid \phi^* q \phi = q\}.$$

The group

$$G = GL(H) \times Sp(q)$$

operates on the vector space

$$L(V, L(H, K))$$

by the following prescription:

$$[(g, \phi) \circ \alpha](v) = \phi \circ \alpha(v) \circ g^{-1}.$$

Let

$$P = \{\alpha \in L(V, L(H, K)) \mid \alpha \text{ has properties (E1), (E2), (E3)}\}.$$

P is a G-invariant subset of $L(V, L(H, K))$. We have

THEOREM 4.1.4. *There is a bijection from the set of isomorphism classes of stable 2-bundles over $\mathbb{P}(V)$ with $c_1 = 0$, $c_2 = n$ onto the orbit space P/G. The isotropy group in each point is $\{\pm(\mathrm{id}_H, \mathrm{id}_K)\}$.*

PROOF. Let $\alpha \in P$ and a the morphism defined by α:

$$a \colon H \otimes \mathcal{O}_{\mathbb{P}}(-1) \to K \otimes \mathcal{O}_{\mathbb{P}}$$

with

$$a(x)(h \otimes v) = \alpha(v)(h)$$

over the point $x = \mathbb{P}(\mathbb{C}v) \in \mathbb{P}$. Let

$$a^* \circ q \colon K \otimes \mathcal{O}_{\mathbb{P}} \to H^* \otimes \mathcal{O}_{\mathbb{P}}(1)$$

be the transposed morphism. From (E1) it follows that a is injective on every fibre of the bundle. Because of (E2) then

$$M(\alpha)\colon \qquad 0 \to H \otimes \mathcal{O}_{\mathbb{P}}(-1) \xrightarrow{a} K \otimes \mathcal{O}_{\mathbb{P}} \xrightarrow{a^* \circ q} H^* \otimes \mathcal{O}_{\mathbb{P}}(1) \to 0$$

is a monad. Let $E(\alpha)$ be the cohomology bundle. $E(\alpha)$ is stable ((E3)!) and has Chern classes

$$c_1(E(\alpha)) = 0, \qquad c_2(E(\alpha)) = n.$$

Thus we have defined a mapping

$$I \colon P \to \{\text{isomorphism classes of stable 2-bundles over } \mathbb{P}_2 \text{ with } c_1 = 0,\ c_2 = n\}$$
$$\alpha \mapsto [E(\alpha)].$$

i) I factors through P/G:
for let $\alpha' = (g, \phi) \circ \alpha$ with

$$(g, \phi) \in GL(H) \times Sp(q).$$

Using g and ϕ we form the following diagram of monads:

$$\begin{array}{lccccccccc}
M(\alpha): & 0 \longrightarrow & H \otimes \mathcal{O}_{\mathbb{P}}(-1) & \xrightarrow{a} & K \otimes \mathcal{O}_{\mathbb{P}} & \xrightarrow{a^*q} & H^* \otimes \mathcal{O}_{\mathbb{P}}(1) & \longrightarrow 0 \\
{\scriptstyle (g,\phi)}\downarrow & & \downarrow{\scriptstyle g\otimes\mathrm{id}} & & \downarrow{\scriptstyle \phi\otimes\mathrm{id}} & & \downarrow{\scriptstyle (g^{-1})^*\otimes\mathrm{id}} & \\
M(\alpha'): & 0 \longrightarrow & H \otimes \mathcal{O}_{\mathbb{P}}(-1) & \xrightarrow{a'} & K \otimes \mathcal{O}_{\mathbb{P}} & \xrightarrow{a'^*q} & H^* \mathcal{O}_{\mathbb{P}}(1) & \longrightarrow 0
\end{array}$$

Over $x = \mathbb{P}(\mathbb{C}v) \in \mathbb{P}$ we have

$$\begin{aligned}
[(\phi \otimes \mathrm{id}) \circ a](x)(h \otimes v) &= \phi(\alpha(v)(h)) \\
&= \alpha'(v)(g(h)) = [a' \circ (g \otimes \mathrm{id})](x)(h \otimes v)
\end{aligned}$$

and

$$\begin{aligned}
[a'^*q \circ (\phi \otimes \mathrm{id})](x)(k) &= [a'(x)^* \circ q \circ \phi](k) \\
&= [a'(x)^* \circ \phi^{*-1} \circ q](k) \\
&= [((g^{-1})^* \otimes \mathrm{id}) \circ a^*q](x)(k).
\end{aligned}$$

Thus the diagram commutes, and by passing to cohomology we get an isomorphism of the bundles $E(\alpha)$, $E(\alpha')$.

ii) I is surjective:
we have already seen that every bundle occurs as the cohomology of a self-dual monad in which the middle term has some symplectic form q'. Such a form is unique up to isomorphism. If one then chooses an isomorphism

$$H \cong H^1(\mathbb{P}, E(-2))$$

and an isomorphism of symplectic spaces $(K, q) \cong (H^1(\mathbb{P}, E \otimes \Omega^1), q')$, it follows that $E \cong E(\alpha)$ for a suitable $\alpha \in P$.

iii) The induced mapping

$$\begin{aligned}
I/G\colon P/G \to \{&\text{isomorphism classes of} \\
&\text{stable 2-bundles over } \mathbb{P}_2 \text{ with } c_1 = 0,\, c_2 = n\}
\end{aligned}$$

is injective:
an isomorphism

$$\psi\colon E(\alpha) \to E(\alpha')$$

can by Lemma 4.1.3 be lifted to an isomorphism of the monads $M(\alpha)$, $M(\alpha')$; we thus get a commutative diagram

$$\begin{array}{lccccccccc}
M(\alpha): & 0 \longrightarrow & H \otimes \mathcal{O}_{\mathbb{P}}(-1) & \xrightarrow{a} & K \otimes \mathcal{O}_{\mathbb{P}} & \xrightarrow{a^*q} & H^* \otimes \mathcal{O}_{\mathbb{P}}(1) & \longrightarrow 0 \\
 & & \downarrow{\scriptstyle g\otimes\mathrm{id}} & & \downarrow{\scriptstyle \phi\otimes\mathrm{id}} & & \downarrow{\scriptstyle h^*\otimes\mathrm{id}} & \\
M(\alpha'): & 0 \longrightarrow & H \otimes \mathcal{O}_{\mathbb{P}}(-1) & \xrightarrow{a'} & K \otimes \mathcal{O}_{\mathbb{P}} & \xrightarrow{a'^*q} & H^* \mathcal{O}_{\mathbb{P}}(1) & \longrightarrow 0
\end{array}$$

with $g, h \in GL(H)$, $\phi \in GL(K)$. Because the triple

$$(h \otimes \mathrm{id}, q^{-1}\phi^* q \otimes \mathrm{id}, g^* \otimes \mathrm{id})$$

defines an isomorphism from $M(\alpha')$ onto $M(\alpha)$, the composition

$$(h \otimes \mathrm{id}, q^{-1}\phi^* q \otimes \mathrm{id}, g^* \otimes \mathrm{id}) \circ (g \otimes \mathrm{id}, \phi \otimes \mathrm{id}, h^* \otimes \mathrm{id})\colon M(\alpha) \to M(\alpha)$$

is an automorphism of $M(\alpha)$.

Now $E(\alpha)$ is stable and thus simple, i.e., $\mathbb{C} \cong \mathrm{Hom}(E(\alpha), E(\alpha)) \cong \mathrm{Hom}(M(\alpha), M(\alpha))$. Hence there is a $c \in \mathbb{C} \setminus \{0\}$ with

$$((h \circ g) \otimes \mathrm{id}, q^{-1}\phi^* q\phi \otimes \mathrm{id}, (g^* \circ h^*) \otimes \mathrm{id}) = c(\mathrm{id}, \mathrm{id}, \mathrm{id}).$$

We now set

$$\phi' = \frac{1}{\sqrt{c}}\phi, \qquad g' = \frac{1}{\sqrt{c}}g, \qquad h' = \frac{1}{\sqrt{c}}h.$$

Then we have

$$h' = (g')^{-1}, \quad \phi' \in Sp(q)$$

and

$$\alpha'(v) = \phi' \circ \alpha(v) \circ (g')^{-1},$$

that is

$$\alpha' = (g', \phi') \cdot \alpha.$$

iv) Let G_α denote the isotropy subgroup of $\alpha \in P$. For $(g, \phi) \in G_\alpha$ we have $\alpha = (g, \phi) \cdot \alpha$. As in iii) one sees that the isomorphism

$$(g \otimes \mathrm{id}, \phi \otimes \mathrm{id}, (g^{-1})^* \otimes \mathrm{id})\colon M(\alpha) \to M(\alpha)$$

from $M(\alpha)$ onto $M(\alpha)$ must be of the form $c \cdot \mathrm{id}$, $c \in \mathbb{C} \setminus \{0\}$. Thus

$$g = c\ \mathrm{id}_H, \qquad \phi = c\ \mathrm{id}_K.$$

Because ϕ is symplectic, it follows that

$$c^2 = 1$$

i.e.,

$$(g, \phi) \in \{\pm(\mathrm{id}_H, \mathrm{id}_K)\}.$$

The converse is clear. □

REMARK 4.1.5. If instead of self-dual monads we consider arbitrary monads

$$0 \to H \otimes \mathcal{O}_{\mathbb{P}}(-1) \xrightarrow{a} K \otimes \mathcal{O}_{\mathbb{P}} \xrightarrow{b} H^* \otimes \mathcal{O}(1) \to 0$$

whose cohomology bundle E is stable, then by taking

$$\alpha(v)(h) = a(x)(h \otimes v) \qquad x = \mathbb{P}(\mathbb{C}v) \in \mathbb{P}$$
$$\beta(v)(k)(h) = (b(x)(k))(h \otimes v)$$

we get linear mappings

$$\alpha \in L(V, L(H, K)), \qquad \beta \in L(V, L(K, H^*))$$

with the following properties:

(E′1) $\alpha(v)$ is injective, $\beta(v)$ is surjective for all $v \neq 0$;
(E′2) $\beta(v) \circ \alpha(v) = 0$ for all $v \in V$;
(E′3) $\hat{\beta}\colon K \to V^* \otimes H^*$ is injective, where $\hat{\beta}(k)(v \otimes h) = (\beta(v)(k))(h)$.

Let

$$P' = \{(\alpha, \beta) \in L(V, L(H, K)) \times L(V, L(K, H^*)) \mid \text{(E}'1\text{), (E}'2\text{), (E}'3\text{) hold}\}.$$

Then the group

$$G' = GL(H) \times GL(K) \times GL(H^*)$$

operates on P' by the prescription

$$[(g, \phi, f) \circ (\alpha, \beta)](v) = (\phi\alpha(v)g^{-1}, f\beta(v)\phi^{-1}).$$

If one defines a homomorphism of groups

$$\gamma\colon G \to G'$$

by $\gamma(g, \phi) = (g, \phi, (g^{-1})^*)$, then the map

$$\Phi\colon P \to P'$$

given by

$$\Phi(\alpha) = (\alpha, \alpha^t)$$

is γ-equivariant. The proof of the above theorem shows that $\Phi\colon P \to P'$ induces a bijection

$$P/G \to P'/G'.$$

In particular

$$\hat{\alpha}\colon V \otimes H \to K$$

is always surjective if (α, β) is a pair in P'. By what was proved above in fact there is a symplectic form $q\colon K \to K^*$ on K with $\hat{\beta} = \hat{\alpha}^* \circ q$. Because $\hat{\beta}$ is injective, $\hat{\alpha}$ must be surjective.

REMARK 4.1.6. Let V be a 4-dimensional vector space. In example 5 of §3.2 we showed that all stable 2-bundles over $\mathbb{P}_3 \cong \mathbb{P}(V)$ with $c_1 = 0$, $c_2 = n$ and $H^1(\mathbb{P}_3, E(-2)) = 0$ are the cohomology bundles of monads of the form

$$0 \to H \otimes \mathcal{O}_{\mathbb{P}}(-1) \to K \otimes \mathcal{O}_{\mathbb{P}} \to H^* \otimes \mathcal{O}_{\mathbb{P}}(1) \to 0.$$

The same considerations as those we have just carried out yield the following

THEOREM: *There is a bijection from the set of isomorphism classes of stable 2-bundles over* $\mathbb{P}_3$ *with* $c_1 = 0$, $c_2 = n$ *and* $H^1(\mathbb{P}_3, E(-2)) = 0$ *to the orbit space*

$$\{\alpha \in L(V, L(H, K)) \mid \alpha \text{ satisfies (E1), (E2), (E3)}\}/GL(H) \times Sp(q).$$

Next, we wish to show that P is a manifold; for this we need the following lemma.

LEMMA 4.1.7. *Let*

$$0 \to A \xrightarrow{a_0} B \xrightarrow{b_0} C \to 0$$

be a monad over a complex manifold X *with cohomology* E. *For the mapping*

$$d_0\colon \operatorname{Hom}(A, B) \oplus \operatorname{Hom}(B, C) \to \operatorname{Hom}(A, C)$$

defined by $d_0(a, b) = b_0 a + b a_0$ *we have*

$$\operatorname{coker} d_0 \cong H^2(X, \operatorname{End} E)$$

if the following groups vanish:

$$H^1(X, B^* \otimes C),\ H^1(X, A^* \otimes B),\ H^1(X, B^* \otimes B),\ H^1(X, C^* \otimes C),$$
$$H^1(X, A^* \otimes A),\ H^2(X, B^* \otimes B),\ H^2(X, A^* \otimes A),\ H^2(X, C^* \otimes C),$$
$$H^2(X, B^* \otimes A),\ H^2(X, C^* \otimes B),\ H^3(X, C^* \otimes A),\ H^3(X, B^* \otimes A),$$
$$H^3(X, C^* \otimes B),\ H^4(X, C^* \otimes A).$$

PROOF. We consider the double complex:

$$\begin{array}{ccccc}
C^* \otimes A & \longrightarrow & C^* \otimes B & \longrightarrow & C^* \otimes C \\
\downarrow & & \downarrow & & \downarrow \\
B^* \otimes A & \longrightarrow & B^* \otimes B & \longrightarrow & B^* \otimes C \\
\downarrow & & \downarrow & & \downarrow \\
A^* \otimes A & \longrightarrow & A^* \otimes B & \longrightarrow & A^* \otimes C
\end{array}$$

The associated chain complex is of the form

$$(*) \qquad 0 \to K^{-2} \to K^{-1} \xrightarrow{S} K^0 \xrightarrow{T} K^1 \to K^2 \to 0,$$

where

$$
\begin{aligned}
K^{-2} &= C^* \otimes A \\
K^{-1} &= (B^* \otimes A) \oplus (C^* \otimes B) \\
K^{0} &= (A^* \otimes A) \oplus (B^* \otimes B) \oplus (C^* \otimes C) \\
K^{1} &= (A^* \otimes B) \oplus (B^* \otimes C) \\
K^{2} &= A^* \otimes C.
\end{aligned}
$$

The complex (*) is exact except at K^0 and there we have

$$\ker T/\operatorname{im} S \cong \operatorname{End} E;$$

this is simply the Künneth theorem (cf. Godement [43], p. 100). The mapping d_0 is the induced homomorphism

$$d_0\colon H^0(K^1) \to H^0(K^2).$$

From (*) we get the following exact sequences:

$$
\begin{aligned}
&0 \to \operatorname{im} S \to \ker T \to \operatorname{End} E \to 0 \\
&0 \to \ker T \to K^0 \to \operatorname{im} T \to 0 \\
&0 \to K^{-2} \to K^{-1} \to \operatorname{im} S \to 0 \\
&0 \to \operatorname{im} T \to K^1 \to K^2 \to 0
\end{aligned}
$$

From these sequences and the vanishing of the cohomology groups in the hypothesis of the lemma we deduce

$$
\begin{aligned}
H^2(X, \operatorname{End} E) &\cong H^2(X, \ker T) \\
&\cong H^1(X, \operatorname{im} T) \\
&\cong \operatorname{coker}(d_0\colon H^0(K^1) \to H^0(K^2)).
\end{aligned}
$$

□

Now we can prove the following theorem.

THEOREM 4.1.8. *P is a complex manifold of dimension $3n^2 + 9n$.*

PROOF. Let $S^2(V, \Lambda^2 H^*)$ be the vector space of all mappings

$$\phi\colon V \to \Lambda^2 H^*$$

for which there is a symmetric bilinear mapping $\tilde{\phi}\colon V \times V \to \Lambda^2 H^*$ with $\phi(v) = \tilde{\phi}(v, v)$. We consider the mapping

$$g\colon L(V, L(H, K)) \to S^2(V, \Lambda^2 H^*)$$

with $g(\alpha)(v) = \alpha(v)^t \alpha(v)$, where we set $\alpha(v)^t = \alpha(v)^* \circ q$. Because $P \subset g^{-1}(0)$ is a Zariski-open subset, it suffices to show that g has a surjective differential

$$d_{\alpha_0} g\colon L(V, L(H, K)) \to S^2(V, \Lambda^2 H^*)$$

in every point $\alpha_0 \in P$.

To determine $d_{\alpha_0} g$ we consider

$$g(\alpha_0 + \alpha) - g(\alpha_0).$$

For $v \in V$ we have

$$\begin{aligned}[g(\alpha_0 + \alpha) - g(\alpha_0)](v) &= (\alpha_0(v)^t + \alpha(v)^t)(\alpha_0(v) + \alpha(v)) - \alpha_0(v)^t \alpha_0(v) \\ &= \alpha_0(v)^t \alpha(v) + \alpha(v)^t \alpha_0(v) + g(\alpha)(v)\end{aligned}$$

and thus

$$g(\alpha_0 + \alpha) - g(\alpha_0) = S_{\alpha_0}(\alpha) + g(\alpha)$$

with the mapping $S_{\alpha_0}(\alpha) = \alpha_0^t \alpha + \alpha^t \alpha_0$, which is linear in α. The map $g(\alpha)$ is homogeneous of degree 2 in α. It follows that

$$dg_{\alpha_0} = S_{\alpha_0}.$$

We consider now the mapping

$$d\colon L(V, L(H, K)) \oplus L(V, L(K, H^*)) \to S^2(V, L(H, H^*))$$

defined by

$$d(\alpha, \beta)(v) = \alpha_0(v)^t \alpha(v) + \beta(v)\alpha_0(v).$$

Under the identifications

$$\begin{aligned}\operatorname{Hom}(H \otimes \mathcal{O}_{\mathbb{P}}(-1), K \otimes \mathcal{O}_{\mathbb{P}}) &\cong L(V, L(H, K)) \\ \operatorname{Hom}(K \otimes \mathcal{O}_{\mathbb{P}}, H^* \otimes \mathcal{O}_{\mathbb{P}}(1)) &\cong L(V, L(K, H^*)) \\ \operatorname{Hom}(H \otimes \mathcal{O}_{\mathbb{P}}(-1), H^* \otimes \mathcal{O}_{\mathbb{P}}(1)) &\cong S^2(V, L(H, H^*))\end{aligned}$$

the map d corresponds to the mapping d_0 of Lemma 4.1.7. The hypotheses of that lemma are satisfied. Moreover

$$H^2(\mathbb{P}, \operatorname{End} E) \cong H^0(\mathbb{P}, (\operatorname{End} E)(-3))$$

vanishes, since if there were a nontrivial section

$$s \in H^0(\mathbb{P}, (\operatorname{End} E)(-3)),$$

then by multiplying by a section $t \in H^0(\mathbb{P}, \mathcal{O}_{\mathbb{P}}(3))$ we would get an endomorphism of E, which must then be of the form

$$t \cdot s = c \cdot \mathrm{id}_E, \quad c \in \mathbb{C} \setminus \{0\}.$$

But t has zeros, so this is a contradiction. Hence $\operatorname{coker} d = 0$, i.e., d is surjective.

Now let some $\gamma \in S^2(V, \Lambda^2 H^*)$ be given. Then $\gamma(v)^* = -\gamma(v)$ for all $v \in V$. Since d is surjective, there is a pair (α, β) in

$$L(V, L(H, K)) \oplus L(V, L(K, H^*))$$

with

$$d(\alpha, \beta) = \gamma.$$

We then have

$$\begin{aligned}S_{\alpha_0}\left(\frac{\alpha + q^{*-1}\beta^*}{2}\right) &= \tfrac{1}{2}[\alpha_0^t(\alpha + q^{*-1}\beta^*) + (\alpha^t + \beta)\alpha_0] \\ &= \tfrac{1}{2}[\alpha_0^t\alpha + \beta\alpha_0 + \alpha_0^t q^{*-1}\beta^* + \alpha^t\alpha_0] \\ &= \tfrac{1}{2}[d(\alpha,\beta) - d(\alpha,\beta)^*] = \tfrac{1}{2}(\gamma - \gamma^*) \\ &= \gamma.\end{aligned}$$

Hence $d_{\alpha_0}g = S_{\alpha_0}$ is surjective.

Thus the fibre $g^{-1}(0)$ is smooth and of dimension

$$\begin{aligned}\dim_{\mathbb{C}}(L(V, L(H,K))) - \dim_{\mathbb{C}}(L(S^2V, \Lambda^2H^*)) \\ = 3n(2n+2) - 6\binom{n}{2} = 3n^2 + 9n.\end{aligned}$$

It follows that $\dim P = 3n^2 + 9n$ also. □

We have already seen that the isotropy subgroups of

$$G = GL(H) \times Sp(q)$$

in points of P always consist of the two elements $\pm(\mathrm{id}_H, \mathrm{id}_K)$. Let

$$G_0 = G/\{\pm(\mathrm{id}_H, \mathrm{id}_K)\}.$$

G_0 operates on P freely in the topological sense. We even have

THEOREM 4.1.9. *P is a holomorphic G_0-principal bundle with a $(4n-3)$-dimensional manifold P/G_0 as base space.*

PROOF. The dimension of P/G_0 is

$$\begin{aligned}\dim P - \dim G_0 &= 3n^2 + 9n - \dim(GL(H) \times Sp(q)) \\ &= 3n^2 + 9n - (n^2 + 2(n+1)^2 + n + 1) \\ &= 4n - 3.\end{aligned}$$

In order to show that P is a G_0-principal bundle it suffices to prove that the mapping

$$\begin{aligned}\gamma\colon P \times G_0 &\to P \times P \\ (\alpha, g) &\mapsto (\alpha, g\alpha)\end{aligned}$$

is an isomorphism onto a closed analytic subspace Γ (cf. Holmann [64] p. 359, [63] p. 433).

We claim that the image Γ of γ is a closed analytic subspace in $P \times P$. A pair $(\alpha, \alpha') \in P \times P$ lies in Γ if and only if

$$h^0(\mathbb{P}, \underline{\mathrm{Hom}}(E(\alpha), E(\alpha'))) = 1.$$

In order to be able to apply the semicontinuity theorem we consider the universal monad over $P \times \mathbb{P}$

$$\mathbb{M}\colon \qquad 0 \to \mathbb{H} \boxtimes \mathcal{O}_{\mathbb{P}}(-1) \to \mathbb{K} \boxtimes \mathcal{O}_{\mathbb{P}} \to \mathbb{H}^* \boxtimes \mathcal{O}_{\mathbb{P}}(1) \to 0$$

with $\mathbb{H} = H \otimes \mathcal{O}_P$, $\mathbb{K} = K \otimes \mathcal{O}_P$, $\mathbb{H}^* = H^* \otimes \mathcal{O}_P$. Over every point $\alpha \in P$ this monad $\mathbb{M}$ induces the monad

$$M(\alpha)\colon \qquad 0 \to H \otimes \mathcal{O}_{\mathbb{P}}(-1) \xrightarrow{a} K \otimes \mathcal{O}_{\mathbb{P}} \xrightarrow{a^*q} H^* \otimes \mathcal{O}_{\mathbb{P}}(1) \to 0.$$

Let $\mathbb{E}$ be the cohomology bundle of $\mathbb{M}$ over $P \times \mathbb{P}$. On

$$P \times P \times \mathbb{P}$$

we have three projection mappings

$$\begin{array}{ccc} P \times P \times \mathbb{P} & \overset{\mathrm{pr}_{13}}{\underset{\mathrm{pr}_{23}}{\rightrightarrows}} & P \times \mathbb{P} \\ {\scriptstyle \mathrm{pr}_{12}}\downarrow & & \\ P \times P. & & \end{array}$$

Let

$$F = \mathrm{pr}_{13}^*(\mathbb{E}^*) \otimes \mathrm{pr}_{23}^*(\mathbb{E}).$$

Then

$$\Gamma = \{(\alpha, \alpha') \in P \times P \mid h^0(\mathrm{pr}_{12}^{-1}(\alpha, \alpha'), F|\mathrm{pr}_{12}^{-1}(\alpha, \alpha')) > 0\}.$$

From the semicontinuity theorem it then follows that Γ is closed and analytic in $P \times P$.

It remains to show that the graph mapping

$$\gamma\colon P \times G_0 \to \Gamma$$

has local holomorphic sections. To this end we consider the bundle

$$F|_{\Gamma \times \mathbb{P}}$$

over $\Gamma \times \mathbb{P}$. Let $p\colon \Gamma \times \mathbb{P} \to \Gamma$ be the projection. The base change theorem shows that

$$L = p_*(F|_{\Gamma \times \mathbb{P}})$$

is a line bundle over Γ.

Now let $W \subset \Gamma$ be a simply connected open subset in Γ,

$$\phi \in H^0(W, L)$$

a section of L over W. Over $W \times \mathbb{P}$ we have the monads

$$\mathbb{M}_i = \mathrm{pr}_{i3}^*(\mathbb{M})|_{W \times \mathbb{P}}$$

with the cohomology bundles

$$\mathbb{E}_i = \mathrm{pr}_{i3}^* \mathbb{E}|_{W \times \mathbb{P}}.$$

ϕ defines a homomorphism

$$\mathbb{E}_1 \to \mathbb{E}_2$$

which is an isomorphism on every fibre. By Lemma 4.1.3 it follows that ϕ is induced by an isomorphism

$$\Phi\colon \mathbb{M}_1 \to \mathbb{M}_2$$

of monads over $W \times \mathbb{P}$. Φ is given by a holomorphic mapping

$$(f, g, h)\colon W \to GL(H) \times GL(K) \times GL(H^*)$$

for which

$$\alpha' = g(\alpha, \alpha')\alpha f(\alpha, \alpha')^{-1}$$

and

$$(\alpha')^* q = h(\alpha, \alpha')\alpha^* q g(\alpha, \alpha')^{-1}.$$

$$(h^*, g^t, f^*)\colon \mathbb{M}_2 \to \mathbb{M}_1$$

is then likewise an isomorphism over $W \otimes \mathbb{P}$. Therefore — since all $E(\alpha)$ are stable —

$$(h^*, g^t, f^*)(f, g, h) = c \cdot \mathrm{id}$$

for some holomorphic function

$$c\colon W \to \mathbb{C}^*.$$

We now choose a square root of c on W and obtain with

$$s\colon W \to P \times G_0$$

$$(\alpha, \alpha') \mapsto (\alpha, (\frac{f}{\sqrt{c}}, \frac{g}{\sqrt{c}}))$$

a holomorphic section of γ over W. □

REMARK 4.1.10. The preceding considerations show that

$$P' = \{(\alpha, \beta) \in L(V, L(H, K)) \times L(V, L(K, H^*)) \mid \text{(E}'\text{1)–(E}'\text{3) hold}\}$$

is also a complex manifold. Let $\mathbb{C}^* \subset G'$ be the subgroup of homotheties $\lambda(\mathrm{id}_H, \mathrm{id}_K, \mathrm{id}_H^*)$. The group

$$G_0' = G'/\mathbb{C}^*$$

operates freely on P' and the canonical map

$$P/G_0 \xrightarrow{\sim} P'/G_0'$$

is an isomorphism of complex manifolds.

We now wish to show that $M_{\mathbb{P}_2}(0,n) = P'/G'$ is a coarse moduli space for $\Sigma_{0,n}^{\mathbb{P}_2}$. An essential tool is the relative version of the Theorem of Beilinson: let S be a complex space, $p\colon \mathbb{P}_n \times S \to S$, $q\colon \mathbb{P}_n \times S \to \mathbb{P}_n$ the projections. We again define $A \boxtimes B$ to be $q^*A \otimes p^*B$ for sheaves A over $\mathbb{P}_n$, B over S. By now replacing the diagram

$$\begin{array}{ccc} \mathbb{P}_n \times \mathbb{P}_n & \xrightarrow{p_1} & \mathbb{P}_n \\ {\scriptstyle p_2}\downarrow & & \downarrow \\ \mathbb{P}_n & \longrightarrow & \{*\} \end{array}$$

and the Euler sequence

$$0 \to \mathcal{O}_{\mathbb{P}_n}(-1) \to \mathcal{O}_{\mathbb{P}_n}^{\oplus(n+1)} \to Q \to 0$$

by the diagram

$$\begin{array}{ccc} \mathbb{P}_n \times \mathbb{P}_n \times S & \xrightarrow{p_1} & \mathbb{P}_n \times S \\ {\scriptstyle p_2}\downarrow & & \downarrow{\scriptstyle p} \\ \mathbb{P}_n \times S & \xrightarrow[p]{} & S \end{array}$$

and the relative Euler sequence

$$0 \to \mathcal{O}_{\mathbb{P}_n}(-1) \boxtimes \mathcal{O}_S \to \mathcal{O}_{\mathbb{P}_n \times S}^{\oplus(n+1)} \to Q \boxtimes \mathcal{O}_S \to 0$$

we obtain quite in analogy to the Theorem of Beilinson in §3 the following

THEOREM 4.1.11 (Beilinson). *For every holomorphic r-bundle E over $\mathbb{P}_n \times S$ there is a spectral sequence with E_1-term*

$$E_1^{s,t} = \mathcal{O}_{\mathbb{P}_n}(s) \boxtimes R^t p_*(E \otimes \Omega^{-s}_{\mathbb{P}_n \times S/S}(-s))$$

which converges to

$$E^i = \begin{cases} E & i = 0 \\ 0 & \textit{otherwise.} \end{cases}$$

We now prove

THEOREM 4.1.12. *$M_{\mathbb{P}_2}(0,n) = P'/G'$ is a coarse moduli space for stable 2-bundles over $\mathbb{P}_2$ with Chern classes $c_1 = 0$, $c_2 = n$.*

PROOF. 1) First we must construct a natural transformation

$$\Phi\colon \Sigma_{0,n}^{\mathbb{P}_2} \to \operatorname{Hom}(-, P'/G')$$

which is a bijection $\Sigma_{0,n}^{\mathbb{P}_2}(*) \xrightarrow{\sim} \operatorname{Hom}(*, P'/G')$ for every reduced point $*$. Let then S be a complex space, $\xi \in \Sigma_{0,n}^{\mathbb{P}_2}(S)$ and E a family of stable 2-bundles over $\mathbb{P}_2$ with $c_1 = 0$, $c_2 = n$ parametrized by S and representing the element ξ. As in §3 Example 3 one obtains E as the cohomology of the monad

$$M(E): \quad 0 \to \mathcal{O}_{\mathbb{P}_2}(-1) \boxtimes R^1 p_* E(-2) \to \mathcal{O}_{\mathbb{P}_2} \boxtimes R^1 p_*(E \otimes \Omega^1_{\mathbb{P}_2 \times S/S}) \\ \to \mathcal{O}_{\mathbb{P}_2}(1) \boxtimes R^1 p_* E(-1) \to 0,$$

which is *canonically* associated to E. To this monad moreover we can canonically associate a holomorphic mapping

$$f\colon S \to P'/G',$$

which only depends on the equivalence class $\xi = [E]$, so that altogether we have a natural transformation

$$\Phi\colon \Sigma^{\mathbb{P}_2}_{0,n} \to \operatorname{Hom}(-, P'/G').$$

The map f is constructed as follows: let (S_i) be an open covering of S, so that $M(E)|\mathbb{P}_2 \times S_i$ is isomorphic to a monad

$$M(a_i, b_i)\colon \; 0 \to \mathcal{O}_{\mathbb{P}_2}(-1) \boxtimes (H \otimes \mathcal{O}_{S_i}) \xrightarrow{a_i} \mathcal{O}_{\mathbb{P}_2} \boxtimes (K \otimes \mathcal{O}_{S_i}) \\ \xrightarrow{b_i} \mathcal{O}_{\mathbb{P}_2}(1) \boxtimes (H^* \otimes \mathcal{O}_{S_i}) \to 0.$$

a_i and b_i define holomorphic mappings

$$\alpha_i\colon S_i \to L(V, L(H, K)) \quad \text{and} \quad \beta_i\colon S_i \to L(V, L(K, H^*)).$$

One readily sees that this gives a holomorphic mapping

$$f_i = (\alpha_i, \beta_i)\colon S_i \to P'.$$

By construction we have

$$f_i(s) \underset{G'}{\sim} f_j(s)$$

for every point $s \in S_i \cap S_j$. The family (f_i) thus defines a holomorphic mapping

$$f\colon S \to P'/G'.$$

$\Phi(\xi) = f$ is the desired natural transformation. By Remark 4.1.5 it follows that

$$\Phi(*)\colon \Sigma^{\mathbb{P}_2}_{0,n}(*) \to \operatorname{Hom}(*, P'/G')$$

is bijective.

2) Now let an arbitrary natural transformation

$$\Psi\colon \Sigma^{\mathbb{P}_2}_{0,n} \to \operatorname{Hom}(-, N)$$

be given. Let $\mathbb{E}$ be the cohomology of the universal monad $\mathbb{M}$ over $\mathbb{P}_2 \times P'$ and

$$\eta = [\mathbb{E}] \in \Sigma^{\mathbb{P}_2}_{0,n}(P').$$

Φ is so constructed that $\Phi(\eta) = \pi\colon P' \to P'/G'$ is precisely the quotient map. The mapping $\tilde{\phi} = \Psi(\eta)\colon P' \to N$ is constant on the fibres of π: for if $\alpha_1, \alpha_2 \in \mathrm{Hom}(*, P')$ are such that $\pi(\alpha_1) = \pi(\alpha_2)$, then we have

$$\Phi(\alpha_1^*\eta) = \Phi(\eta)(\alpha_1) = \Phi(\eta)(\alpha_2) = \Phi(\alpha_2^*\eta)$$

and thus $\alpha_1^*\eta = \alpha_2^*\eta$, i.e.,

$$\tilde{\phi}(\alpha_1) = \Psi(\eta)(\alpha_1) = \Psi(\alpha_1^*\eta) = \Psi(\alpha_2^*\eta) = \Psi(\eta)(\alpha_2) = \tilde{\phi}(\alpha_2).$$

Since π has holomorphic sections locally, $\tilde{\phi}$ induces a holomorphic mapping

$$\phi\colon P'/G' \to N \quad \text{with} \quad \phi \circ \Phi(\eta) = \Psi(\eta).$$

If S is an arbitrary complex space and $\xi \in \Sigma_{0,n}^{\mathbb{P}_2}(S)$, then we have

$$\phi \circ \Phi(\xi) = \Psi(\xi).$$

For we can choose an open covering (S_i) of S and holomorphic mappings $g_i\colon S_i \to P'$ so that

$$\xi | S_i = g_i^*(\eta).$$

It then follows that

$$\begin{aligned}\Psi(\xi)|S_i = \Psi(\xi|S_i) = \Psi(g_i^*\eta) &= g_i^*\Psi(\eta)\\ &= g_i^*\phi\Phi(\eta) = \phi\Phi(g_i^*\eta) = \phi\Phi(\xi|S_i) = \phi\Phi(\xi)|S_i\end{aligned}$$

and thus

$$\Psi(\xi) = \phi\Phi(\xi).$$

Since P'/G' is reduced, ϕ is uniquely determined. Hence the theorem is proved. □

REMARK 4.1.13. Let E_0 be a stable 2-bundle over $\mathbb{P}_2$ with $c_1 = 0$, $c_2 = n$. Let $x_0 \in M_{\mathbb{P}_2}(0,n)$ be the point defined by E_0 in the coarse moduli space $M_{\mathbb{P}_2}(0,n) = M$. Let $M' \subset M$ be an open neighborhood of x_0 with a section $f\colon M' \to P'$ of $\pi\colon P' \twoheadrightarrow M$. Then $U = f^*\mathbb{E}$ is a family of 2-bundles over $\mathbb{P}_2$ with $U(x_0) \cong E_0$.

Let $\underline{\mathrm{An}}_0$ be the category of germs of complex spaces. For a germ (S, s_0) of a complex space we consider the set $D_{E_0}(S, s_0)$ of all isomorphism classes $[E]$ of local deformations E of the bundle E_0 over (S, s_0):

$$\begin{aligned}D_{E_0}(S, s_0) = \{[E] \mid &E \text{ is a holomorphic vector bundle over}\\ &\mathbb{P}_2 \times S' \text{ with } E(s_0) \cong E_0, \text{ where } S' \text{ is a representative of } (S, s_0)\}.\end{aligned}$$

Let (M, x_0) denote the germ of the space M in x_0. U then defines an element in $D_{E_0}(M, x_0)$. We then have the following

THEOREM. *U is the* universal *local deformation of the bundle E_0, i.e., the mapping*

$$\mathrm{Hom}_{\underline{\mathrm{An}}_0}((S,s_0),(M,x_0)) \to D_{E_0}(S,s_0)$$
$$\phi \mapsto (\mathrm{id}_{\mathbb{P}_2} \times \phi)^* U$$

is bijective for every germ (S,s_0) of a complex space.

The local universal deformations do not in general define a global universal family over M. We wish to show that in the case n odd M has a universal family $U \in \Sigma_{0,n}^{\mathbb{P}_2}(M)$. Le Potier has shown that this is not the case when n is even.

THEOREM 4.1.14. *If n is odd then $M = M_{\mathbb{P}_2}(0,n)$ is a fine moduli space for $\Sigma_{0,n}^{\mathbb{P}_2}$.*

PROOF. Here we employ the representation $M = P/G$ of the coarse moduli space $M = M_{\mathbb{P}_2}(0,n)$. Let

$$\mathbb{M}\colon \qquad \mathcal{O}_{\mathbb{P}_2}(-1)\boxtimes(H\otimes\mathcal{O}_P) \to \mathcal{O}_{\mathbb{P}_2}\boxtimes(K\otimes\mathcal{O}_P) \to \mathcal{O}_{\mathbb{P}_2}(1)\boxtimes(H^*\otimes\mathcal{O}_P)$$

be the universal monad over $\mathbb{P}_2\times P$ and let $\mathbb{E}$ be its cohomology bundle. The group G operates on $\mathbb{M}$ and induces a G-operation on $\mathbb{E}$. If we define a G-operation on $\mathbb{P}_2\times P$ by the prescription $g\cdot(x,\alpha) = (x, g\cdot\alpha)$, then $\mathbb{E}$ becomes a G-vector bundle over $\mathbb{P}_2 \times P$.

In order to get a vector bundle $\mathbb{E}/G$ over $\mathbb{P}_2\times(P/G)$ it is necessary and sufficient that the isotropy group $G_{(x,\alpha)} = \{\pm 1\}$ of every point $(x,\alpha) \in \mathbb{P}_2 \times P$ operates trivially on the vector bundle fibre $\mathbb{E}(x,\alpha)$. In our case however $-1 \in G$ operates as $-\mathrm{id}$ on the fibre $\mathbb{E}(x,\alpha)$. We attempt to eliminate this difficulty by tensoring $\mathbb{E}$ with a suitable G-line bundle p^*L ($p\colon \mathbb{P}_2 \times P \to P$). This does not alter the equivalence class $[\mathbb{E}] \in \Sigma_{0,n}^{\mathbb{P}_2}(P)$. The direct image $R^1p_*(\mathbb{E}(-1))$ is a holomorphic vector bundle of rank n over P, and the fibre over a point $\alpha \in P$ is canonically isomorphic to $H^1(\mathbb{P}_2, E(\alpha)(-1)) \simeq H^*$. Thus $R^1p_*\mathbb{E}(-1)$ is a G-vector bundle on which $-1 \in G$ also operates as $-\mathrm{id}$. We define the G-line bundle L by

$$L = \Lambda^n R^1 p_* \mathbb{E}(-1).$$

If n is odd, then $-1 \in G$ operates as $-\mathrm{id}_L$ on L and thus as $+\mathrm{id}$ on the fibres of $\mathbb{E}\otimes p^*L$.

We thus get a 2-bundle $U = (\mathbb{E}\otimes p^*L)/G$ over $\mathbb{P}_2 \times M$. U is a universal family: for each complex space S the mapping

$$\mathrm{Hom}(S,M) \to \Sigma_{0,n}^{\mathbb{P}_2}(S)$$
$$\phi \mapsto \phi^*[U] = [(\mathrm{id}_{\mathbb{P}_2}\times\phi)^*U]$$

is bijective. □

Before proceeding in the next section to prove the irreducibility of $M_{\mathbb{P}_2}(0,n)$ we wish to indicate how one can describe the moduli space $M_{\mathbb{P}_2}(-1,n)$ — in a manner similar to that for $M_{\mathbb{P}_2}(0,n)$ — as the orbit space of a holomorphic principal bundle.

Construction of $M_{\mathbb{P}_2}(-1,n)$: Let V again be a 3-dimensional $\mathbb{C}$-vector space, $\mathbb{P} = \mathbb{P}(V)$. Let E be a stable 2-bundle over $\mathbb{P}$ with Chern classes

$$c_1(E) = -1, \qquad c_2(E) = n.$$

In §3 Example 4 we saw that E is the cohomology bundle of a monad of the form

$$M: \qquad 0 \to H \otimes \mathcal{O}_{\mathbb{P}}(-1) \xrightarrow{a} K \otimes \Omega^1_{\mathbb{P}}(1) \xrightarrow{b} H' \otimes \mathcal{O}_{\mathbb{P}} \to 0,$$

where $\mathcal{O}_{\mathbb{P}}(-1) \cong \Omega^2_{\mathbb{P}}(2)$ is the determinant bundle of $\Omega^1_{\mathbb{P}}(1)$.

The monomorphism $a\colon H \otimes \mathcal{O}_{\mathbb{P}}(-1) \to K \otimes \Omega^1_{\mathbb{P}}(1)$ corresponds to an element in

$$H^* \otimes K \otimes V$$

which we can regard as a linear mapping

$$\alpha \in L(V^*, L(H,K)).$$

If we identify $\mathcal{O}_{\mathbb{P}}(-1)$ with

$$\Omega^2_{\mathbb{P}}(2) = \Lambda^2 Q^*, \qquad Q^* \subset V^* \otimes \mathcal{O}_{\mathbb{P}},$$

then for $z', z'' \in V^*$ and $x = \mathbb{P}(\mathbb{C}v) \in \mathbb{P}$ with $z'(v) = z''(v) = 0$ we have the equation

$$a(x)(h \otimes z' \wedge z'') = \alpha(z')(h) \otimes z'' - \alpha(z'')(h) \otimes z'.$$

Because $c_1(E) = -1$, the bundle E has a symplectic structure f, i.e., there is an isomorphism

$$f\colon E \to E^*(-1)$$

with $f^*(-1) = -f$. f is induced by the mapping

$$\hat{f}\colon E \otimes E \to \Lambda^2 E \cong \mathcal{O}(-1)$$
$$e \otimes e' \mapsto e \wedge e'.$$

Because E is simple, there is up to multiplication by a constant precisely one symplectic structure on E. We consider the monad

$$M^*(-1)\colon \quad 0 \to H'^* \otimes \mathcal{O}_{\mathbb{P}}(-1) \xrightarrow{b^*(-1)} K^* \otimes \Omega^1_{\mathbb{P}}(1)^*(-1) \xrightarrow{a^*(-1)} H^* \otimes \mathcal{O}_{\mathbb{P}} \to 0.$$

As in Corollary 2 to Lemma 4.1.3 one sees that the symplectic structure on E is induced by an isomorphism of the monads M and $M^*(-1)$.

The isomorphism

$$K \otimes \Omega^1_{\mathbb{P}}(1) \to K^* \otimes \Omega^1_{\mathbb{P}}(1)^*(-1)$$

is of the form $q \otimes g$ for a symplectic structure

$$g\colon \Omega^1_{\mathbb{P}}(1) \to \Omega^1_{\mathbb{P}}(1)^*(-1)$$

on $\Omega^1_{\mathbb{P}}(1)$ and an isomorphism

$$q\colon K \to K^*.$$

From the equation

$$-(q \otimes g) = (q \otimes g)^*(-1) = q^* \otimes g^*(-1) = -q^* \otimes g$$

one sees that q is symmetric:

$$q^* = q.$$

Thus we have shown: every stable 2-bundle E over $\mathbb{P} = \mathbb{P}(V)$ with $c_1(E) = -1$ is the cohomology of a self-dual monad

$$M(\alpha)\colon \quad 0 \to H \otimes \mathcal{O}_{\mathbb{P}}(-1) \xrightarrow{a} K \otimes \Omega^1_{\mathbb{P}}(1) \xrightarrow{a^t(-1)} H^* \otimes \mathcal{O}_{\mathbb{P}} \to 0$$

with $a(x)(h \otimes (z' \wedge z'')) = \alpha(z')(h) \otimes z'' - \alpha(z'')(h) \otimes z'$ and

$$a^t(-1)(x)(k \otimes z) = (\alpha(z)^* \circ q)(k),$$

where $x = \mathbb{P}(\mathbb{C}v) \in \mathbb{P}(V)$, $z', z'' \in V^*$, $z'(v) = z''(v) = z(v) = 0$. We set $\alpha(z)^t = \alpha(z)^* q$.

Because $M(\alpha)$ is a monad we have

(F1) The map $\alpha_h\colon V^* \to K$ defined by $\alpha_h(z) = \alpha(z)(h)$ has rank ≥ 2 for $h \neq 0$.

(F2) $\alpha(z')^t \alpha(z'') = \alpha(z'')^t \alpha(z')$ for all $z', z'' \in V^*$.

Conversely suppose given vector spaces H, K of dimensions $n-1$ resp. n and

$$q\colon K \to K^*$$

a fixed non-degenerate quadratic form on K. We then consider the set

$$P = \{\alpha \in L(V^*, L(H, K)) \mid \alpha \text{ satisfies (F1) and (F2)}\}.$$

Let $O(q)$ be the orthogonal group of (K, q):

$$O(q) = \{\phi \in GL(K) \mid \phi^* q \phi = q\}.$$

$G = GL(H) \times O(q)$ operates on P by the prescription

$$[(g, \phi) \cdot \alpha](z) = \phi \circ \alpha(z) \circ g^{-1}.$$

To each element $\alpha \in P$ we can associate the cohomology bundle $E(\alpha)$ of the monad $M(\alpha)$. From the display of $M(\alpha)$ one deduces that $E(\alpha)$ is stable with Chern classes

$$c_1(E(\alpha)) = -1, \qquad c_2(E(\alpha)) = n.$$

As in the case $c_1 = 0$ we have

THEOREM 4.1.15. *The assignment* $[\alpha] \mapsto [E(\alpha)]$ *is a bijective mapping of the orbit space* P/G *onto the set of isomorphism classes of stable 2-bundles over* $\mathbb{P}$ *with* $c_1 = -1$, $c_2 = n$. *The isotropy subgroup of* G *in* $\alpha \in P$ *is always* $\{\pm(\mathrm{id}_H, \mathrm{id}_K)\}$. *If* G_0 *denotes* $G/\{\pm 1\}$, *then* G_0 *operates freely on* P.

THEOREM 4.1.16. P *is a smooth holomorphic* G_0*-principal bundle over the* $(4n-4)$*-dimensional complex manifold* P/G_0.

To prove this theorem one considers the mapping

$$g\colon L(V^*, L(H,K)) \to L(\Lambda^2 V^*, \Lambda^2 H^*)$$

defined by

$$\alpha \mapsto \alpha^t \wedge \alpha$$

$$(\alpha^t \wedge \alpha)(z' \wedge z'') = \alpha(z')^t \alpha(z'') - \alpha(z'')^t \alpha(z').$$

P is a Zariski-open subset of the fibre $g^{-1}(0)$. The differential $d_{\alpha_0} g$ of g in a point $\alpha_0 \in P$ is given by

$$d_{\alpha_0} g(\alpha) = \alpha_0^t \wedge \alpha + \alpha^t \wedge \alpha_0.$$

The cokernel of the map

$$d_0\colon L(V^*, L(H,K)) \oplus L(V^*, L(K,H^*)) \to L(\Lambda^2 V^*, L(H,H^*))$$

defined by

$$d_0(\alpha, \beta) = \alpha_0^t \wedge \alpha + \beta \wedge \alpha_0$$

is isomorphic to $H^2(\mathbb{P}, \operatorname{End} E)$. Because

$$H^2(\mathbb{P}, \operatorname{End} E) \cong H^0(\mathbb{P}, (\operatorname{End} E)(-3))$$

vanishes, d_0 — and thus also $d_{\alpha_0} g$ — is surjective, and hence $g^{-1}(0)$ is smooth in the point α_0. Thus P is a complex manifold of dimension

$$\begin{aligned}\dim_{\mathbb{C}} L(V^*, L(H,K)) &- \dim_{\mathbb{C}} L(\Lambda^2 V^*, \Lambda^2 H^*) \\ &= 3(n-1)n - \tfrac{3(n-1)(n-2)}{2} = \tfrac{3}{2}(n^2 + n - 2).\end{aligned}$$

With the help of the universal monad

$$\mathbb{M}\colon \qquad 0 \to \mathcal{O}_{\mathbb{P}}(-1) \boxtimes \mathbb{H} \to \Omega^1_{\mathbb{P}}(1) \boxtimes \mathbb{K} \to \mathcal{O}_{\mathbb{P}} \otimes \mathbb{H}^* \to 0$$

over $\mathbb{P} \times P$ one can just as in the case $c_1 = 0$ show that the graph mapping

$$\gamma\colon G_0 \times P \to P \times P$$

is a closed embedding. P is thus a holomorphic G_0-principal bundle. The dimension of the base space P/G_0 is

$$\dim P/G_0 = \dim P - \dim G = \tfrac{3}{2}(n^2+n-2) - (n-1)^2 - \tfrac{n(n-1)}{2} = 4n-4.$$

In the case $c_1 = -1$ we have for arbitrary $n \geq 1$:

THEOREM 4.1.17. *$M_{\mathbb{P}_2}(-1,n) = P/G$ is a fine moduli space for stable 2-bundles over $\mathbb{P}_2$ with Chern classes $c_1 = -1$, $c_2 = n$.*

PROOF. The cohomology bundle $\mathbb{E}$ of the universal monad $\mathbb{M}$ over $\mathbb{P} \times P$ is again a G-vector bundle on which $-1 \in G$ operates as $-\mathrm{id}$. The bundle

$$L = \Lambda^n R^1 p_* \mathbb{E}(-1) \otimes \Lambda^{n-1} R^1 p_* \mathbb{E} \qquad (p\colon \mathbb{P} \times P \to P)$$

is a G-line bundle over P on which $-1 \in G$ also acts as $-\mathrm{id}$. Thus one gets as in the proof of Theorem 4.1.14 a universal family

$$U = (\mathbb{E} \otimes p^* L)/G \quad \text{over} \quad M_{\mathbb{P}_2}(-1,n). \qquad \square$$

4.2. Irreducibility of $M_{\mathbb{P}_2}(0,n)$. The goal of this section is the following

THEOREM 4.2.1 (Barth, Maruyama). *The moduli space $M_{\mathbb{P}_2}(0,n)$ is irreducible.*

To show this, we consider the model (cf. Remark 4.1.5)

$$M_{\mathbb{P}_2}(0,n) = P'/G'$$

with

$$P' = \{(\alpha,\beta) \in L(V, L(H,K)) \oplus L(V, L(K,H^*)) \mid (\alpha,\beta) \text{ satisfy (E}'1)\text{–(E}'3)\}$$

and $G' = GL(H) \times GL(K) \times GL(H^*)$.

Let $G'_0 = G'/\mathbb{C}^*$. The monomorphism

$$GL(K) \to G'_0, \qquad \phi \mapsto (\mathrm{id}_H, \phi, \mathrm{id}_{H^*})/\mathbb{C}^*$$

makes P' into a free $GL(K)$-space. Thus P' is a $GL(K)$-principal bundle over the complex manifold

$$P'' = P'/GL(K).$$

It therefore suffices to show that P'' is connected.

To do this we associate the pairs $(\alpha,\beta) \in P'$ with certain Kronecker modules

$$\gamma \in L(\Lambda^2 V, L(H,H^*)).$$

DEFINITION. A *stable Kronecker module of rank* 2 is a linear mapping

$$\gamma\colon \Lambda^2 V \to L(H,H^*),$$

so that for the associated linear mapping

$$\hat{\gamma}\colon V \otimes H \to V^* \otimes H^*$$

with

$$\hat{\gamma}(v \otimes h)(v' \otimes h') = [\gamma(v \wedge v')(h)](h')$$

the following hold:

(K1) $\hat{\gamma}(v \otimes -): H \to V^* \otimes H^*$ is injective for $v \neq 0$.

(K2) If $v^{**}: V^* \otimes H^* \to H^*$ is the evaluation mapping associated to $v \in V$, then

$$v^{**} \circ \hat{\gamma}: V \otimes H \to H^*$$

is surjective for $v \neq 0$.

(K3) $\operatorname{rk} \hat{\gamma} = 2n + 2$.

Let N be the set of all stable Kronecker modules of rank 2.

If $(\alpha, \beta) \in P'$, then we can define a bilinear mapping

$$\gamma': V \times V \to L(H, H^*)$$

alternating in V by setting

$$\gamma'(v_1, v_2) = \beta(v_2) \circ \alpha(v_1);$$

it then defines an element

$$\gamma \in L(\Lambda^2 V, L(H, H^*)).$$

We wish to verify that γ is a stable Kronecker module of rank 2.

Let $\hat{\alpha}: V \otimes H \to K$, $\hat{\beta}: K \to V^* \otimes H^*$ and $\hat{\gamma}: V \otimes H \to V^* \otimes H^*$ be the linear maps associated to α, β, γ. By the definition of γ we have

$$\hat{\gamma} = \hat{\beta} \circ \hat{\alpha},$$

for with $v \otimes h, v' \otimes h' \in V \otimes H$ we have

$$\begin{aligned}
\hat{\beta} \circ \hat{\alpha}(v \otimes h)(v' \otimes h') &= [\hat{\beta}(\alpha(v)(h))](v' \otimes h') \\
&= [\beta(v')(\alpha(v)(h))](h') \\
&= [(\beta(v') \circ \alpha(v))(h)](h') \\
&= [\gamma(v \wedge v')(h)](h') \\
&= \hat{\gamma}(v \otimes h)(v' \otimes h').
\end{aligned}$$

Because $\hat{\beta}$ is injective and $\hat{\alpha}$ is surjective it follows that $\hat{\gamma}$ is of rank $2n + 2$. The properties (K1) and (K2) follow immediately from (E′1).

We thus get a mapping

$$\begin{aligned}
\Psi: P' &\to N \\
(\alpha, \beta) &\mapsto \gamma \qquad \text{with } \hat{\gamma} = \hat{\beta} \circ \hat{\alpha}.
\end{aligned}$$

Ψ is $GL(K)$-invariant and we have

LEMMA 4.2.2. *$\Psi/GL(K): P'' = P'/GL(K) \to N$ is a homeomorphism.*

PROOF. 1) Injectivity: let $(\alpha,\beta),(\alpha',\beta') \in P'$ be such that $\hat{\gamma} = \hat{\beta}\circ\hat{\alpha} = \hat{\beta}'\circ\hat{\alpha}' = \hat{\gamma}'$; then one can define a mapping

$$\phi\colon K \to K$$

by

$$\phi(k) = \hat{\alpha}'(u) \quad \text{with} \quad \hat{\alpha}(u) = k \quad (\hat{\alpha} \text{ is surjective}).$$

ϕ is well defined since $\hat{\beta}$ and $\hat{\beta}'$ are injective and $\hat{\beta}\circ\hat{\alpha} = \hat{\beta}'\circ\hat{\alpha}'$. By definition we then have

$$(*) \qquad \phi\circ\hat{\alpha} = \hat{\alpha}';$$

hence ϕ is surjective and therefore an isomorphism. From (*) one further gets

$$\phi^{-1}\circ\hat{\alpha}' = \hat{\alpha};$$

thus

$$\hat{\beta}\circ\phi^{-1}\circ\hat{\alpha}' = \hat{\beta}\circ\hat{\alpha} = \hat{\beta}'\circ\hat{\alpha}'.$$

Since $\hat{\alpha}'$ is surjective, it then follows that

$$\hat{\beta}' = \hat{\beta}\circ\phi^{-1},$$

i.e.,

$$\phi\cdot(\alpha,\beta) = (\alpha',\beta').$$

2) Surjectivity: let $\gamma \in N$ be a stable Kronecker module of rank 2. Because $\hat{\gamma}$ has a $(2n+2)$-dimensional image, there are linear mappings

$$\hat{\alpha}\colon V\otimes H \to K$$
$$\hat{\beta}\colon K \to V^*\otimes H^*$$

with

$$\hat{\gamma} = \hat{\beta}\circ\hat{\alpha}.$$

Clearly $\hat{\alpha}$ is surjective and $\hat{\beta}$ is injective. The associated pair (α,β) lies in P': (E′1) follows immediately from (K1) and (K2), (E′2) from the equation

$$\begin{aligned}[\beta(v)\circ\alpha(v)(h)](h') &= \hat{\beta}\circ\hat{\alpha}(v\otimes h)(v\otimes h')\\ &= \hat{\gamma}(v\otimes h)(v\otimes h') = \gamma(v\wedge v)(h)(h') = 0.\end{aligned}$$

The map $\Psi/GL(K)$ is thus bijective; since P'' is smooth N must be locally irreducible. But then $\Psi/GL(K)$ is open (cf. Fischer [39] p. 143) and hence a homeomorphism. □

If we can now show that the set N of stable Kronecker modules of rank 2 is irreducible, the irreducibility of the moduli space $M_{\mathbb{P}_2}(0,n)$ will follow.

We need some preparation:

LEMMA 4.2.3. *Let E be the cohomology bundle belonging to a pair $(\alpha, \beta) \in P'$, $\gamma \in N$ the Kronecker module associated to (α, β). For two linearly independent vectors $v_1, v_2 \in V$ we have: $\gamma(v_1 \wedge v_2)$ is an isomorphism if and only if the restriction $E|L$ to the line*

$$L = \mathbb{P}(\{v \mid v \wedge v_1 \wedge v_2 = 0\}) \subset \mathbb{P}(V)$$

is trivial.

With the help of the Theorem of Grauert and Mülich (§2) we deduce from this lemma the following

COROLLARY. *For every Kronecker module $\gamma \in N$ we have:*
(K4) *$\gamma(v_1 \wedge v_2)$ is an isomorphism for almost all vectors $v_1 \wedge v_2$.*

PROOF OF LEMMA. We consider the monad

$$0 \to H \otimes \mathcal{O}_{\mathbb{P}}(-1) \xrightarrow{a} K \otimes \mathcal{O}_{\mathbb{P}} \xrightarrow{b} H^* \otimes \mathcal{O}_{\mathbb{P}}(1) \to 0$$

defined by $(\alpha, \beta) \in P'$. Suppose $v_1, v_2 \in V$ are such that $v_1 \wedge v_2 \neq 0$,

$$W = \mathbb{C}v_1 + \mathbb{C}v_2$$

the subspace of V spanned by v_1, v_2. The restriction of the monad to $L = \mathbb{P}(W)$ is the monad

$$0 \to H \otimes \mathcal{O}_L(-1) \xrightarrow{a|L} K \otimes \mathcal{O}_L \xrightarrow{b|L} H^* \otimes \mathcal{O}_L(1) \to 0.$$

Therefore we have

$$H^0(L, E|L) \cong H^0(L, \ker(b|L)) \cong \ker(K \to H^* \otimes W^*).$$

But $E|L$ is trivial if and only if no section

$$s \in H^0(L, E|L) \setminus \{0\}$$

has zeros. We must show that this is the case if and only if $\gamma(v_1 \wedge v_2)$ is invertible.

Let $F = \ker(b|L)$. We have inclusions

$$H \otimes \mathcal{O}_L(-1) \overset{i}{\hookrightarrow} F \overset{j}{\hookrightarrow} K \otimes \mathcal{O}_L.$$

Let $s \in H^0(L, F)$ be a section; there is then some $k \in K$ with

$$j \circ s(x) = k \qquad \text{for all } x \in L.$$

The section

$$s' \in H^0(L, E|L)$$

defined by s has a zero at $x = \mathbb{P}(\mathbb{C}v) \in L$ if and only if $s(x)$ lies in the image of the inclusion

$$i(x) \colon H \otimes \mathcal{O}_L(-1)(x) \to F(x),$$

i.e., if and only if there is an $h \in H$ with

$$\alpha(v)(h) = k.$$

But s is a section in $F = \ker(b|L)$, so for every $v' \in W$ we must have

$$\beta(v')(k) = 0.$$

We have thus demonstrated the following: $E|L$ has a nontrivial section with a zero if and only if

$$\operatorname{im} \alpha(v) \subset \bigcap_{v' \in W} \ker \beta(v')$$

for at least one vector $v \in W \setminus \{0\}$. This is equivalent to the statement that for any basis $v, v' \in W$ of W

$$\gamma(v \wedge v') = \beta(v') \circ \alpha(v)$$

is not an isomorphism. □

REMARK 4.2.4. Lemma 4.2.3 can be formulated and is correct for Kronecker modules of higher rank over $\mathbb{P}_n$. The corollary remains true for Kronecker modules of rank 2 over $\mathbb{P}_n$, $n \geq 2$.

Now we can begin the proof of the irreducibility of the set N of stable Kronecker modules of rank 2. To this end we consider the open set

$$N_z = \{\gamma \in N \mid \gamma(z) \text{ is an isomorphism}\}$$

for $z \in \Lambda^2 V \setminus \{0\}$. Because every Kronecker module in N has the property (K4), the family

$$(N_z)_{z=v_1 \wedge v_2}, \quad v_1 \wedge v_2 \in \Lambda^2 V \setminus \{0\}$$

forms an open covering of N. Furthermore for any two vectors

$$z, z' \in \Lambda^2 V \setminus \{0\}$$

we have

$$N_z \cap N_{z'} \neq \varnothing.$$

For let $\gamma \in N$ be an arbitrary stable Kronecker module of rank 2, and

$$D(\gamma) = \{z \in \Lambda^2 V \mid \gamma(z) \text{ is not an isomorphism}\}$$

be the set of the points in $\Lambda^2 V$ in which γ is not invertible.

Then for every automorphism ϕ of $\Lambda^2 V$ with $\phi(z), \phi(z') \notin D(\gamma)$ (Lemma 4.2.3, Corollary!) we see that $\gamma \circ \phi$ is an element in the intersection

$$N_z \cap N_{z'}.$$

We thus have a covering of N by pairwise intersecting open sets N_z. N is therefore certainly irreducible if all the sets

$$N_z, \quad z = v_1 \wedge v_2 \in \Lambda^2 V \setminus \{0\}$$

are irreducible.

Let then $v_1 \wedge v_2$ be in $\Lambda^2 V \setminus \{0\}$. We choose a $v_0 \in V$ so that $\{v_0, v_1, v_2\}$ is a basis for V and set

$$z_0 = v_1 \wedge v_2, \quad z_1 = v_0 \wedge v_2, \quad z_2 = v_0 \wedge v_1.$$

$\gamma \in N_{z_0}$ is then determined by the triple

$$(\gamma_0, \gamma_1, \gamma_2) = (\gamma(z_0), \gamma(z_1), \gamma(z_2)).$$

Because γ has the property (K3), the matrix

$$(\gamma(v_i \wedge v_j)) = \begin{pmatrix} 0 & \gamma_2 & \gamma_1 \\ -\gamma_2 & 0 & \gamma_0 \\ -\gamma_1 & -\gamma_0 & 0 \end{pmatrix}$$

has rank $2n+2$. By elementary row and column operations one sees that this is equivalent to the following condition:

$$\operatorname{rk}(\gamma_1 \gamma_0^{-1} \gamma_2 - \gamma_2 \gamma_0^{-1} \gamma_1) = 2.$$

Because (K1) and (K2) are Zariski-open conditions, N_{z_0} is a Zariski-open subset of the set

$$N_0 = \{(\gamma_0, \gamma_1, \gamma_2) \in L(H, H^*)^{\oplus 3} \mid \\ \gamma_0 \text{ is an isomorphism}, \operatorname{rk}(\gamma_1 \gamma_0^{-1} \gamma_2 - \gamma_2 \gamma_0^{-1} \gamma_1) = 2\}.$$

It thus suffices to prove the irreducibility of N_0.

We now fibre N_0 over the connected complex manifold $\operatorname{Iso}(H, H^*)$ of isomorphisms from H to H^* by the mapping

$$p_0 \colon N_0 \to \operatorname{Iso}(H, H^*)$$
$$(\gamma_0, \gamma_1, \gamma_2) \mapsto \gamma_0.$$

With respect to the canonical $GL(H) \times GL(H^*)$-operation on N_0 and $\operatorname{Iso}(H, H^*)$ the map p_0 is equivariant. Because $\operatorname{Iso}(H, H^*)$ is a transitive $GL(H) \times GL(H^*)$-space, we have only to prove that the fibre

$$F_0 = p_0^{-1}(\gamma_0)$$

is irreducible for a fixed $\gamma_0 \in \operatorname{Iso}(H, H^*)$.

We identify F_0 with the set

$$F = \{(A, B) \in \operatorname{End} H \oplus \operatorname{End} H \mid \operatorname{rk}[A, B] = 2\}$$

by means of the mapping

$$(\gamma_0, \gamma_1, \gamma_2) \mapsto (\gamma_1 \gamma_0^{-1}, \gamma_2 \gamma_0^{-1}).$$

Thus we have reduced the proof of the irreducibility of $M_{\mathbb{P}_2}(0,n)$ to the following theorem.

THEOREM 4.2.5 (Hulek). *For* $2 \le r \le n = \dim H$ *we have that*

$$F = \{(A,B) \in \operatorname{End} H \oplus \operatorname{End} H \mid \operatorname{rk}(AB - BA) = r\}$$

is irreducible.

PROOF. An endomorphism $A \in \operatorname{End} H$ is *regular* if its minimal polynomial $m_A(t)$ has degree n. Let

$$F_{\text{reg}} = \{(A,B) \in F \mid A \text{ is regular}\}$$

be the set of all pairs $(A,B) \in F$ such that the minimal polynomial of A has maximal degree. $F_{\text{reg}} \subset F$ is Zariski-open in F.

Claim. $F_{\text{reg}} \subset F$ is dense with respect to the usual topology.

Proof. If $(A,B) \in F$, then there is a regular endomorphism $C \in \operatorname{End} H$ which commutes with B (one sees this by bringing B into Jordan normal form). Let $A(t) = A + tC$. The set

$$\{t \in \mathbb{C} \mid A(t) \text{ is regular}\}$$

is non-empty and Zariski-open in $\mathbb{C}$, so there is an $\varepsilon > 0$ with $A(t)$ regular for all t with $0 < |t| < \varepsilon$.

By the choice of C we have

$$[A(t), B] = [A, B],$$

and thus $(A(t), B) \in F_{\text{reg}}$ for $0 < |t| < \varepsilon$. Hence one can approximate $(A,B) = (A(0), B)$ arbitrarily closely by elements of F_{reg}.

We now want to show that F_{reg} is irreducible.

Claim. F_{reg} is irreducible.

Since the proof is rather complicated, we carry it out in several steps. The first two consist in reducing the proof of the irreducibility of F_{reg} to showing that a certain intersection C_{A,I_r} of quadrics in $\operatorname{End} H \oplus \operatorname{End} H$ is irreducible (Steps 1 and 2). Then we show that C_{A,I_r} is a product

$$C_{A,I_r} \cong C^1_{A,I_r} \times \cdots \times C^m_{A,I_r}$$

of intersections C^i_{A,I_r} of quadrics in certain subspaces M_i of $\operatorname{End} H \oplus \operatorname{End} H$ (Step 3). Then we introduce coordinates, compute the codimension of C^i_{A,I_r} in M_i and determine the singular locus $\operatorname{Sing}(C^I_{A,I_r})$ of these intersections C^i_{A,I_r} (Steps 4 and 5). We then prove a lemma (Step 6) with which finally we can show that the sets C^i_{A,I_r} are irreducible. The claim then follows.

STEP 1: In every point $(A_0, B_0) \in F$ of the locally analytic set $F \subset \operatorname{End} H \oplus \operatorname{End} H$ we have

$$(*) \qquad \dim_{(A_0,B_0)} F \geq 2n^2 - (n-r)^2.$$

PROOF. Choose a basis for H so that the upper left r-minor $D_1([A_0, B_0])$ of $[A_0, B_0]$ does not vanish.

$$U = \{(A,B) \in \operatorname{End} H \oplus \operatorname{End} H \mid D_1([A,B]) \neq 0\}$$

is an open neighborhood of (A_0, B_0) in $\operatorname{End} H \oplus \operatorname{End} H$. A pair (A, B) is in $U \cap F$ if and only if all $(r+1)$-minors of $[A, B]$ which "contain" the minor $D_1([A, B])$ vanish. Thus $F \cap U$ is given by $(n-r)^2$ equations in U; the inequality (*) follows.

Now let

$$p\colon F_{\text{reg}} \to \operatorname{End}_{\text{reg}} H, \qquad (A,B) \mapsto A$$

be the projection into the set of regular endomorphisms of H. If we show that all p-fibres are $(2rn - r^2)$-dimensional, then we also have (cf. Fischer [39], p. 142)

$$\dim F_{\text{reg}} \leq 2rn - r^2 + n^2 = 2n^2 - (n-r)^2.$$

Together with (*) it follows then that F_{reg} is equidimensional.

If we further show that all p-fibres are irreducible, then the irreducibility of F_{reg} can be deduced as follows: we only need to show that $F'_{\text{reg}} = F_{\text{reg}} \setminus \operatorname{Sing}(F_{\text{reg}})$ is connected (cf. Narasimhan [91], p. 68). $p' = p|F'_{\text{reg}}\colon F'_{\text{reg}} \to \operatorname{End} H$ is open (cf. Fischer [39], p. 143). Since F'_{reg} is an (algebraically) constructible set in $\operatorname{End} H \oplus \operatorname{End} H$, its image $p'(F'_{\text{reg}}) \subset \operatorname{End} H$ is also (algebraically) constructible (see Mumford [90]), p. 37); in particular $p'(F'_{\text{reg}})$ contains a Zariski-open subset of $\operatorname{End} H$. The set $p'(F'_{\text{reg}})$ is thus connected because $\operatorname{End} H$ is. If the p-fibres and thus the p'-fibres are irreducible, then it follows that F'_{reg} is also connected.

Thus it suffices to show that all p-fibres are irreducible and $(2rn - r^2)$g-dimensional. We consider an arbitrary point $A \in \operatorname{End}_{\text{reg}} H$ and set

$$F_A = p^{-1}(A).$$

Let $C_A = \{[A,B] \mid B \in \operatorname{End} H,\ \operatorname{rk}[A,B] = r\}$. The mapping

$$q\colon F_A \to C_A, \qquad B \mapsto [A,B]$$

makes F_A into an affine bundle over C_A with fibre $\mathbb{C}^n$. We are thus finished if we know that C_A is *irreducible* and $(2rn - r^2 - n)$-*dimensional.*

STEP 2: Let $C_1, \ldots, C_n$ be a basis for the centralizer

$$Z(A) = \{B \mid [A,B] = 0\}$$

of A. Because A is regular, the vector space

$$\{D \in \operatorname{End} H \mid D = [A, B] \text{ for some } B \in \operatorname{End} H\}$$

is n-codimensional in $\operatorname{End} H$, and for all elements $D = [A, B]$ we have

$$\operatorname{Trace}(DC_i) = 0,$$

and hence

$$\{D \in \operatorname{End} H \mid D = [A, B],\ B \in \operatorname{End} H\} \\ \subset \{D \in \operatorname{End} H \mid \operatorname{Tr}(DC_i) = 0,\ i = 1, \dots, n\}.$$

For reasons of dimension we must have equality, so that we can also describe C_A as follows:

$$C_A = \{D \in \operatorname{End} H \mid \operatorname{rk} D = r,\ \operatorname{Tr}(DC_i) = 0,\ i = 1, \dots, n\}.$$

Now let $I_r \in \operatorname{End} H$ be an endomorphism of rank r. The mapping

$$h\colon GL(H) \times GL(H) \to \{D \in \operatorname{End} H \mid \operatorname{rk} D = r\} \\ (X, Y) \mapsto XI_rY$$

is surjective with $(2n^2 - 2rn + r^2)$-dimensional fibres. We set

$$C_{A,I_r} = \{(X, Y) \in \operatorname{End} H \times \operatorname{End} H \mid \operatorname{Tr}(XI_rYC_i) = 0,\ i = 1, \dots, n\}.$$

If we can now show that C_{A,I_r} is *irreducible* and $(2n^2 - n)$-*dimensional*, then it follows that also the image C_A of the Zariski-open subset

$$C_{A,I_r} \cap (GL(H) \times GL(H))$$

under the mapping h is irreducible and $(2rn - r^2 - n)$-dimensional.

STEP 3: Let $H = H_1 \oplus \cdots \oplus H_m$ be a decomposition of H into A-invariant subspaces, $A_i = A|H_i \in \operatorname{End} H_i$ the restriction of A to H_i; we set

$$M_i = L(H, H_i) \times L(H_i, H)$$

and for each i choose a basis

$$C_{ij} \in \operatorname{End} H_i, \quad j = 1, \dots, n_i$$

for the centralizer $Z(A_i)$ of A_i in $\operatorname{End} H_i$. The C_{ij} give a basis for $Z(A)$. Let

$$C^i_{A,I_r} = \{(X, Y) \in M_i \mid \operatorname{Tr}(XI_rYC_{ij}) = 0 \text{ for } j = 1, \dots, n_i\}.$$

C_{A,I_r} is then a product

$$C_{A,I_r} \cong C^1_{A,I_r} \times \cdots \times C^m_{A,I_r},$$

and it suffices to show that each factor C^i_{A,I_r} is *irreducible* and $(2n_in - n_i)$-*dimensional.*

STEP 4: Let then $H = H' \oplus H''$ be an A-invariant decomposition of H such that with respect to a suitable basis of H' the matrix of $A' = A|H' \in \operatorname{End} H'$ is a Jordan matrix: we choose a basis $\{h_1, \dots, h_s\}$ for H' and complete it to a basis $\{h_1, \dots, h_s, h_{s+1}, \dots, h_n\}$ of H such that A' is a Jordan matrix

$$J_s(\lambda) = \begin{pmatrix} \lambda & 1 & & 0 \\ & \ddots & \ddots & \\ & & \ddots & 1 \\ 0 & & & \lambda \end{pmatrix}$$

with respect to $\{h_1, \dots, h_s\}$. Then the matrices

$$C_k = (\delta_{i,j-k+1})_{i,j=1,\dots,s} = \begin{pmatrix} 0 & \dots & 1 & & 0 \\ & \ddots & & \ddots & \\ & & \ddots & & 1 \\ & & & \ddots & \vdots \\ 0 & & & & 0 \end{pmatrix}$$

form a basis of the centralizer of A' in $\operatorname{End} H'$.

Let $M(s,n)$ be the space of all $(s \times n)$-matrices,

$$M_{s,n} = M(s,n) \times M(n,s) \cong L(H,H') \times L(H',H)$$

and

$$I_r = \left(\begin{array}{ccc|c} 1 & & 0 & \\ & \ddots & & 0 \\ 0 & & 1 & \\ \hline & 0 & & 0 \end{array}\right).$$

Then $g_k(X,Y) = \operatorname{Tr}(X I_r Y C_k)$ defines a mapping

$$g = (g_1, \dots, g_s) \colon M_{s,n} \to \mathbb{C}^s.$$

We must show that the fibre

$$Q^r_{s,n} = g^{-1}(0) \subset M_{s,n}$$

is $(2sn - s)$-dimensional and irreducible.

By definition

$$(*) \qquad g_k(X,Y) = \sum_{i=k}^{s} \Big(\sum_{j=1}^{r} x_{ij} y_{j,i-k+1} \Big), \quad k = 1, \dots, s.$$

The differential

$$d_{(X,Y)} g \colon M_{s,n} \to \mathbb{C}^s$$

of g at (X,Y) is thus easy to compute. With respect to the coordinates

$$((x_{1\nu}, \dots, x_{s\nu})_{\nu=1,\dots,n};\ (y_{\mu 1}, \dots, y_{\mu s})_{\mu=1,\dots,n})$$

it is given by the $(2sn \times s)$-matrix

$$\left[\left(\left(\frac{\partial g_k}{\partial x_{l\nu}}\right)_{1\le k,l\le s}\right)_{1\le\nu\le n}, \left(\left(\frac{\partial g_k}{\partial y_{\mu l}}\right)_{1\le k,l\le s}\right)_{1\le\mu\le n}\right]$$

$$= \left[\left.\begin{pmatrix} y_{\nu 1} & \cdots & y_{\nu s} \\ & \ddots & \vdots \\ 0 & & y_{\nu 1}\end{pmatrix}_{1\le\nu\le r}\right| 0 \left| \begin{pmatrix} x_{1\mu} & \cdots & x_{s\mu} \\ \vdots & \cdot^{\cdot^{\cdot}} & \\ x_{s\mu} & & 0\end{pmatrix}_{1\le\mu\le r}\right| 0 \right].$$

One sees that $d_{(X,Y)}g$ is surjective if and only if the vector

$$(x_{s\,1}, \dots, x_{s\,r}, y_{1\,1} \dots, y_{r\,1})$$

is not zero.

Let $S(M_{s,n}) \subset M_{s,n}$ be the linear subspace given by

$$(**) \qquad x_{s\,1} = \dots = x_{s\,r} = y_{1\,1} = \dots = y_{r\,1} = 0.$$

We define

$$S(Q^r_{s,n}) = S(M_{s,n}) \cap Q^r_{s,n}.$$

STEP 5: In this step we show: for $1 \le s \le n$ the subset $Q^r_{s,n}$ is purely s-codimensional in $M_{s,n}$ with singular locus $S(Q^r_{s,n})$ and we have

$$\operatorname{codim}(S(Q^r_{s,n}), Q^r_{s,n}) = \begin{cases} 2r-2 & \text{if } s \ge 2 \\ 2r-1 & \text{if } s = 1. \end{cases}$$

One sees this directly for $s = 1, 2$.

Therefore let $s \ge 3$ and suppose the statement is already proved for $s-2$. We already know that

$$Q^r_{s,n} \setminus S(Q^r_{s,n})$$

is smooth of codimension s in $M_{s,n}$. Let $M_{s-2,n}$ be the set of pairs of matrices

$$(X', Y') = \left((x_{i\,j})_{\substack{i=2,\dots,s-1\\ j=1,\dots,n}}, (y_{i\,j})_{\substack{i=1,\dots,n\\ j=2,\dots,s-1}}\right) \in M(s-2, n) \times M(n, s-2)$$

and $Q^r_{s-2,n} \subset M_{s-2,n}$ be given by the $s-2$ equations

$$0 = g'_k(X', Y') = \sum_{i=k+1}^{s-1}\left(\sum_{j=1}^{r} x_{i\,j}y_{j,i-k+1}\right), \quad k = 1, \dots, s-2.$$

From (*) and (**) one then gets

$$S(Q^r_{s,n}) = Q^r_{s-2,n} \times \mathbb{C}^{2n} \times \mathbb{C}^{2(n-r)}$$

where the factor $\mathbb{C}^{2n} \times \mathbb{C}^{2(n-r)}$ is given by

$$(x_{1\,1}, \dots, x_{1\,n}, y_{1\,s}, \dots, y_{n\,s}, x_{s,r+1}, \dots, x_{s\,n}, y_{r+1,1}, \dots, y_{n\,1}).$$

By the induction hypothesis

$$\operatorname{codim}(Q^r_{s-2,n}, M_{s-2,n}) = s-2,$$

and hence

$$\begin{aligned}\dim S(Q^r_{s,n}) &= 2(s-2)n - (s-2) + 2n + 2(n-r) \\ &= 2sn - s - 2r + 2,\end{aligned}$$

i.e.,

$$\operatorname{codim}(S(Q^r_{s,n}), M_{s,n}) = s + 2r - 2.$$

The claim follows.

STEP 6: In order to see that

$$Q = Q^r_{s,n} \subset M = M_{s,n}$$

is irreducible we consider the analytic set

$$X = \mathbb{P}(Q) \subset \mathbb{P}(M) = \mathbb{P}_N \quad (N = 2sn - 1)$$

in $\mathbb{P}_N$ and show that X is irreducible. We know that X is a complete intersection of codimension s in $\mathbb{P}_N$ with at least 2-codimensional singular locus $S(X) = X \cap \mathbb{P}(S(M_{s,n}))$; thus X is irreducible, as the following general theorem of Hartshorne [54] shows. □

LEMMA 4.2.6. *Every s-codimensional complete intersection X in $\mathbb{P}_N$ with $\dim X \geq 1$ and at least 2-codimensional singular locus S ($\operatorname{codim}(S, X) \geq 2$) is irreducible.*

PROOF. Assume X is reducible. Then $X \setminus S$ is not connected and therefore there are two non empty Zariski-open sets $U_1, U_2 \subset X$ such that

$$U_1 \cap U_2 = \varnothing \quad \text{and} \quad U_1 \cup U_2 = X \setminus S.$$

Now let $H \subset \mathbb{P}_N$ be a general $(s+1)$-dimensional linear subspace with the properties

i) $H \cap S = \varnothing$
ii) $H \cap X$ is a 1-dimensional complete intersection.

(The existence of such a subspace follows from the hypotheses about the codimension.) $H \cap X$ is connected (cf. Hartshorne [56], p. 231), and thus $H \cap U_1 = \varnothing$ or $H \cap U_2 = \varnothing$ — say $H \cap U_1 = \varnothing$. Because the Zariski closure $\bar{U}_1$ of U_1 lies in $U_1 \cup S$ and $H \cap S = \varnothing$, it follows that also $H \cap \bar{U}_1 = \varnothing$. But that is impossible. Hence the assumption is disproved and the lemma proved. □

REMARK 4.2.7. With the help of theorem 4.2.5 one can with some additional effort show that the moduli space $M_{\mathbb{P}_2}(0, n, r)$ for stable r-bundles over $\mathbb{P}_2$ with Chern classes $c_1 = 0$, $c_2 = n$ is irreducible (cf. Hulek [73]).

4.3. Examples. In this section we shall describe the moduli spaces $M_{\mathbb{P}_2}(-1,2)$, $M_{\mathbb{P}_2}(0,2)$ and $M_{\mathbb{P}_3}(0,1)$ more explicitly.

EXAMPLE 1. Let V be a 3-dimensional $\mathbb{C}$-vector space, $(\ ,\ )$ the usual symmetric bilinear form on $\mathbb{C}^2$, $O(2)$ the associated orthogonal group. According to 4.1 we can describe the moduli space $M_{\mathbb{P}_2}(-1,2)$ as

$$P/\mathbb{C}^* \times O(2)$$

with $P = \{\alpha\colon V^* \to \mathbb{C}^2 \mid \alpha \text{ linear and surjective}\}$; for $c_2 = 2$ the space H is 1-dimensional; the condition (F2) is thus automatically fulfilled and (F1) is equivalent to the surjectivity of α.

If in V^* we introduce a basis z_0, z_1, z_2, then we can identify P with the set

$$M_{2,3}^2$$

of all (2×3)-matrices of rank 2. $\mathbb{C}^*$ and $O(2)$ operate by left multiplication on this set.

Let S_n be the set of symmetric $(n \times n)$-matrices, $S_n^r \subset S_n$ the set of matrices in S_n of rank r. We consider the mapping

$$\phi\colon M_{2,3}^2 \to S_3^2$$

with $\phi(A) = A^t A$. If $S = (s_{ij}) \in S_n$, then we regard the entries

$$s_{ij}, \quad 1 \le i \le j \le n,$$

as linear coordinates and denote by $\mathbb{P}(S_n)$ the associated projective space. One defines the (nonlinear!) subspace $\mathbb{P}(S_n^r) \subset \mathbb{P}(S_n)$ analogously.

CLAIM. ϕ induces an isomorphism

$$\Phi\colon M_{\mathbb{P}_2}(-1,2) \to \mathbb{P}(S_3^2).$$

PROOF. ϕ induces a mapping

$$\phi/O(2)\colon M_{2,3}^2/O(2) \to S_3^2.$$

$\phi/O(2)$ is surjective, for every symmetric (3×3)-matrix S of rank 2 is congruent to

$$I_2 = \begin{pmatrix} 1 & 0 & 0 \\ 0 & 1 & 0 \\ 0 & 0 & 0 \end{pmatrix},$$

i.e., there is a matrix $X \in GL(3,\mathbb{C})$ with

$$X^{-1} S X^{-1t} = I_2.$$

The first two rows of X^t thus define a matrix $A \in M_{2,3}^2$ with $A^t A = S$.

To show that $\phi/O(2)$ is also injective, it suffices to investigate the matrices A with $\phi(A) = I_2$. The rank 2 matrices A with $\phi(A) = A^t A = I_2$ are however precisely the matrices of the form

$$\begin{pmatrix} A' & \begin{matrix} 0 \\ 0 \end{matrix} \end{pmatrix}$$

with $A' \in O(2)$; thus $\phi/O(2)$ is bijective and the induced mapping

$$\Phi\colon M_{\mathbb{P}_2}(-1,2) \to \mathbb{P}(S_3^2)$$

hence an isomorphism. □

In particular it follows from this description of $M_{\mathbb{P}_2}(-1,2)$ that all stable bundles over $\mathbb{P}_2$ with the Chern classes $c_1 = -1$, $c_2 = 2$ are projectively equivalent — say to the bundle defined by the matrix

$$\alpha = \begin{pmatrix} 1 & 0 & 0 \\ 0 & 1 & 0 \end{pmatrix} \in M_{2,3}^2.$$

To compactify $M_{\mathbb{P}_2}(-1,2)$ we consider the set

$$\mathbb{P}(S_3^{\text{sing}}) = \{x \in \mathbb{P}(S_3) \mid \det x = 0\}.$$

We have

$$\mathbb{P}(S_3^{\text{sing}}) = \overline{\mathbb{P}(S_3^2)} = \mathbb{P}(S_3^2) \cup \mathbb{P}(S_3^1).$$

As above one sees that the mapping

$$\Psi\colon \mathbb{P}_2 \to \mathbb{P}(S_3^1)$$

$$\Psi(a_0 : a_1 : a_2) = \begin{pmatrix} a_0 \\ a_1 \\ a_2 \end{pmatrix} (a_0\ a_1\ a_2)$$

is an isomorphism.

$\mathbb{P}(S_3^{\text{sing}})$ is a hypersurface of degree 3 in $\mathbb{P}_5$ and a compactification of $M_{\mathbb{P}_2}(-1,2)$. The complement $\mathbb{P}(S_3^1) = \mathbb{P}(S_3^{\text{sing}}) \setminus M_{\mathbb{P}_2}(-1,2)$ is precisely the singular locus of $\mathbb{P}(S_3^{\text{sing}})$.

Remark. The bundle given by the matrix $\alpha = \begin{pmatrix} 1 & 0 & 0 \\ 0 & 1 & 0 \end{pmatrix}$ has exactly one jump line, for the exact cohomology sequence of

$$0 \to E(-2) \to E(-1) \to E(-1)|L \to 0$$

shows that the line L defined by a $z \in V^*$ is a jump line if and only if the mapping

$$H^1(\mathbb{P}, E(-2)) \xrightarrow{\alpha(z)} H^1(\mathbb{P}, E(-1))$$

is injective. If one writes $z = \sum t_i z_i$, then this means that $\alpha(z) = (t_1, t_2)$ must be non-zero. The only jump line is thus the one defined by $(0:0:1) \in \mathbb{P}(V^*)$.

Hulek has shown that for almost all lines $L \subset \mathbb{P}$ the restriction of the bundle E to the first infinitesimal neighborhood L^2 of L has no sections, i.e.,

$$H^0(L^2, E|L^2) = 0$$

for general L. This is equivalent to the statement that the mapping

$$H^1(\mathbb{P}, E(-2)) \to H^1(\mathbb{P}, E)$$

induced by the sequence

$$0 \to E(-2) \to E \to E|L^2 \to 0$$

is injective. This mapping is just the composition

$$\alpha(z)^t \alpha(z)$$

if z defines the line L. In our example with $\alpha = \begin{pmatrix} 1 & 0 & 0 \\ 0 & 1 & 0 \end{pmatrix}$ we have

$$\alpha(z)^t \alpha(z) = t_1^2 + t_2^2.$$

The jump lines of the second kind, i.e., the lines L with

$$h^0(L^2, E|L^2) \neq 0,$$

are in this case the two lines

$$\{(t : it : s)\}, \quad \{(t : -it : s)\} \text{ in } \mathbb{P}(V^*);$$

They intersect in the point $(0:0:1)$.

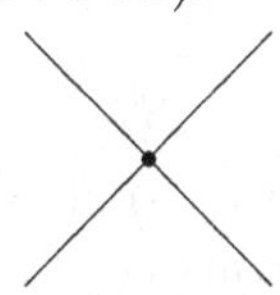

EXAMPLE 2. Let V and H be 3- respectively 2-dimensional vector spaces, $\mathbb{P} = \mathbb{P}(V)$, (K, q) a 6-dimensional vector space with a non-degenerate symplectic form q. The moduli space $M_{\mathbb{P}_2}(0,2)$ is according to 4.1 the orbit space

$$P/GL(H) \times \mathrm{Sp}(q)$$

with

$$P = \{\alpha \in L(V, L(H,K)) \mid \hat{\alpha}\colon V \otimes H \to K \text{ is an isomorphism and } \alpha(v)^* q \alpha(v) = 0 \text{ for all } v \in V\}.$$

In order to give a simpler description of this space, we identify $P/\mathrm{Sp}(q)$ with the space of *symmetric* stable Kronecker modules

$$\gamma \in L(\Lambda^2 V, L(H, H^*))$$

of rank 2: if $\alpha \in P$ then the Kronecker module γ associated to (α, α^t) by

$$\gamma(v_1 \wedge v_2) = \alpha(v_2)^* q \alpha(v_1) \in L(H, H^*)$$

is symmetric, i.e.,

$$\gamma(v_1 \wedge v_2)^* = \gamma(v_1 \wedge v_2),$$

i.e., $\gamma \in L(\Lambda^2 V, S^2 H^*)$.

The conditions (K1)–(K3) for Kronecker modules (cf. 4.2) are in the case $c_2 = 2$ equivalent to the bijectivity of the mapping

$$\hat{\gamma}\colon V \otimes H \to V^* \otimes H^*$$

and in our case this is equivalent to the bijectivity of the mapping $\gamma\colon \Lambda^2 V \to S^2 H^*$, as one can easily verify.

With the assignment $\alpha \mapsto \gamma$ we get a bijective mapping

$$P/\mathrm{Sp}(q) \to \mathrm{Iso}(\Lambda^2 V, S^2 H^*).$$

The injectivity is proved as in Lemma 4.2.2. The surjectivity can be seen as follows: let $\gamma \in \mathrm{Iso}(\Lambda^2 V, S^2 H^*)$. For the associated isomorphism $\hat{\gamma}\colon V \otimes H \to V^* \otimes H^*$ we then have $\hat{\gamma}^* = -\hat{\gamma}$ and thus there is an isomorphism $\hat{\alpha}\colon V \otimes H \to K$ with $\hat{\alpha}^* q \hat{\alpha} = \hat{\gamma}$. Then $\hat{\alpha}$ determines an element $\alpha \in P$, whose associated Kronecker module is precisely γ.

Thus for $M_{\mathbb{P}_2}(0,2)$ we get the description

$$M_{\mathbb{P}_2}(0,2) \cong \mathrm{Iso}(\Lambda^2 V, S^2 H^*)/GL(H).$$

After choosing a basis for H we can identify the vector space $S^2 H^*$ with the space

$$S_2$$

of all symmetric (2×2)-matrices. Let $S_2^{\mathrm{sing}} \subset S_2$ be the singular matrices in S_2. In the homogeneous coordinates s_{ij}, $1 \leq i \leq j \leq 2$, we have that

$$\mathbb{P}(S_2^{\mathrm{sing}}) \subset \mathbb{P}(S_2) \cong \mathbb{P}_2$$

is the conic

$$\mathbb{P}(S_2^{\mathrm{sing}}) = \{(s_{11} : s_{12} : s_{22}) \mid s_{11}s_{22} = s_{12}^2\}.$$

If $\gamma\colon \Lambda^2 V \to S^2 H^* \cong S_2$ is an isomorphism, then the preimage C_γ of $\mathbb{P}(S_2^{\mathrm{sing}})$ under the mapping

$$\mathbb{P}(\gamma)\colon \mathbb{P}(\Lambda^2 V) \to \mathbb{P}(S_2)$$

induced by γ is a non-singular conic in $\mathbb{P}^* = \mathbb{P}(\Lambda^2 V)$. The points $z \in \Lambda^2 V$ with $\gamma(z) \in S_2^{\mathrm{sing}}$ characterize according to Lemma 4.2.3

precisely the jump lines of the bundle $E(\gamma)$ belonging to γ. Thus C_γ is the curve of jump lines of $E(\gamma)$ in $\mathbb{P}^*$; $z \in C_\gamma$ defines the jump line $L_z = \{x \in \mathbb{P} \mid x \wedge z = 0\}$ in $\mathbb{P}^*$.

CLAIM: The mapping

$$C\colon M_{\mathbb{P}_2}(0,2) \to \mathbb{P}(H^0(\mathbb{P}^*, \mathcal{O}_{\mathbb{P}^*}(2))) \cong \mathbb{P}(S_3)$$
$$[E(\gamma)] \mapsto C_\gamma$$

maps $M_{\mathbb{P}_2}(0,2)$ bijectively onto the set $\mathbb{P}(S_3^3)$ of non-singular conics in $\mathbb{P}^*$.

PROOF. The mapping $C\colon M_{\mathbb{P}_2}(0,2) \to \mathbb{P}(S_3^3)$ is well defined and surjective because all non-singular conics are projectively equivalent.

Let $\gamma, \gamma'\colon \Lambda^2 V \to S_2$ be two isomorphisms such that

$$C_\gamma = C_{\gamma'}.$$

Then $\sigma = \mathbb{P}(\gamma') \circ \mathbb{P}(\gamma)^{-1}$ is an automorphism of $\mathbb{P}(S_2)$ which leaves $\mathbb{P}(S_2^{\text{sing}})$ invariant. σ is then of the form

$$\sigma(S) = A^t S A \text{ with a matrix } A \in GL(2, \mathbb{C}).$$

This we show as follows: let $f\colon \mathbb{P}_1 \xrightarrow{\sim} \mathbb{P}(S_2^{\text{sing}})$ be the isomorphism $f(z_0 : z_1) = \begin{pmatrix} z_0 \\ z_1 \end{pmatrix} (z_0, z_1)$. The homomorphism

$$\Phi_f\colon \operatorname{Aut}(\mathbb{P}(S_2), \mathbb{P}(S_2^{\text{sing}})) \to \operatorname{Aut}\mathbb{P}_1$$
$$\sigma \mapsto f^{-1} \circ \sigma | \mathbb{P}(S_2^{\text{sing}}) \circ f$$

is injective, because $\mathbb{P}(S_2^{\text{sing}})$ meets every general line L in $\mathbb{P}(S_2)$ in two points, i.e., σ is determined by $\sigma|\mathbb{P}(S_2^{\text{sing}})$.

For the homomorphism

$$\Psi\colon \operatorname{Aut}\mathbb{P}_1 \to \operatorname{Aut}(\mathbb{P}(S_2), \mathbb{P}(S_2^{\text{sing}}))$$
$$((z_0 : z_1) \mapsto (z_0, z_1)A) \mapsto (S \mapsto A^t S A)$$

we have

$$\Phi_f \circ \Psi = \mathrm{id}_{\operatorname{Aut}\mathbb{P}_1}.$$

Since Φ_f is injective, Ψ is an isomorphism. In particular then two isomorphisms $\gamma, \gamma'\colon \Lambda^2 V \to S^2 H^* \cong S_2$ with $C_\gamma = C_{\gamma'}$ are equivalent modulo $GL(H)$, i.e., $E(\gamma) \cong E(\gamma')$. The mapping C is therefore bijective. □

In particular all stable 2-bundles E over $\mathbb{P}_2$ with $c_1 = 0$, $c_2 = 2$ are projectively equivalent to the (up to isomorphism uniquely determined) bundle E_0 with curve of jump lines

$$\{(z_0 : z_1 : z_2) \in \mathbb{P}^* \mid z_0^2 + z_1^2 + z_2^2 = 0\}.$$

REMARK. Let $C \subset \mathbb{P}^*$ be the curve of jump lines of a stable 2-bundle E with $c_1 = 0$, $c_2 = 2$.

The curve C is a rational normal curve in $\mathbb{P}^*$ and hence given by an embedding

$$f \colon \mathbb{P}_1 \hookrightarrow \mathbb{P}^*$$

of degree 2. Every line in $\mathbb{P}^*$ determines a point in $\mathbb{P}$,

$$(\mathbb{P}^*)^* \cong \mathbb{P}.$$

Let $f^* \colon \mathbb{P}_1 \hookrightarrow \mathbb{P}$ be the curve in $\mathbb{P}$ dual to C. The map f^* associates to every point $t \in \mathbb{P}_1$ the point $f^*(t)$ given by the tangent to C in the point $f(t) \in C$. We have

$$(f^*)^* = f.$$

The tangents to the dual curve $C^* = f^*(\mathbb{P}_1) \subset \mathbb{P}$ are thus simply the lines determined by points in $C \subset \mathbb{P}^*$, hence exactly the jump lines of the bundle E.

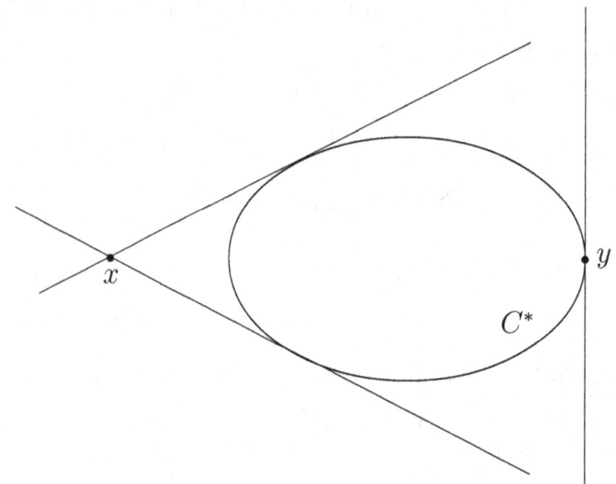

From this one sees that through every point $x \in \mathbb{P}$ not on the curve C^* exactly two jump lines pass, whereas through points $y \in C^*$ only one jump line passes — namely the tangent to C^* at y.

Let $\mathbb{F}_2 = \{(x, \ell) \in \mathbb{P} \times \mathbb{P}^* \mid x \in \ell\}$ be the flag manifold, p and q the projections onto $\mathbb{P}$ respectively $\mathbb{P}^*$. We consider the preimage in $\mathbb{F}_2$

$$q^{-1}(C) \subset \mathbb{F}_2$$

of the curve of jump lines. Let

$$q' \colon q^{-1}(C) \setminus p^{-1}(C^*) \to C^*$$

be the mapping which to a point $(x, \ell) \in q^{-1}(C)$ with $x \notin C^*$ associates the uniquely determined point $q'(x, \ell) \in C^*$ in which the jump line through x other than L touches C^*.

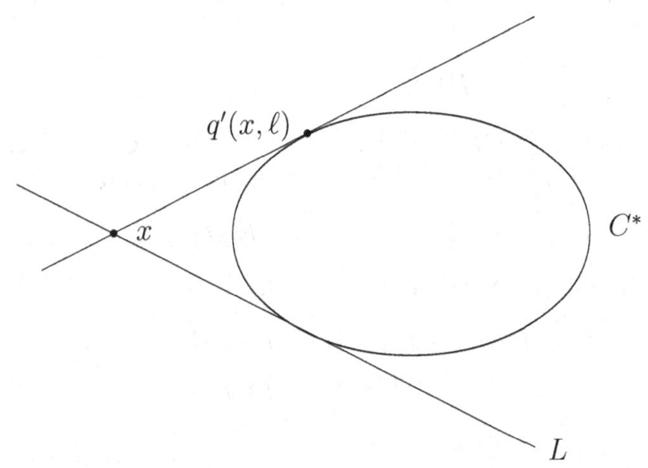

q' can be extended to a mapping

$$\bar{q}\colon q^{-1}(C) \to C^*;$$

with

$$q \times \bar{q}\colon q^{-1}(C) \to C \times C^*$$

one gets a biholomorphic mapping of $q^{-1}(C)$ onto the quadric $Q = C \times C^*$. Let $p'\colon Q \to \mathbb{P}$ be the restriction of the projection $p\colon \mathbb{F}_2 \to \mathbb{P}$ to Q. p' is a 2-fold covering, which is branched exactly in the points over C^*.

In §2 we defined the curve of jump lines C with the help of the sheaf

$$F = R^1 q_* p^* E(-1).$$

In our case $F \simeq F|C$ is a line bundle over C. One can show (cf. Barth [13]) that for the bundle $L = F(1)|C$ one has

$$(p'_* q^* L)(-1) \cong E.$$

This is the original description due to Schwarzenberger [108] of a 2-bundle as direct image of a line bundle under a branched 2-fold covering

$$p'\colon Q \to \mathbb{P}$$

of $\mathbb{P}_2$ by a quadric. The tangents to the discriminant $C^* \subset \mathbb{P}$ of this covering are the jump lines of E.

Example 3. The moduli space $M_{\mathbb{P}_3}(0,1)$ for stable 2-bundles over $\mathbb{P}_3$ with Chern classes $c_1 = 0$, $c_2 = 1$ can be constructed as follows. We begin with a lemma.

LEMMA 4.3.1. *For a stable 2-bundle E over $\mathbb{P}_3$ with Chern classes $c_1 = 0$, $c_2 = 1$ we have*

$$H^1(\mathbb{P}_3, E(-2)) = 0.$$

PROOF. Take a line L such that $E|L = \mathcal{O}_L \oplus \mathcal{O}_L$ and choose a plane $H \subset \mathbb{P}_3$ containing L. It follows immediately that

$$h^0(E(-1)|H) = 0, \qquad h^0(E|H) \leq 1.$$

By Riemann–Roch and Serre–Duality we have

$$\begin{aligned} h^0(E|H) &= 1 + h^1(E|H) - h^2(E|H) \\ &= 1 + h^1(E|H) - h^0(E(-3)|H) \\ &= 1 + h^1(E|H). \end{aligned}$$

This shows that

$$h^0(E|H) = 1, \qquad h^1(E|H) = 0.$$

From the exact sequence

$$0 \to E(k-1)|H \to E(k)|H \to E(k)|L \to 0$$

one deduces that

$$h^1(E(k)|H) = 0 \quad \text{for} \quad k \geq 0.$$

This implies, using the exact sequence

$$0 \to E(k-1) \to E(k) \to E(k)|H \to 0$$

that

$$h^2(E(k)) = 0 \quad \text{for} \quad k \geq -1.$$

By Serre-Duality this gives

$$h^1(E(k)) = 0 \quad \text{for} \quad k \leq -3.$$

Using Riemann–Roch we can write down the $h^q(E(p))$-diagram, $0 \leq q \leq 3$, $-3 \leq p \leq 0$:

$$\begin{array}{cccc|l} 0 & 0 & 0 & 0 & \uparrow q \\ 1 & a & 0 & 0 & \\ 0 & a & 1 & 0 & \\ 0 & 0 & 0 & 0 & \\ \hline & & & & \to p \end{array}$$

where $a = h^1(E(-2)) = h^2(E(-2))$.

From the exact sequence

$$\cdots \to H^1(E(-2)|H) \to H^2(E(-3)) \to H^2(E(-2)) \to H^2(E(-2)|H) = 0$$

we deduce that

$$a = h^2(E(-2)) \leq h^2(E(-3)) = 1.$$

Suppose that $a = 1$. The theorem of Beilinson (3.1.3) then shows (after a short calculation) the existence of an isomorphism

$$\Omega^2_{\mathbb{P}_3}(2) \xrightarrow{\sim} \Omega^1_{\mathbb{P}_3}(1)$$

which is absurd. Hence $a = 0$ and the lemma is proved. □

In §3 we showed that every stable 2-bundle E over $\mathbb{P}_3$ with $c_1(E) = 0$, $c_2(E) = 1$, $H^1(\mathbb{P}_3, E(-2)) = 0$ occurs as quotient of a bundle monomorphism

$$a\colon \mathcal{O}_{\mathbb{P}_3}(-1) \to \Omega^1_{\mathbb{P}_3}(1).$$

Together with the lemma just proved this yields the following.

LEMMA 4.3.2. *Every stable 2-bundle E over $\mathbb{P}_3$ with $c_1(E) = 0$, $c_2(E) = 1$ is given by an exact sequence*

$$0 \to \mathcal{O}_{\mathbb{P}_3}(-1) \to \Omega^1_{\mathbb{P}_3}(1) \to E \to 0.$$

Each of these bundles E is a null correlation bundle.

In order to be able to describe the moduli space $M_{\mathbb{P}_3}(0,1)$ of the null correlation bundles, we must investigate the question, which bundle monomorphisms

$$a\colon \mathcal{O}_{\mathbb{P}_3}(-1) \to \Omega^1_{\mathbb{P}_3}(1)$$

define isomorphic quotient bundles.

LEMMA 4.3.3. *Let $a, a'\colon \mathcal{O}_{\mathbb{P}_3}(-1) \to \Omega^1_{\mathbb{P}_3}(1)$ be two bundle monomorphisms, E, E' the associated quotient bundles. E and E' are isomorphic if and only if there is a constant $c \in \mathbb{C} \setminus \{0\}$ with $a' = ca$.*

PROOF. Of course the quotients of a and ca are isomorphic. Conversely suppose

$$a, a'\colon \mathcal{O}_{\mathbb{P}_3}(-1) \to \Omega^1_{\mathbb{P}_3}(1)$$

are bundle monomorphisms and

$$\psi\colon E \to E'$$

an isomorphism of the quotients. It is easy to see that ψ is induced by a homomorphism of the associated monads (cf. 4.1). Thus we get a commutative diagram

$$\begin{array}{ccccccccc} 0 & \longrightarrow & \mathcal{O}_{\mathbb{P}_3}(-1) & \xrightarrow{a} & \Omega^1_{\mathbb{P}_3}(1) & \longrightarrow & E & \longrightarrow & 0 \\ & & \downarrow{\scriptstyle \Psi'} & & \downarrow{\scriptstyle \Psi} & & \downarrow{\scriptstyle \psi} & & \\ 0 & \longrightarrow & \mathcal{O}_{\mathbb{P}_3}(-1) & \xrightarrow{a'} & \Omega^1_{\mathbb{P}_3}(1) & \longrightarrow & E' & \longrightarrow & 0. \end{array}$$

But $\mathcal{O}_{\mathbb{P}_3}(-1)$ and $\Omega^1_{\mathbb{P}_3}(1)$ are simple, i.e., Ψ and Ψ' are homotheties:

$$\Psi = \lambda \mathrm{id}_{\Omega^1_{\mathbb{P}_3}(1)}, \qquad \lambda \in \mathbb{C} \setminus \{0\}$$
$$\Psi' = \lambda' \mathrm{id}_{\Omega^1_{\mathbb{P}_3}(1)}, \qquad \lambda' \in \mathbb{C} \setminus \{0\}.$$

Hence we have $a' = ca$ with $c = \lambda/\lambda'$. □

We can now describe the moduli space of the null correlation bundles.

THEOREM 4.3.4. *The moduli space $M_{\mathbb{P}_3}(0,1)$ for the null correlation bundles over $\mathbb{P}_3$ is isomorphic to the complement*

$$\mathbb{P}_5 \setminus G(1,3)$$

in $\mathbb{P}_5$ of the Grassmann manifold of lines in $\mathbb{P}_3$.

PROOF. Let V be a 4-dimensional $\mathbb{C}$-vector space, $\mathbb{P}_3 \cong \mathbb{P}(V)$. With the help of the dual Euler sequence

$$0 \to \Omega^1_{\mathbb{P}(V)}(1) \to V^* \otimes \mathcal{O}_{\mathbb{P}(V)} \to \mathcal{O}_{\mathbb{P}(V)}(1) \to 0$$

one can identify the bundle monomorphisms

$$a\colon \mathcal{O}_{\mathbb{P}(V)}(-1) \to \Omega^1_{\mathbb{P}(V)}(1)$$

with certain sections in $V^* \otimes \mathcal{O}_{\mathbb{P}(V)}(1)$. If one regards elements in

$$H^0(\mathbb{P}(V), V^* \otimes \mathcal{O}_{\mathbb{P}(V)}(1)) \simeq V^* \otimes V^*$$

as linear mappings

$$\alpha\colon V \to V^*,$$

then the non-vanishing sections in $\Omega^1_{\mathbb{P}(V)}(2)$ are given — after choosing a basis in V and the dual basis in V^* — precisely by the skew symmetric non-singular matrices

$$A \in GL(4, \mathbb{C}).$$

Hence

$$M_{\mathbb{P}_3}(0,1) = \{A \in PGL(4,\mathbb{C}) \mid A = -A^t\}$$

gives a coarse moduli space.

If one associates to each skew symmetric matrix

$$A = \begin{bmatrix} 0 & a & b & c \\ -a & 0 & d & e \\ -b & -d & 0 & f \\ -c & -e & -f & 0 \end{bmatrix} \in GL(4, \mathbb{C})$$

the point $(a : b : c : d : e : f) \in \mathbb{P}_5$, then one gets an isomorphism

$$M_{\mathbb{P}_3}(0, 1) \xrightarrow{\sim} \mathbb{P}_5 \setminus G(1, 3),$$

where $G(1, 3)$ denotes the support of the (non-reduced) divisor in $\mathbb{P}_5$ defined by the equation

$$\det A = (af - be + cd)^2 = 0.$$

$G(1, 3)$ is precisely the image under the Plücker embedding of the Grassmann manifold $G(1, 3)$ of lines in $\mathbb{P}_3$. □

4.4. Historical remarks, further results, and open problems. Maruyama has shown that for stable vector bundles with fixed Hilbert polynomial over a projective-algebraic manifold $X \hookrightarrow \mathbb{P}_N$ there is a coarse moduli space and it is algebraic. For surfaces this was also proved by Gieseker [42]. The coarse moduli space is even quasi-projective (cf. [40]).

The description of the moduli space for stable 2-bundles over $\mathbb{P}_2$ with $c_1 = 0$ by means of monads was carried out by Barth in his important paper [13]. Hulek [72] then dealt with the case $c_1 = -1$. With this "explicit" construction Barth and Hulek then succeeded in proving that these moduli spaces are irreducible and rational. Independently Maruyama [85] has proved the irreducibility and the unirationality (in some cases the rationality). Furthermore one finds in his paper a sufficient number-theoretic criterion for the moduli space to be fine. In the concrete case of 2-bundles over $\mathbb{P}_2$ this gives precisely the statements in §4.1:

$$M_{\mathbb{P}_2}(-1, n) \text{ is fine,}$$
$$M_{\mathbb{P}_2}(0, n) \text{ is fine if } n \text{ is odd.}$$

In our presentation in 4.1 we mainly followed Le Potier [97]. He proves that $M_{\mathbb{P}_2}(0, n)$ is *not* a fine moduli space for n even. To see this he calculates

$$\pi_1(M_{\mathbb{P}_2}(0,n)) = \begin{cases} \mathbb{Z}/3\mathbb{Z} & \text{for } n=2 \\ 0 & \text{otherwise} \end{cases}$$

$$\pi_2(M_{\mathbb{P}_2}(0,n)) = \begin{cases} \mathbb{Z}/2\mathbb{Z} & \text{for } n=2 \\ \mathbb{Z}/2\mathbb{Z}\oplus\mathbb{Z} & \text{for } n>2,\ n \text{ even} \\ \mathbb{Z} & \text{for } n \text{ odd.} \end{cases}$$

PROBLEM 4.4.1. Determine the topological invariants of $M_{\mathbb{P}_2}(0,n)$ and $M_{\mathbb{P}_2}(-1,n)$.

With an entirely different approach Ellingsrud and Strømme [36] have shown that $M_{\mathbb{P}_2}(-1,n)$ is irreducible and rational. Furthermore Le Potier and Ellingsrud–Strømme [37] have computed the algebraic Picard groups:

$$\mathrm{Pic}_{\mathrm{alg}}(M_{\mathbb{P}_2}(0,n)) = \begin{cases} \mathbb{Z} & \text{for } n>2 \\ \mathbb{Z}/3\mathbb{Z} & \text{for } n=2 \end{cases}$$

$$\mathrm{Pic}_{\mathrm{alg}}(M_{\mathbb{P}_2}(-1,n)) = \begin{cases} \mathbb{Z} & \text{for odd } n \\ \mathbb{Z}/2\mathbb{Z} & \text{for even } n. \end{cases}$$

If one wishes to compactify the moduli spaces for stable bundles, one must admit semistable torsion-free coherent sheaves (cf. Langton [79]). To get Hausdorff compactifications it is necessary to identify semistable sheaves if their associated graded sheaves are isomorphic, i.e., their Jordan–Hölder filtrations have isomorphic stable factors. Moreover here it is essential to use the Gieseker concept of semistability (cf. [85]). The resulting moduli spaces are then even projective.

In Theorem 2.2.4 we saw that for a (semi-)stable 2-bundle E with $c_1=0$ over $\mathbb{P}_n$ the set S_E of jump lines can in a natural way be regarded as a divisor of degree c_2 in the Grassmann manifold G_n. One can ask to what extent S_E determines the bundle E. We saw in 4.3 that over $\mathbb{P}_2$ for $c_2=2$ this is true. Barth [13] has shown quite generally that the curve of jump lines together with a "θ-characteristic" determines the bundle uniquely. This θ-characteristic is an $\mathcal{O}_{S_E}$-module sheaf θ, which satisfies the equation

$$\theta^2 = \omega_{S_E}$$

if S_E is reduced and θ is locally free. With the notation of 2.2.4 one has

$$\theta = (R^1 q_* p^*(E(-1))) \otimes \mathcal{O}_{G_n}(1).$$

In particular Barth shows that for $c_2 = 3$ the curve of jump lines alone no longer determines the bundle uniquely. He moreover investigates for $c_2 \leq 5$ which curves can occur as curves of jump lines.

Hulek [72] carries out analogous investigations for the case $c_1 = -1$. Here however — as we indicated in 2.3 — one has to consider jump lines of the second kind. It turns out that the bundles with $c_2 = 2$ and $c_2 = 3$ are still uniquely determined by their curves of jump lines of the second kind. Barth and Hulek further show that for the bundles constructed by Hulsbergen [74] (which form a Zariski-open subset of the moduli space) the curves of jump lines of the first kind are smooth and that the singularities of the curve of jump lines of the second kind are precisely the jump lines of the first kind.

In section 4.3 we determined the moduli space $M_{\mathbb{P}_3}(0,1)$. This had already been done by other means by Barth [12] (cf. also Hartshorne [58]). For higher second Chern class little is known. Hartshorne [58] has investigated $M_{\mathbb{P}_3}(0,2)$ and shown that it is smooth and connected. On the other hand Barth–Hulek [15] and Hartshorne [58] have shown that $M_{\mathbb{P}_3}(0,2n+1)$ for $n \geq 2$ (apart from the splitting into 2 components by the α-invariant) is no longer irreducible.

We shall follow now established usage in calling stable 2-bundles E over $\mathbb{P}_3$ with $c_1(E) = 0$ and $H^1(\mathbb{P}_3, E(-2)) = 0$ *complex instanton bundles* of rank 2. In Section 3.2 (Example 5) we saw that these bundles can be described by means of self-dual monads and that one thus gets a description of the moduli space (cf. 4.1.6). Lemma 4.3.1 shows that stable 2-bundles over $\mathbb{P}_3$ with $c_1 = 0$ and $c_2 = 1$ are instanton bundles. Barth and Elencwajg [14] have proved the following general vanishing theorem for stable 2-bundles E over $\mathbb{P}_n$, $n \geq 3$, with $c_1(E) = 0$:

$$H^1(\mathbb{P}_n, E(-i)) = 0 \quad \text{for} \quad i > \left[\frac{c_2+1}{2}\right].$$

From this it follows that also for $c_2 = 2$ all stable 2-bundles over $\mathbb{P}_3$ are instanton bundles. One can further show that for $c_2 = 3$ and $c_2 = 4$ stable 2-bundles over $\mathbb{P}_3$ whose α-invariant vanishes are also instanton bundles.

PROBLEM 4.4.2. What sort of vanishing theorem holds for stable bundles over $\mathbb{P}_3$ with $c_1 = -1$?

It would be very interesting to know more about the moduli spaces of stable 2-bundles over $\mathbb{P}_3$. For example $M_{\mathbb{P}_3}(-1,2)$ is unknown. One also knows nothing about $M_{\mathbb{P}_3}(0,3)$, $M_{\mathbb{P}_3}(0,4)$. In studying $M_{\mathbb{P}_3}(0,n)$ it is useful to restrict attention to the Zariski-open set of instanton bundles, since here a beautiful description with monads is available. The following problem is still open.

PROBLEM 4.4.3. Is the space of complex instanton bundles over $\mathbb{P}_3$ irreducible and rational? Does it have singularities?

A possible point of attack on this problem might be to study the restrictions of these bundles to quadrics (cf. [60]).

We wish to conclude with some remarks about the connection between stable vector bundles and real instantons in physics. On $\mathbb{P}_3$ one has a real structure given by the involution

$$\sigma\colon (x_0 : x_1 : x_2 : x_3) \mapsto (-\bar{x}_1 : -\bar{x}_0 : -\bar{x}_3 : -\bar{x}_2).$$

The lines in $\mathbb{P}_3$ which are left invariant by σ will be called *real.* Atiyah and Ward [4] have shown that the euclidean self-dual $SU(2)$–Yang–Mills fields on S^4 (modulo gauge transformations) correspond precisely to the isomorphism classes of holomorphic 2-bundles over $\mathbb{P}_3$ with

i) E is trivial on real lines;
ii) E has a symplectic structure lying over σ, i.e., one has an antilinear isomorphism $\tau\colon E \to E$ with $\tau^2 = -\mathrm{id}_E$ such that the diagram

$$\begin{array}{ccc} E & \xrightarrow{\tau} & E \\ \downarrow & & \downarrow \\ \mathbb{P}_3 & \xrightarrow{\sigma} & \mathbb{P}_3 \end{array}$$

commutes. Here we must regard two symplectic structures on E as equal if they differ by a multiplicative constant $z \in \mathbb{C}$, $|z| = 1$.

We shall call such bundles E *real instanton bundles* (of rank 2). From the conditions i) and ii) it follows that real instanton bundles are stable and have vanishing first Chern class. Atiyah–Hitchin–Drinfeld–Manin [6], [30] have shown that for real instanton bundles E we have

$$H^1(\mathbb{P}_3, E(-2)) = 0$$
$$H^1(\mathbb{P}_3, (\operatorname{End} E)(-2)) = 0.$$

Thus real instanton bundles are also complex instanton bundles and one can describe them by monads of the form

$$0 \to H \otimes \mathcal{O}_{\mathbb{P}_3}(-1) \to K \otimes \mathcal{O}_{\mathbb{P}_3} \to H^* \otimes \mathcal{O}_{\mathbb{P}_3}(1) \to 0.$$

In particular one can see that the space of real instanton bundles with second Chern class n has real dimension $8n - 3$ [6]. Because of

$H^1(\mathbb{P}_3, (\operatorname{End} E)(-2)) = 0$ it follows that $H^2(\mathbb{P}_3, \operatorname{End}(E)) = 0$. Therefore this moduli space is also smooth. It is an unsolved problem whether these moduli spaces are connected.

A reader interested in these questions is referred to the literature [3], [4], [5], [6], [7], [27], [28], [29], [30], [57], [100], [131], [138]. In Hartshorne's problem list one finds still further literature.

Bibliography

[1] Altman, A. B., Kleiman, S. L.: Introduction to Grothendieck duality theory. Lecture Notes in Math. 146. Springer 1970.

[2] Atiyah, M. F., Rees, E.: Vector bundles on projective 3-space. Invent. Math. **35**, 131–153 (1976).

[3] Atiyah, M. F., Hitchin, N. J., Singer, I. M.: Deformations of instantons. Proc. Nat. Acad. Sci. USA **74**, 2662–2663 (1977).

[4] Atiyah, M. F., Ward, R. S.: Instantons and algebraic geometry. Comm. Math. Phys. **55**, 117–124 (1977).

[5] Atiyah, M. F., Hitchin, N. J., Singer, I. M.: Self-duality in four-dimensional Riemannian geometry. Proc. Roy. Soc. London A **362**, 425–461 (1978).

[6] Atiyah, M. F., Hitchin, N. J., Drinfeld, V. G., Manin, Y. I.: Construction of instantons. Phys. Lett. **65** A, 185–187 (1978).

[7] Atiyah, M. F., Jones, J. D. S.: Topological aspects of Yang–Mills theory. Comm. Math. Phys. **61**, 97–118 (1978).

[8] Bănică, C., Stănăşilă, O.: Algebraic methods in the global theory of complex spaces, Wiley 1976.

[9] Barth, W., Van de Ven, A.: A decomposability criterion for algebraic 2-bundles on projective spaces. Invent. Math. **25**, 91–106 (1974).

[10] Barth, W., Van de Ven, A.: On the geometry in codimesion 2 of Grassmann manifolds. In Classification of algebraic varieties and compact complex manifolds, Lecture Notes in Math. 412, 1–35. Springer 1974.

[11] Barth, W.: Submanifolds of low codimension in projective space. Proc. I.C.M. Vancouver, 409–413 (1975).

[12] Barth, W.: Some properties of stable rank-2 vector bundles on $\mathbb{P}_n$. Math. Ann. **226**, 125–150 (1977).

[13] Barth, W.: Moduli of vector bundles on the projective plane. Invent. Math. **42**, 63–91 (1977).

[14] Barth, W., Elencwajg, G.: Concernant la cohomologie des fibrés algébriques stables sur $\mathbb{P}_n(\mathbb{C})$. In Variétés analytiques compactes, Nice 1977, Lecture Notes in Math. 683, 1–24. Springer 1978.

[15] Barth, W., Hulek, K.: Monads and moduli of vector bundles. Manuscripta math. **25**, 323–347 (1978).

[16] Barth, W.: Kummer surfaces associated with the Horrocks–Mumford bundle. To appear in the Proceedings of the Angers conference 1979 on algebraic geometry.

[17] Barth, W.: Counting singularities of quadratic forms on vector bundles. To appear in the Proceedings of the Nice Conference 1979 on Vector bundles and Differential equations.

[18] Beilinson, A.: Coherent sheaves on P^N and problems of linear algebra. Functional Anal. Appl. **12**, 214–216 (1978). English translation.
[19] Birkhoff, G. D.: Singular points of ordinary linear differential equations. TAMS **10**, 436–470 (1909).
[20] Birkhoff, G. D.: A theorem on matrices of analytic functions. Math. Ann. **74**, 122–133 (1913).
[21] Bogomolov, F. A.: Holomorphic tensors and vector bundles on projective manifolds. In Russian. Isvestija Akademii Nauk SSr, Ser Mat. **42**, 1227–1287 (1978).
[22] Borel, A., Serre, J. P.: Le théorème de Riemann–Roch. Bull. Soc. math. France **86**, 97–136 (1958).
[23] Bott, R.: Homogeneous vector bundles. Ann. of Math. **66**, 203–248 (1957).
[24] Daoudy, M.: Fibrés bi-uniformes sur $\mathbb{P}_2(\mathbb{C})$. Preprint, Nice 1979.
[25] Dedekind, R., Weber, H.: Theorie der algebraischen Funktionen einer Veränderlichen. Crelle Journal **92**, 181–290 (1882).
[26] Deschamps, M.: Courbes de genre géométrique borné sur une surface de type général (D'après F. A. Bogomolov) Sém. Bourbaki 1977/78, exposé 519. Lecture Notes in Math. 710. Springer 1979.
[27] Douady, A., Verdier, J. L.: Séminaire E.N.S. 1977/78.
[28] Drinfeld, V. G., Manin, Ju.: On locally free sheaves over CP^3 connected with Yang–Mills fields. Uspehi Mat. Nauk **33**, 165–166 (1978).
[29] Drinfeld, V. G., Manin, Ju.: Self-dual Yang–Mills fiels over a sphere. Functional Anal. Appl. **12**, 140–142 (1978).
[30] Drinfeld, V. G., Manin, Ju.: A description of instantons. Comm. Math. Phys. **63**, 177–192 (1978).
[31] Elencwajg, G.: Les fibrés uniformes de rang 3 sur $\mathbb{P}_2(\mathbb{C})$ sont homogènes. Math. Ann. **231**, 217–227 (1978).
[32] Elencwajg, G.: Des fibrés uniformes non homogènes. Math. Ann. **239**, 185–192 (1979).
[33] Elencwajg, G.: Concernant les fibrés uniformes de rang 4 sur $\mathbb{P}_2$. Preprint, Nice 1979.
[34] Elencwajg, G., Forster, O.: Bounding cohomology groups of vector bundles over $\mathbb{P}_n$. Math. Ann. **246**, 251–270 (1980).
[35] Elencwajg, G., Hirschowitz, A., Schneider, M.: Les fibrés uniformes de rang au plus n sur $\mathbb{P}_n(\mathbb{C})$ sont ceux qu'on croit. To appear in the Proceedings of the Nice Conference 1979 on Vector bundles and Differential equations.
[36] Ellingsrud, G., Strømme, S. A.: On the moduli space for stable rank-2 vector bundles on $\mathbb{P}_2$ with odd first Chern class. Preprint, Oslo 1979.
[37] Ellingsrud, G., Strømme, S. A.: The Picard group of the moduli space for stable rank-2 vector bundles on $\mathbb{P}_2$ with odd first Chern class. Preprint, Oslo 1979.
[38] Ferrand, D.: Courbes gauches et fibrés de rang 2. CRAS Paris **281**, A 345–347 (1975).
[39] Fischer, G.: Complex analytic geometry. Lecture Notes in Math. 538. Springer 1976.
[40] Forster, O., Hirschowitz, A., Schneider, M.: Type de scindage généralisé pour les fibrés stable. To appear in the Proceedings of the Nice Conference 1979 on Vector bundles and Differential equations.

[41] Fulton, W.: Ample vector bundles, Chern classes and numerical criteria. Invent. Math. **32**, 171–178 (1976).

[42] Gieseker, D.: On the moduli of vector bundles on an algebraic surface. Ann. of Math. **106**, 45–60 (1977).

[43] Godement, R.: Topologie algébrique et théorie des faisceaux. Hermann 1958.

[44] Grauert, H., Remmert, R.: Analytische Stellenalgebren. Springer 1971.

[45] Grauert, H., Mülich, G.: Vektorbündel vom Rang 2 über dem n-dimensionalen komplex projektiven Raum. Manuscripta math. **16**, 75–100 (1975).

[46] Grauert, H., Remmert, R.: Zur Spaltung lokalfreier Garben über Riemannschen Flächen. Math. Z., **144**, 35–43 (1975).

[47] Grauert, H., Remmert, R.: Theorie der Steinschen Räume. Springer 1977.

[48] Grauert, H., Schneider, M.: Komplexe Unterräume und holomorphe Vektorraumbündel vom Rang zwei. Math. Ann. **230**, 75–90 (1977).

[49] Griffiths, Ph., Harris, J.: Principles of algebraic geometry. Wiley 1978.

[50] Grothendieck, A.: Sur la classification des fibrés holomorphes sur la sphère de Riemann. Amer. J. Math. **79**, 121–138 (1956).

[51] Grothendieck, A.: La théorie des classes de Chern. Bull. Soc. math. France **86**, 137–154 (1958).

[52] Grothendieck, A.: Techniques de construction et théorèmes d'existence en géométrie algébrique : Les schémas de Hilbert. Sém. Bourbaki 1960/61, exposé 221.

[53] Grothendieck, A.: EGA III, 1 IHES Publ. Math. no **11**, Paris 1961.

[54] Hartshorne, R.: Cohomological dimension of algebraic varieties. Ann. Math. **88** 403–450 (1968).

[55] Hartshorne, R.: Varieties of small codimension in projective space. Bull. Amer. Math. Soc. **80**, 1017–1032 (1974).

[56] Hartshorne, R.: Algebraic Geometry. Springer 1977.

[57] Hartshorne, R.: Stable vector bundles and instantons. Comm. Math. Phys. **59**, 1–15 (1978).

[58] Hartshorne, R.: Stable vector bundles of rank 2 on $\mathbb{P}^3$. Math. Ann. **238**, 229–280 (1978).

[59] Hartshorne, R.: Algebraic vector bundles on projective spaces: a problem list. Topology **18**, 117–128 (1979).

[60] Hartshorne, R.: Restricting bundles from $\mathbb{P}^3$ to a quadric. Preprint, 1979.

[61] Hilbert, D.: Grundzüge einer allgemeinen Theorie der linearen Integralgleichungen. Nachr. Wiss. Göttingen, math. nat. Klasse, 307–338 (1905).

[62] Hirzebruch, F.: Topological methods in algebraic geometry. Springer 1966.

[63] Holmann, H.: Quotienten komplexer Räume. Math. Ann. **142**, 407–440 (1961).

[64] Holmann, H.: Komplexe Räume mit komplexen Transformationsgruppen. Math. Ann. **150**, 327–360 (1963).

[65] Horrocks, G.: Vector bundles on the punctured spectrum of a local ring. Proc. London Math. Soc. (3), **14**, 689–713 (1964).

[66] Horrocks, G.: A construction for locally free sheaves. Topology **7**, 117–120 (1968).

[67] Horrocks, G.: Letter to Mumford (1971).

[68] Horrocks, G., Mumford, D.: A rank 2 vector bundle on P^4 with $15,000$ symmetries. Topology **12**, 63–81 (1973).

[69] Horrocks, G.: Examples of rank three vector bundles on five-dimensional projective space. J. London Math. Soc. (2), **18**, 15–27 (1978).
[70] Hosoh, T.: Ample vector bundles on a rational surface. Nagoya Math. J. **59**, 135–148 (1975).
[71] Hosoh, T.: Ample vector bundles on a rational surface (higher rank). Nagoya Math. J. **66**, 77–88 (1977).
[72] Hulek, K.: Stable rank-2 vector bundles on $\mathbb{P}_2$ with c_1 odd. Math. Ann. **242**, 241–266 (1979).
[73] Hulek, K.: On the classification of stable rank-r vector bundles over the projective plane. To appear in the Proceedings of the Nice Conference 1979 on Vector bundles and Differential equations.
[74] Hulsbergen, W.: Vector bundles on the complex projective plane. Thesis, Leiden 1976.
[75] Kleiman, S. L.: Geometry on Grassmannians and applications to splitting bundles and smoothing cycles. IHES Publ. math. no **36**, 281–297 (1969).
[76] Lange, H.: The stability degree of a rank-2 vector bundle on $\mathbb{P}^2$. Notas e comunicações de matemática no 91, Pernambuco, Recife (1978).
[77] Lange, H.: On stable and uniform rank-2 vector bundles on $\mathbb{P}^2$ in characteristic p. Manuscripta math. 29, 11–28 (1979). Correction, Manuscripta math. **32**, 407–408 (1980).
[78] Lange, H.: Invertierbare Untergarben maximalen Grades von Rang-2 Vektorbündeln auf der projektiven Ebene. Arch. Math. (Basel) **34**, 313–321 (1980).
[79] Langton, S. G.: Valuative criteria for families of vector bundles on algebraic varieties. Ann. of Math. **101**, 88–110 (1975).
[80] Larsen, M. E.: On the topology of projective manifolds. Invent. math. **19**, 251–260 (1973).
[81] Maruyama, M.: On a family of algebraic vector bundles. Number theory, algebraic geometry and commutative algebra, in honor of Y. Akizuki, Kinokuniya, Tokyo, 95–146 (1973).
[82] Maruyama, M.: Stable vector bundles on an algebraic surface. Nagoya Math. J. **58**, 25–68 (1975).
[83] Maruyama, M.: Openness of a family of torsion-free sheaves. J. Math. Kyoto Univ. **16**, 627–637 (1976).
[84] Maruyama, M.: Moduli of stable sheaves, I. J. Math. Kyoto Univ. **17**, 91–126 (1977).
[85] Maruyama, M.: Moduli of stable sheaves, II. J. Math. Kyoto Univ. **18**, 557–614 (1978).
[86] Maruyama, M.: Boundedness of semi-stable sheaves of small ranks. Nagoya Math. J. **78**, 65–94 (1980).
[87] Miyanishi, M.: Some remarks on algebraic homogeneous vector bundles. Number theory, algebraic geometry and commutative algebra, in honor of Y. Akizuki, Kinokuniya, Tokyo, 95–146 (1973).
[88] Mülich, G.: Familien holomorpher Vektorraumbündel über $\mathbb{P}_1$ und unzerlegbare holomorphe 2-Bündel über der projektiven Ebene. Thesis, Göttingen 1974.
[89] Mumford, D.: Geometric invariant theory. Springer 1965.
[90] Mumford, D.: Algebraic geometry I, Complex projective varieties. Springer 1976.

[91] Narasimhan, R.: Introduction to the theory of analytic spaces. Lecture Notes in Math. 25. Springer 1966.
[92] Norton, V. A.: Nonseparation in the moduli of complex vector bundles. Math. Ann. **235**, 1–16 (1978).
[93] Norton, V. A.: Analytic moduli of complex vector bundles. Indiana Univ. Math. J. **28**, 365–387 (1979).
[94] Oda, T.: Vector bundles on abelian surfaces. Invent. Math. **13**, 247–260 (1971).
[95] Ogus, A.: On the formal neighborhood of a subvariety of projective space. Amer. J. Math. **97**, 1085–1107 (1976).
[96] Plemelj, J.: Riemannsche Funktionenscharen mit gegebener Monodromiegruppe. Monatsh. Math. **19**, 211–245 (1908).
[97] le Potier, J.: Fibrés stables de rang 2 sur $\mathbb{P}_2(\mathbb{C})$. Math. Ann. **241**, 217–256 (1979).
[98] le Potier, J.: Stabilité et amplitude sur $\mathbb{P}_2(\mathbb{C})$. To appear in the Proceedings of the Nice Conference 1979 on Vector bundles and Differential equations.
[99] Raynaud, M.: Séminaire de géométrie algébrique, Orsay 1977/78.
[100] Rawnsley, J. H.: On the Atiyah–Hitchin–Drinfeld–Manin vanishing theorem for cohomology groups of instanton bundles. Math. Ann. **241**, 43–56 (1979).
[101] Rees, E.: Some rank two bundles on $\mathbb{P}_n(\mathbb{C})$ whose Chern classes vanish. In Variétés analytiques compactes. Lecture Notes in Math. 683, 25–28. Springer 1978.
[102] Sato, E.: Uniform vector bundles on a projective space. J. Math. Soc. Japan **28**, 123–132 (1976).
[103] Sato, E.: On the decomposability of infinitely extendable vector bundles on projective spaces and Grassmann varieties. J. Math. Kyoto Univ. **17**, 127–150 (1977).
[104] Schneider, M.: Stabile Vektorraumbündel vom Rang 2 auf der projektiven Ebene. Nachr. Akad. Wiss. Göttingen. Math. Phys. Klasse 1976, 83–86.
[105] Schneider, M.: Holomorphic vector bundles on $\mathbb{P}_n$. Sém. Bourbaki 1978/79, exposé 520.
[106] Schneider, M.: Chernklassen semi-stabiler Vektorraumbündel vom Rank 3 auf dem komplex-projektiven Raum. J. Reine Angew. Math. **315**, 211–220 (1980).
[107] Schwarzenberger, R. L. E.: Vector bundles on algebraic surfaces. Proc. London Math. Soc. **11**, 601–622 (1961).
[108] Schwarzenberger, R. L. E.: Vector bundles on the projective plane. Proc. London Math. Soc. **11**, 623–640 (1961).
[109] Schwarzenberger, R. L. E.: The secant bundle of a projective variety. Proc. London Math. Soc. **14**, 369–384 (1964).
[110] Serre, J.-P.: Faisceaux algébriques cohérents. Ann. of Math. **61**, 197–278 (1955).
[111] Serre, J.-P.: Géométrie algébrique et géométrie analytique. Ann. Inst. Fourier **6**, 1–42 (1956).
[112] Serre, J.-P.: Sur les modules projectifs. Sém. Dubreil-Pisot 1960/1, exposé 2.
[113] Serre, J.-P.: Algèbre locale, multiplicités. Lecture Notes in Math. 11. Springer 1975.
[114] Seshadri, C. S.: Generalized multiplicative meromorphic functions on a complex analytic manifold. J. Indian Math. Soc. **21**, 149–178 (1957).

[115] Shatz, S. S.: On subbundles of vector bundles over $\mathbb{P}^1$. J. Pure Appl. Algebra **10**, 315–322 (1977).

[116] Shatz, S. S.: Fibre equivalence of vector bundles on ruled surfaces. J. Pure Appl. Algebra **12**, 201–205 (1978).

[117] Simonis, J.: A class of indecomposable algebraic vector bundles. Math. Ann. **192**, 262–278 (1971).

[118] Smith, L.: Complex 2-plane bundles over $CP(n)$. Manuscripta math. **24**, 221–228 (1978).

[119] Spindler, H.: Der Satz von Grauert–Mülich für beliebige semistabile holomorphe Vektorraumbündel über dem n-dimensionalen komplex-projektiven Raum. Math. Ann. **243**, 131–141 (1979).

[120] Switzer, R. M.: Algebraic topology—homotopy and homology. Springer 1975.

[121] Switzer, R. M.: Complex 2-plane bundles over complex projective space. Math. Z. **168**, 275–287 (1979).

[122] Takemoto, F.: Stable vector bundles on algebraic surfaces. Nagoya Math. J. **47**, 29–48 (1972).

[123] Takemoto, F.: Stable vector bundles on algebraic surfaces II. Nagoya Math. J. **52**, 173–195 (1973).

[124] Tango, H.: On $(n-1)$-dimensional projective spaces contained in the Grassmann variety $\mathrm{Gr}(n,1)$. J. Math. Kyoto Univ. **14**, 415–460 (1974).

[125] Tango, H.: An example of indecomposable vector bundle of rank $n-1$ on $\mathbb{P}^n$. J. Math. Kyoto Univ. **16**, 137–141 (1976).

[126] Tango, H.: On morphisms from projective space $\mathbb{P}_n$ to the Grassmann variety $\mathrm{Gr}(n,d)$. J. Math. Kyoto Univ. **16**, 201–207 (1976).

[127] Thomas, A.: Almost complex structures on complex projective spaces. TAMS **193**, 123–132 (1974).

[128] Tjurin, A. N.: Finite dimensional vector bundles over infinite varieties. Math. U.S.S.R. Isvestija **10**, 1187–1204 (1976).

[129] Trautmann, G.: Darstellungen von Vektorraumbündeln über $\mathbb{C}^n \setminus 0$. Arch. Math. **24**, 303–313 (1973).

[130] Trautmann, G.: Moduli for vector bundles on $\mathbb{P}_n(\mathbb{C})$. Math. Ann. **237**, 167–186 (1978).

[131] Trautmann, G.: Zur Berechnung von Yang–Mills Potentialen durch holomorphe Vektorbündel. To appear in the Proceedings of the Nice Conference 1979 on Vector bundles and Differential equations.

[132] Umemura, H.: Some results in the theory of vector bundles. Nagoya Math. J. **52**, 97–128 (1973).

[133] Umemura, H.: Stable vector bundles with numerically trivial Chern classes over a hyperelliptic surface. Nagoya Math. J. **59**, 107–134 (1975).

[134] Van de Ven, A.: On uniform vector bundles. Math. Ann. **195**, 245–248 (1972).

[135] Van de Ven, A.: Some recent results on surfaces of general type. Sém. Bourbaki 1976/77, exposé 500. Lecture Notes in Math. 677. Springer 1978.

[136] Vetter, U.: Zu einem Satz von G. Trautmann über den Rang gewisser kohärenter analytischer Moduln. Arch. Math. **24**, 158–161 (1973).

[137] Vogelaar, J. A.: Constructing vector bundles from codimension-two subvarieties. Thesis, Leiden 1978.

[138] Wells, R. O.: Complex manifolds and mathematical physics. Bull. Amer. Math. Soc. (new series) **1**, 296–336 (1979).

Supplemental Bibliography

[1] Y. Akimoto, *On the stable vector bundles of rank 2 on P^3*, TRU Math. **16** (1980), no. 1, 1–12.

[2] S. Arima, F. Hidaka, and S. Ishimura, *An example of simple vector bundles on projective plane which are not stable*, Bull. Sci. Engrg. Res. Lab. Waseda Univ. (1977), no. 79, 68–70.

[3] E. Arrondo and I. Sols, *Classification of smooth congruences of low degree*, J. Reine Angew. Math. **393** (1989), 199–219.

[4] M.F. Atiyah, *On the Krull-Schmidt theorem with application to sheaves*, Bull. Soc. Math. Fr. **84** (1956), 307–317.

[5] ———, *Geometry of Yang-Mills fields*, Mathematical problems in theoretical physics (Proc. Internat. Conf., Univ. Rome, Rome, 1977), Lecture Notes in Phys., vol. 80, Springer, Berlin, 1978, pp. 216–221.

[6] ———, *Vector bundles on algebraic varieties*, Tata Institute of Fundamental Research Studies in Mathematics, vol. 11, Published for the Tata Institute of Fundamental Research, Bombay, 1987, Papers presented at the international colloquium held in Bombay, January 9–16, 1984.

[7] M.F. Atiyah and R. Bott, *Yang-Mills and bundles over algebraic curves*, Proc. Indian Acad. Sci., Math. Sci. **90** (1981), 11–20.

[8] ———, *The Yang-Mills equations over Riemann surfaces*, Philos. Trans. R. Soc. Lond., A **308** (1983), 523–615.

[9] E. Ballico, *Uniform vector bundles of rank $(n+1)$ on P_n*, Tsukuba J. Math. **7** (1983), 215–226.

[10] E. Ballico and L. Chiantini, *On smooth subcanonical varieties of codimension 2 in P^n, $n \geq 4$*, Ann. Mat. Pura Appl., IV. Ser. **135** (1983), 99–117.

[11] E. Ballico and P. Ellia, *Fibrés homogènes sur $\mathbf{P}^n$*, C. R. Acad. Sci. Paris Sér. I Math. **294** (1982), no. 12, 403–406.

[12] ———, *Fibrés uniformes de rang 5 sur P^3*, Bull. Soc. Math. Fr. **111** (1983), 59–87.

[13] E. Ballico and P.E. Newstead, *Uniform bundles on quadric surfaces and some related varieties*, J. Lond. Math. Soc., II. Ser. **31** (1985), 211–223.

[14] C. Bănică, *Fibrés stables de rang 2 et classes de Chern $c_1 = -1$, $c_2 = 4$ sur $P^3(C)$*, Inst. Elie Cartan, Univ. Nancy I **8** (1983), 1–11.

[15] ———, *Topologisch triviale holomorphe Vektorbündel auf $P^n(C)$*, J. Reine Angew. Math. **344** (1983), 102–119.

[16] ———, *Sur les fibrés instables de rang 2 sur $\mathbf{P}^3(C)$*, Arch. Math. **43** (1984), 250–257.

[17] ———, *On surfaces associated to reflexive sheaves on a projective threefold*, Rev. Roum. Math. Pures Appl. **31** (1986), 479–488.

[18] C. Bănică and J. Coandă, *Existence of rank 3 vector bundles with given Chern classes on homogeneous rational 3-folds*, Manuscr. Math. **51** (1985), 121–143.

[19] C. Bănică and O. Forster, *Multiplicity structures on space curves*, The Lefschetz centennial conference, Part I (Mexico City, 1984), Contemp. Math., vol. 58, Amer. Math. Soc., Providence, RI, 1986, pp. 47–64.

[20] C. Bănică and J. Le Potier, *Sur l'existence des fibrés vectoriels holomorphes sur les surfaces non- algébriques*, J. Reine Angew. Math. **378** (1987), 1–31.

[21] C. Bănică and N.. Manolache, *Remarks on rank 2 stable vector bundles on* $\mathbf{P}^3$ *with Chern classes* $c_1 = -1$, $c_2 - 4$, Preprint Series in Math. INCREST No. 104, Bucaresti.

[22] C. Bănică and N. Manolache, *Moduli space* $M_{\mathbf{P}^3}(-1, 4)$*: Minimal spectrum*, Preprint Series in Math. INCREST No. 19, Bucaresti (1983).

[23] ______, *Rank 2 stable vector bundles on* $\mathbf{P}^3(C)$ *with Chern classes* $c_1 = -1$, $c_2 = 4$, Math. Z. **190** (1985), 315–339.

[24] C. Bănică and M. Putinar, *On complex vector bundles on rational threefolds*, Math. Proc. Camb. Philos. Soc. **97** (1985), 279–288.

[25] ______, *On complex vector bundles on projective threefolds*, Invent. Math. **88** (1987), 427–438.

[26] C. Bănică, M. Putinar, and G. Schumacher, *Variation der globalen Ext in Deformationen kompakter komplexer Räume*, Math. Ann. **250** (1980), 135–155.

[27] W. Barth, *Lectures on mathematical instanton bundles*, Gauge theories: fundamental interactions and rigorous results (Poiana Brasov, 1981), Progr. Phys., vol. 5, Birkhäuser Boston, Boston, MA, pp. 177–206.

[28] ______, *Stable vector bundles on* $\mathbf{P}^3$*, some experimental data, In: les équations de Yang-Mills*, Astérique **71-72** (1908), 205–218.

[29] ______, *Algebraische Vektorbündel*, Jahresber. Dtsch. Math.-Ver. **83** (1981), 106–118.

[30] ______, *Irreducibility of the space of mathematical instanton bundles with rank 2 and* $c_2 = 4$, Math. Ann. **258** (1981), 81–106.

[31] ______, *Report on vector bundles*, Proceedings of the International Congress of Mathematicians, Vol. 1, 2 (Warsaw, 1983) (Warsaw), PWN, 1984, pp. 783–789.

[32] W. Barth and K. Hulek, *Projective models of Shioda modular surfaces*, Manuscr. Math. **50** (1985), 73–132.

[33] W. Barth, K. Hulek, and R. Moore, *Degenerations of Horrocks-Mumford surfaces*, Math. Ann. **277** (1987), 735–755.

[34] ______, *Shioda's modular surface* $S(5)$ *and the Horrocks-Mumford bundle*, Vector bundles on algebraic varieties (Bombay, 1984), Tata Inst. Fund. Res. Stud. Math., vol. 11, Tata Inst. Fund. Res., Bombay, 1987, pp. 35–106.

[35] W. Barth, C. Peters, and A. Van de Ven, *Compact complex surfaces*, Ergebnisse der Mathematik und ihrer Grenzgebiete (3), vol. 4, Springer-Verlag, Berlin, 1984.

[36] I. N. Bernšteĭn, I. M. Gel′fand, and S. I. Gel′fand, *Algebraic vector bundles on* $\mathbf{P}^n$ *and problems of linear algebra*, Funktsional. Anal. i Prilozhen. **12** (1978), no. 3, 66–67.

[37] J. Bertin and G. Elencwajg, *Symétries des fibrés vectoriels sur* P^n *et nombre d'Euler*, Duke Math. J. **49** (1982), 807–831.

[38] J. Bertin and I. Sols, *Quelques formules énumératives concernant les fibrés vectoriels sur P^r*, C. R. Acad. Sci., Paris, Sér. I **294** (1982), 197–200.

[39] F.A. Bogomolov, *Unstable vector bundles and curves on surfaces*, Proceedings of the International Congress of Mathematicians (Helsinki, 1978) (Helsinki), Acad. Sci. Fennica, 1980, pp. 517–524.

[40] W. Böhmer, *Monads and matrices for instantons and Horrocks-Mumford-examples*, Preprint, Kaiserlautern.

[41] ———, *Geometric and computational interpretation of intanton bundles on* $\mathbf{P}_3$ *with* $h^0E(1) = 2.$, Singularities, Representation of Algebras, and Vector Bundles, Proceedings of a Symposium held in Lambrecht/Pfalz, Fed.Rep. of Germany, Dec. 13-17, 1985, Lecture Notes in Math., vol. 1273, Springer, Berlin, 1987.

[42] G. Bolondi, *Seminatural cohomology and stability*, C. R. Acad. Sci., Paris, Sér. I **301** (1985), 407–410.

[43] ———, *Arithmetically normal sheaves*, Bull. Soc. Math. Fr. **115** (1987), 71–95.

[44] V. Brînzănescu and M. Stoia, *Topologically trivial algebraic 2-vector bundles on ruled surfaces. I*, Rev. Roumaine Math. Pures Appl. **29** (1984), no. 8, 661–673.

[45] ———, *Topologically trivial algebraic 2-vector bundles on ruled surfaces. II*, Algebraic geometry, Bucharest 1982 (Bucharest, 1982), Lecture Notes in Math., vol. 1056, Springer, Berlin, 1984, pp. 34–46.

[46] V. Brînzănescu and P. Flondor, *Holomorphic 2-vector bundles on nonalgebraic 2-tori*, J. Reine Angew. Math. **363** (1985), 47–58.

[47] J.E. Brosius, *Rank-2 vector bundles on a ruled surface. I*, Math. Ann. **265** (1983), 155–168.

[48] ———, *Rank-2 vector bundles on a ruled surface. II*, Math. Ann. **266** (1983), 199–214.

[49] J. Brun, *Les fibrés de rang deux sur P_2 et leurs sections*, Bull. Soc. Math. Fr. **108** (1980), 457–473.

[50] ———, *Des fibrés instantons ayant beaucoup de plans instables*, C. R. Acad. Sci., Paris, Sér. I **296** (1983), 203–204.

[51] J. Brun and A. Hirschowitz, *Droites de saut des fibrés stables de rang élevé sur P_2*, Math. Z. **181** (1982), 171–178.

[52] ———, *Variété des droites sauteuses du fibré instanton général*, Compos. Math. **53** (1984), 325–336.

[53] N.P. Buchdahl, *Instantons on* $\mathbf{CP}_2$, J. Differential Geom. **24** (1986), no. 1, 19–52.

[54] ———, *Stable 2-bundles on Hirzebruch surfaces*, Math. Z. **194** (1987), no. 1, 143–152.

[55] ———, *Hermitian-Einstein connections and stable vector bundles over compact complex surfaces*, Math. Ann. **280** (1988), no. 4, 625–648.

[56] M.-C. Chang, *A bound on the order of jumping lines*, Math. Ann. **262** (1983), 511–516.

[57] ———, *Stable rank 2 bundles on P^3 with $c_1 = 0$, $c_2 = 4$, and $\alpha = 1$*, Math. Z. **184** (1983), 407–415.

[58] ———, *Stable rank 2 reflexive sheaves on P^3 with large c_3*, J. Reine Angew. Math. **343** (1983), 99–107.

[59] ———, *Stable rank 2 reflexive sheaves on P^3 with small c_2 and applications*, Trans. Am. Math. Soc. **284** (1984), 57–89.

[60] I. Coandă, *About a theorem of Elencwajg-Forster*, Stud. Cercet. Mat. **35** (1983), 3–9.

[61] ———, *The Chern classes of the stable rank 3 vector bundles on P^3*, Math. Ann. **273** (1985), 65–79.

[62] ———, *On the duals of the stable rank 3 reflexive sheaves on P^3 with c_3 maximal*, Ann. Univ. Ferrara, Nuova Ser., Sez. VII **32** (1986), 71–78.

[63] ———, *On the spectrum of a stable rank 3 reflexive sheaf on P^3*, J. Reine Angew. Math. **367** (1986), 155–171.

[64] W. Decker, *Über das Horrocks-Mumford-Bündel*, Preprint No. 61, Kaiserslautern (1983).

[65] ———, *Über den Modul-Raum für stabile 2-Vektorbündel über $\mathbf{P}_3$ mit $c_1 = -1$, $c_2 = 2$*, Manuscripta Math. **42** (1983), no. 2-3, 211–219.

[66] ———, *Über den Modul-Raum für stabile 2-Vektorbündel über $\mathbf{P}_4$ mit $c_1 = -1$, $c_2 = -4$ und $H^1(F(-2)) = 0$*, Preprint No. 62, Kaiserslautern (1983).

[67] ———, *Das Horrocks-Mumford-Bündel und das Modulschema für stabile 2-Vektorbündel über $\mathbf{P}_4$ mit $c_1 = -1$, $c_2 = 4$*, Math. Z. **188** (1984), no. 1, 101–110.

[68] ———, *Über stabile 2 Vektorbündel mit Chern-Klassen $c_1 = -1$, $c_2 = 4$*, Preprint No. 84, Kaiserslautern (1984).

[69] ———, *Stable rank 2 vector bundles with Chern-classes $c_1 = -1$, $c_2 = 4$*, Math. Ann. **275** (1986), no. 3, 481–500.

[70] W. Decker and F.O. Schreyer, *On the uniqueness of the Horrocks-Mumford bundle*, Math. Ann. **273** (1986), no. 3, 415–443.

[71] ———, *Pullbacks of the Horrocks-Mumford bundle*, J. Reine Angew. Math. **382** (1987), 215–220.

[72] S.K. Donaldson, *Instantons and geometric invariant theory*, Commun. Math. Phys. **93** (1984), 453–460.

[73] ———, *Anti self-dual Yang Mills connections over complex algebraic surfaces and stable vector bundles*, Proc. Lond. Math. Soc., III. Ser. **50** (1985), 1–26.

[74] ———, *La topologie différentielle des surfaces complexes*, C. R. Acad. Sci., Paris, Sér. I **301** (1985), 317–320.

[75] ———, *Vector Bundles on the Flag Manifold and the Ward correspondence*, Geometry Today (Rome, 1984), Progr. Math., vol. 60, Birkhäuser Boston, Boston, MA, 1985, pp. 109–119.

[76] ———, *Gauge theory and smooth structures on 4-manifolds*, Geometry and topology (Athens, Ga., 1985), Lecture Notes in Pure and Appl. Math., vol. 105, Dekker, New York, 1987, pp. 89–98.

[77] ———, *Infinite determinants, stable bundles and curvature*, Duke Math. J. **54** (1987), 231–247.

[78] ———, *Irrationality and the h-cobordism conjecture*, J. Differ. Geom. **26** (1987), 141–168.

[79] A. (ed.) Douady and J.-L. (ed.) Verdier, *Les équations de Yang-Mills. Seminaire E.N.S. 1977-1978*, Asterisque, vol. 71-72, Société mathématique de france, 1980.

[80] J.M. Drezet, *Exemples de fibrés uniformes non homogènes sur P^n*, C. R. Acad. Sci., Paris, Sér. A **291** (1980), 125–128.

[81] ______, *Fibres uniformes de type (0,...,0,-1,...,-1) sur P_2*, J. Reine Angew. Math. **325** (1981), 1–27.

[82] ______, *Fibres uniformes de type (1,0,...,0,-1) sur P_2*, Ann. Inst. Fourier **31** (1981), no. 1, 99–134.

[83] ______, *Fibrés exceptionnels et suite spectrale de Beilinson généralisée sur $P_2(C)$*, Math. Ann. **275** (1986), 25–48.

[84] ______, *Fibrés exceptionnels et variétés de modules de faisceaux semi- stables sur $P_2(C)$*, J. Reine Angew. Math. **380** (1987), 14–58.

[85] ______, *Groupe de Picard des variétés de modules de faisceaux semi-stables sur $\mathbf{P}_2$*, Singularities, representation of algebras, and vector bundles (Lambrecht, 1985), Lecture Notes in Math., vol. 1273, Springer, Berlin, 1987, pp. 337–362.

[86] ______, *Cohomologie des variétés de modules de hauteur nulle*, Math. Ann. **281** (1988), no. 1, 43–85.

[87] ______, *Groupe de Picard des variétés de modules de faisceaux semi-stable sur $P_2(C)$*, Ann. Inst. Fourier **38** (1988), no. 3, 105–168.

[88] J.M. Drezet and J. Le Potier, *Fibrés stables et fibrés exceptionnels sur P_2*, Ann. Sci. Éc. Norm. Supér. (4) **18** (1985), 193–243.

[89] ______, *Conditions d'existence des fibrés stables de rang élevé sur $\mathbf{P}_2$*, Vector bundles on algebraic varieties (Bombay, 1984), Tata Inst. Fund. Res. Stud. Math., vol. 11, Tata Inst. Fund. Res., Bombay, 1987, pp. 133–158.

[90] V.G. Drinfel'd and Y.I. Manin, *Instantons and sheaves on CP^3*, Funct. Anal. Appl. **13** (1979), 124–134.

[91] J.A. Eagon and D.G. Northcott, *Ideals defined by matrices and a certain complex associated with them*, Proc. R. Soc. Lond., Ser. A **269** (1962), 188–204.

[92] L. Ein, *Stable vector bundles on projective spaces in char $p > 0$*, Math. Ann. **254** (1980), 53–72.

[93] ______, *Some stable vector bundles on P^4 and P^5*, J. Reine Angew. Math. **337** (1982), 142–153.

[94] ______, *Rank 2 vector bundles on projective spaces*, The curves seminar at Queen's, Vol. III (Kingston, Ont., 1983), Queen's Papers in Pure and Appl. Math., vol. 67, Queen's Univ., Kingston, ON, 1984, pp. Exp. No. N, 6.

[95] L. Ein, R. Hartshorne, and H. Vogelaar, *Restriction theorems for stable rank 3 vector bundles on P^n*, Math. Ann. **259** (1982), 541–569.

[96] L. Ein and I. Sols, *Stable vector bundles on quadric hypersurfaces*, Nagoya Math. J. **96** (1984), 11–22.

[97] G. Elencwajg, *Fibrés uniformes de rang élevé sur $\mathbf{P}_2$*, Ann. Inst. Fourier (Grenoble) **31** (1981), no. 4, vii, 89–114.

[98] G. Elencwajg and O. Forster, *Vector bundles on manifolds without divisors and a theorem on deformations*, Ann. Inst. Fourier (Grenoble) **32** (1982), no. 4, 25–51 (1983).

[99] P. Ellia, *Des fibrés uniformes non homogènes et indécomposables de rang $(2n+1)$ sur $\mathbf{P}^n(\mathbf{C})$, $n \geq 3$*, J. Reine Angew. Math. **321** (1981), 113–119.

[100] ______, *Sur les fibrés uniformes de rang $(n+1)$ sur $\mathbf{P}^n$*, Mém. Soc. Math. France (N.S.) (1982), no. 7, 60.

[101] ______, *Faisceaux réflexifs stables: le spectre n'est pas toujours constant*, C. R. Acad. Sci. Paris Sér. I Math. **296** (1983), no. 1, 55–58.

[102] G. Ellingsrud, *Sur l'irréductibilité du module des fibrés stables sur* $\mathbf{P}^2$, Math. Z. **182** (1983), no. 2, 189–192.

[103] G. Ellingsrud and S.A. Strømme, *Stable rank-2 vector bundles on* $\mathbf{P}^3$ *with* $c_1 = 0$ *and* $c_2 = 3$, Math. Ann. **255** (1981), no. 1, 123–135.

[104] H. Flenner, *Eine Bemerkung über relative* Ext-*Garben*, Math. Ann. **258** (1981/82), no. 2, 175–182.

[105] ———, *Restrictions of semistable bundles on projective varieties*, Comment. Math. Helv. **59** (1984), no. 4, 635–650.

[106] ———, *Babylonian tower theorems on the punctured spectrum*, Math. Ann. **271** (1985), no. 1, 153–160.

[107] O. Forster, *Holomorphic vector bundles on tori*, Several complex variables (Hangzhou, 1981), Birkhäuser Boston, Boston, MA, 1984, pp. 143–149.

[108] D.S. Freed and K.K. Uhlenbeck, *Instantons and four-manifolds*, Mathematical Sciences Research Institute Publications, vol. 1, Springer-Verlag, New York, 1984.

[109] K. Fritzsche, *Linear-uniforme Bündel auf Quadriken*, Ann. Scuola Norm. Sup. Pisa Cl. Sci. (4) **10** (1983), no. 2, 313–339.

[110] A. Fujiki and G. Schumacher, *The moduli space of Hermite-Einstein bundles on a compact Kähler manifold*, Proc. Japan Acad. Ser. A Math. Sci. **63** (1987), no. 3, 69–72.

[111] W. Fulton, *Intersection theory*, Ergebnisse der Mathematik und ihrer Grenzgebiete (3), vol. 2, Springer-Verlag, Berlin, 1984.

[112] D. Gieseker, *On a theorem of Bogomolov on Chern classes of stable bundles*, Amer. J. Math. **101** (1979), no. 1, 77–85.

[113] H. Grauert and R. Remmert, *Coherent analytic sheaves*, Grundlehren der Mathematischen Wissenschaften, vol. 265, Springer-Verlag, Berlin, 1984.

[114] P. Griffiths and J. Harris, *Residues and zero-cycles on algebraic varieties*, Ann. of Math. (2) **108** (1978), no. 3, 461–505.

[115] M. Guyot, *Caractérisation par l'uniformité des fibrés universels sur la grassmanienne*, Math. Ann. **270** (1985), no. 1, 47–62.

[116] G. Harder and M. S. Narasimhan, *On the cohomology groups of moduli spaces of vector bundles on curves*, Math. Ann. **212** (1974/75), 215–248.

[117] R. Hartshorne, *Algebraic vector bundles on projective spaces, with applications to the Yang-Mills equation*, Complex manifold techniques in theoretical physics (Proc. Workshop, Lawrence, Kan., 1978), Res. Notes in Math., vol. 32, Pitman, Boston, Mass., 1979, pp. 35–44.

[118] ———, *Four years of algebraic vector bundles: 1975–1979*, Journées de Géometrie Algébrique d'Angers, Juillet 1979/Algebraic Geometry, Angers, 1979, Sijthoff & Noordhoff, Alphen aan den Rijn, 1980, pp. 21–27.

[119] ———, *Stable reflexive sheaves*, Math. Ann. **254** (1980), no. 2, 121–176.

[120] ———, *Stable reflexive sheaves. II*, Invent. Math. **66** (1982), no. 1, 165–190.

[121] ———, *On the classification of algebraic space curves. II*, Algebraic geometry, Bowdoin, 1985 (Brunswick, Maine, 1985), Proc. Sympos. Pure Math., vol. 46, Amer. Math. Soc., Providence, RI, 1987, pp. 145–164.

[122] ———, *Stable reflexive sheaves. III*, Math. Ann. **279** (1988), no. 3, 517–534.

[123] R. Hartshorne and A. Hirschowitz, *Cohomology of a general instanton bundle*, Ann. Sci. École Norm. Sup. (4) **15** (1982), no. 2, 365–390.

[124] ———, *Droites en position générale dans l'espace projectif*, Algebraic geometry (La Rábida, 1981), Lecture Notes in Math., vol. 961, Springer, Berlin, 1982, pp. 169–188.

[125] ———, *Courbes rationnelles et droites en position générale*, Ann. Inst. Fourier (Grenoble) **35** (1985), no. 4, 39–58.

[126] ———, *Nouvelles courbes de bon genre dans l'espace projectif*, Math. Ann. **280** (1988), no. 3, 353–367.

[127] R. Hartshorne and I. Sols, *Stable rank* 2 *vector bundles on* $\mathbf{P}^3$ *with* $c_1 = -1$, $c_2 = 2$, J. Reine Angew. Math. **325** (1981), 145–152.

[128] R. Hernández, *On Harder-Narasimhan stratification over Quot schemes*, J. Reine Angew. Math. **371** (1986), 115–124.

[129] R. Hernández and I. Sols, *On a family of rank* 3 *bundles on* $\mathrm{Gr}(1,3)$, J. Reine Angew. Math. **360** (1985), 124–135.

[130] A. Hirschowitz, *Sur la restriction des faisceaux semi-stables*, Ann. Sci. École Norm. Sup. (4) **14** (1981), no. 2, 199–207.

[131] ———, *Rank techniques and jump stratifications*, Vector bundles on algebraic varieties (Bombay, 1984), Tata Inst. Fund. Res. Stud. Math., vol. 11, Tata Inst. Fund. Res., Bombay, 1987, pp. 159–205.

[132] ———, *Existence de faisceaux réflexifs de rang deux sur* $\mathbf{P}^3$ *à bonne cohomologie*, Inst. Hautes Études Sci. Publ. Math. (1988), no. 66, 105–137.

[133] A. Hirschowitz and K. Hulek, *Complete families of stable vector bundles over* $\mathbf{P}_2$, Complex analysis and algebraic geometry (Göttingen, 1985), Lecture Notes in Math., vol. 1194, Springer, Berlin, 1986, With an appendix by K. Hulek and S. A. Strømme, pp. 19–40.

[134] A. Hirschowitz and R. Marlin, *Nouvelles surfaces à nœuds dans* $\mathbf{P}^3$, Math. Ann. **267** (1984), no. 1, 83–89.

[135] A. Hirschowitz and M.S. Narasimhan, *Fibrés de 't Hooft spéciaux et applications*, Enumerative geometry and classical algebraic geometry (Nice, 1981), Progr. Math., vol. 24, Birkhäuser Boston, Mass., 1982, pp. 143–164.

[136] A. Holme and M. Schneider, *A computer aided approach to codimension* 2 *subvarieties of* $\mathbf{P}_n$, $n \geq 6$, J. Reine Angew. Math. **357** (1985), 205–220.

[137] H.J. Hoppe, *Modulräume stabiler Vektorraumbündel vom Rang* 2 *auf rationalen Regelflächen*, Math. Ann. **264** (1983), no. 2, 227–239.

[138] ———, *Generischer Spaltungstyp und zweite Chernklasse stabiler Vektorraumbündel vom Rang* 4 *auf* $\mathbf{P}_4$, Math. Z. **187** (1984), no. 3, 345–360.

[139] H.J. Hoppe and H. Spindler, *Modulräume stabiler* 2*-Bündel auf Regelflächen*, Math. Ann. **249** (1980), no. 2, 127–140.

[140] G Horrocks, *Construction of bundles in* $\mathbf{P}^n$, Les équations de Yang-Mills, A. Douady - J.L. Verdier séminaire E.N.S, Astérisque, vol. 71-72, 1977-1978, pp. 197–203.

[141] G. Horrocks, *Vector bundles on the punctured spectrum of a local ring. II*, Vector bundles on algebraic varieties (Bombay, 1984), Tata Inst. Fund. Res. Stud. Math., vol. 11, Tata Inst. Fund. Res., Bombay, 1987, pp. 207–216.

[142] T. Hosoh and S. Ishimura, *Stable vector bundles of rank* 2 *on a* 3*-dimensional rational scroll*, J. Math. Soc. Japan **37** (1985), no. 4, 557–568.

[143] K. Hulek, *On the deformation of orthogonal bundles over the projective line*, J. Reine Angew. Math. **329** (1981), 52–57.

[144] ———, *Complete intersection curves, the splitting of the normal bundle and the Veronese surface*, Algebraic geometry, Sitges (Barcelona), 1983, Lecture Notes in Math., vol. 1124, Springer, Berlin, 1985, pp. 132–145.

[145] ———, *Geometry of the Horrocks-Mumford bundle*, Algebraic geometry, Bowdoin, 1985 (Brunswick, Maine, 1985), Proc. Sympos. Pure Math., vol. 46, Amer. Math. Soc., Providence, RI, 1987, pp. 69–85.

[146] K. Hulek and H. Lange, *The Hilbert modular surface for the ideal* $(\sqrt{5})$ *and the Horrocks-Mumford bundle*, Math. Z. **198** (1988), no. 1, 95–116.

[147] K. Hulek and S.A. Strømme, *Appendix to the paper "Complete Families of Stable Vector Bundles over* P_2*".*, Complex analysis and algebraic geometry (Göttingen, 1985), Lecture Notes in Math., vol. 1194, Springer, Berlin, 1986, pp. 34–40.

[148] K. Hulek and A. Van de Ven, *The Horrocks-Mumford bundle and the Ferrand construction*, Manuscripta Math. **50** (1985), 313–335.

[149] S. Ishimura, *On* π*-uniform vector bundles*, Tokyo J. Math. **2** (1979), no. 2, 337–342.

[150] ———, *A descent problem of vector bundles and its applications*, J. Math. Kyoto Univ. **23** (1983), no. 1, 73–83.

[151] M. Itoh, *Moduli of anti-self-dual connections on Kähler manifolds*, Proc. Japan Acad. Ser. A Math. Sci. **57** (1981), no. 3, 176–180.

[152] M. M. Kapranov, *The derived category of coherent sheaves on Grassmann varieties*, Funktsional. Anal. i Prilozhen. **17** (1983), no. 2, 78–79.

[153] H.J. Kim, *Moduli of Hermite-Einstein vector bundles*, Math. Z. **195** (1987), no. 1, 143–150.

[154] J.O. Kleppe, *Deformation of reflexive sheaves of rank 2 on* $\mathbf{P}^3$, Preprint No. 4 (1982).

[155] M.-A. Knus, *Bundles on* $\mathbf{P}^2$ *with a quaternionic structure*, Vector bundles on algebraic varieties (Bombay, 1984), Tata Inst. Fund. Res. Stud. Math., vol. 11, Tata Inst. Fund. Res., Bombay, 1987, pp. 225–250.

[156] S. Kobayashi, *The first Chern class and holomorphic symmetric tensor fields*, J. Math. Soc. Japan **32** (1980), no. 2, 325–329.

[157] ———, *First Chern class and holomorphic tensor fields*, Nagoya Math. J. **77** (1980), 5–11.

[158] ———, *Curvature and stability of vector bundles*, Proc. Japan Acad. Ser. A Math. Sci. **58** (1982), no. 4, 158–162.

[159] ———, *Einstein-Hermitian vector bundles and stability*, Global Riemannian geometry (Durham, 1983), Ellis Horwood Ser. Math. Appl., Horwood, Chichester, 1984, pp. 60–64.

[160] ———, *Differential geometry of holomorphic vector bundles*, PAM-3154, Berkeley March (1986).

[161] ———, *Homogeneous vector bundles and stability*, Nagoya Math. J. **101** (1986), 37–54.

[162] ———, *Simple vector bundles over symplectic Kähler manifolds*, Proc. Japan Acad. Ser. A Math. Sci. **62** (1986), no. 1, 21–24.

[163] S. Kosarew and C. Okonek, *Global moduli spaces and simple holomorphic bundles*, Publ. Res. Inst. Math. Sci. **25** (1989), no. 1, 1–19.

[164] H. Kurke and H. Theel, *Some examples of vector bundles on the flag variety* $\mathbf{F}(1,2)$, Algebraic geometry, Bucharest 1982 (Bucharest, 1982), Lecture Notes in Math., vol. 1056, Springer, Berlin, 1984, pp. 187–254.

[165] H. Lange, *Invertierbare Untergarben maximalen Grades von Rang-2 Vektorbündeln auf der projektiven Ebene*, Arch. Math. (Basel) **34** (1980), no. 4, 313–321.

[166] ———, *Universal families of extensions*, J. Algebra **83** (1983), no. 1, 101–112.

[167] R. Lazarsfeld, *Some applications of the theory of positive vector bundles*, Complete intersections (Acireale, 1983), Lecture Notes in Math., vol. 1092, Springer, Berlin, 1984, pp. 29–61.

[168] J. Le Potier, *Sur le groupe de Picard de l'espace de modules des fibrés stables sur* $\mathbf{P}_2$, Ann. Sci. École Norm. Sup. (4) **14** (1981), no. 2, 141–155.

[169] ———, *Fibrés vectoriels sur les surfaces* $K3$, P. Lelong-P. Dolbeault-H. Skoda analysis seminar, 1981/1983, Lecture Notes in Math., vol. 1028, Springer, Berlin, 1983, pp. 225–238.

[170] ———, *Sur l'espace de modules des fibrés de Yang et Mills*, Mathematics and physics (Paris, 1979/1982), Progr. Math., vol. 37, Birkhäuser Boston, Boston, MA, 1983, pp. 65–137.

[171] ———, *Variétés de modules de faisceaux semi-stables de rang élevé sur* $\mathbf{P}_2$, Algebraic geometry, Bowdoin, 1985 (Brunswick, Maine, 1985), Proc. Sympos. Pure Math., vol. 46, Amer. Math. Soc., Providence, RI, 1987, pp. 87–100.

[172] J. Le Potier and J. Verdier (eds.), *Module des fibrés stables sur les courbes algébriques*, Progress in Mathematics, vol. 54, Birkhäuser Boston Inc., Boston, MA, 1985, Papers from the conference held at the École Normale Supérieure, Paris, 1983.

[173] J. Leiterer, *The Penrose transform for bundles nontrivial on the general line*, Math. Nachr. **112** (1983), 35–67.

[174] ———, *Subsheaves in bundles on* $\mathbf{P}_n$ *and the Penrose transform*, Analytic functions, Błażejewko 1982 (Błażejewko, 1982), Lecture Notes in Math., vol. 1039, Springer, Berlin, 1983, pp. 332–345.

[175] ———, *On holomorphic vector bundles over linearly concave manifolds*, Math. Ann. **274** (1986), no. 3, 391–417.

[176] M. Lübke, *Chernklassen von Hermite-Einstein-Vektorbündeln*, Math. Ann. **260** (1982), no. 1, 133–141.

[177] ———, *Stability of Einstein-Hermitian vector bundles*, Manuscripta Math. **42** (1983), no. 2-3, 245–257.

[178] M. Lübke and C. Okonek, *Differentiable structures of elliptic surfaces with cyclic fundamental group*, Compositio Math. **63** (1987), no. 2, 217–222.

[179] ———, *Moduli spaces of simple bundles and Hermitian-Einstein connections*, Math. Ann. **276** (1987), no. 4, 663–674.

[180] ———, *Stable bundles on regular elliptic surfaces*, J. Reine Angew. Math. **378** (1987), 32–45.

[181] M. Manaresi, *Families of homogeneous vector bundles on* $\mathbf{P}^2$, J. Pure Appl. Algebra **35** (1985), no. 3, 297–304.

[182] N. Manolache, *Rank* 2 *stable vector bundles on* $\mathbf{P}^3$ *with Chern classes* $c_1 = -1$, $c_2 = 2$, Rev. Roumaine Math. Pures Appl. **26** (1981), no. 9, 1203–1209.

[183] ———, *Cohen-Macaulay nilpotent structures*, Rev. Roumaine Math. Pures Appl. **31** (1986), no. 6, 563–575.

[184] ———, *On the normal bundle to abelian surfaces embedded in* $\mathbf{P}^4(\mathbf{C})$, Manuscripta Math. **55** (1986), no. 1, 111–119.

[185] ———, *Syzygies of abelian surfaces embedded in* $\mathbf{P}^4(\mathbf{C})$, J. Reine Angew. Math. **384** (1988), 180–191.

[186] M. Maruyama, *On boundedness of families of torsion free sheaves*, J. Math. Kyoto Univ. **21** (1981), no. 4, 673–701.

[187] ———, *The theorem of Grauert-Mülich-Spindler*, Math. Ann. **255** (1981), no. 3, 317–333.

[188] ———, *Elementary transformations in the theory of algebraic vector bundles*, Algebraic geometry (La Rábida, 1981), Lecture Notes in Math., vol. 961, Springer, Berlin, 1982, pp. 241–266.

[189] ———, *Algebraic vector bundles*, Recent progress of algebraic geometry in Japan, North-Holland Math. Stud., vol. 73, North-Holland, Amsterdam, 1983, pp. 106–151.

[190] ———, *Moduli of stable sheaves—generalities and the curves of jumping lines of vector bundles on* $\mathbf{P}^2$, Algebraic varieties and analytic varieties (Tokyo, 1981), Adv. Stud. Pure Math., vol. 1, North-Holland, Amsterdam, 1983, pp. 1–27.

[191] ———, *Singularities of the curve of jumping lines of a vector bundle of rank* 2 *on* $\mathbf{P}^2$, Algebraic geometry (Tokyo/Kyoto, 1982), Lecture Notes in Math., vol. 1016, Springer, Berlin, 1983, pp. 370–411.

[192] ———, *The equations of plane curves and the moduli spaces of vector bundles on* $\mathbf{P}^2$, Algebraic and topological theories (Kinosaki, 1984), Kinokuniya, Tokyo, 1986, pp. 430–466.

[193] ———, *The rationality of the moduli spaces of vector bundles of rank* 2 *on* $\mathbf{P}^2$, Algebraic geometry, Sendai, 1985, Adv. Stud. Pure Math., vol. 10, North-Holland, Amsterdam, 1987, With an appendix by Isao Naruki, pp. 399–414.

[194] ———, *Vector bundles on* $\mathbf{P}^2$ *and torsion sheaves on the dual plane*, Vector bundles on algebraic varieties (Bombay, 1984), Tata Inst. Fund. Res. Stud. Math., vol. 11, Tata Inst. Fund. Res., Bombay, 1987, pp. 275–339.

[195] ———, *On a compactification of a moduli space of stable vector bundles on a rational surface*, Algebraic geometry and commutative algebra, Vol. I, Kinokuniya, Tokyo, 1988, pp. 233–260.

[196] Y. Matsushima, *Fibrés holomorphes sur un tore complexe*, Nagoya Math. J. **14** (1959), 1–24.

[197] V.B. Mehta, *On some restriction theorems for semistable bundles*, Invariant theory (Montecatini, 1982), Lecture Notes in Math., vol. 996, Springer, Berlin, 1983, pp. 145–153.

[198] V.B. Mehta and M.V. Nori, *Tranlation-invariant and semi-stable sheaves on abelian varieties*, Preprint (1983).

[199] ———, *Semistable sheaves on homogeneous spaces and abelian varieties*, Proc. Indian Acad. Sci. Math. Sci. **93** (1984), no. 1, 1–12.

[200] V.B. Mehta and A. Ramanathan, *Semistable sheaves on projective varieties and their restriction to curves*, Math. Ann. **258** (1981/82), no. 3, 213–224.

[201] ———, *An analogue of Langton's theorem on valuative criteria for vector bundles*, Proc. Roy. Soc. Edinburgh Sect. A **96** (1984), no. 1-2, 39–45.

[202] ———, *Restriction of stable sheaves and representations of the fundamental group*, Invent. Math. **77** (1984), no. 1, 163–172.

[203] J. Meseguer and I. Sols, *Faisceaux semi-stables de rang* 2 *sur* $\mathbf{P}^3$, C. R. Acad. Sci. Paris Sér. I Math. **298** (1984), no. 20, 525–528.

[204] J. Meseguer, I. Sols, and S.A. Strømme, *Compactification of a family of vector bundles on* $\mathbf{P}^3$, 18th Scandinavian Congress of Mathematicians (Aarhus, 1980), Progr. Math., vol. 11, Birkhäuser Boston, Mass., 1981, pp. 474–494.

[205] N. Mestrano, *Poincaré bundles for projective surfaces*, Ann. Inst. Fourier (Grenoble) **35** (1985), no. 2, 217–249.

[206] R.M. Miró-Roig, *Gaps in Chern classes of rank* 2 *stable reflexive sheaves*, Math. Ann. **270** (1985), no. 3, 317–323.

[207] ———, *Haces reflexives sobre espacios proyectivos*, Thesis Barcelona (1985).

[208] ———, *Chern classes of rank-*2 *reflexive sheaves*, J. London Math. Soc. (2) **33** (1986), no. 3, 421–429.

[209] ———, *Faisceaux réflexifs stables de rang* 2 *sur* $\mathbf{P}^3$ *non obstrués*, C. R. Acad. Sci. Paris Sér. I Math. **303** (1986), no. 14, 711–713.

[210] ———, *A sharp bound for the number of points where a rank* 3 *stable reflexive sheaf is not free*, Manuscripta Math. **56** (1986), no. 1, 11–17.

[211] ———, *Chern classes of rank* 3 *stable reflexive sheaves*, Math. Ann. **276** (1987), no. 2, 291–302.

[212] ———, *The moduli space* $M^S_{\mathbf{P}^3}(-1, c_2, c_2^2 - 2c_2 + 4)$, Géométrie algébrique et applications, II (La Rábida, 1984), Travaux en Cours, vol. 23, Hermann, Paris, 1987, pp. 177–186.

[213] ———, *Some moduli spaces for rank* 2 *stable reflexive sheaves on* $\mathbf{P}^3$, Trans. Amer. Math. Soc. **299** (1987), no. 2, 699–717.

[214] Y. Miyaoka, *The Chern classes and Kodaira dimension of a minimal variety*, Algebraic geometry, Sendai, 1985, Adv. Stud. Pure Math., vol. 10, North-Holland, Amsterdam, 1987, pp. 449–476.

[215] R. Moore, $M(-1, 2)$ *is a homogeneous space*, Math. Research Report 4, Canberra (1982).

[216] ———, *Monads for vector bundles of rank 2 on* $\mathbf{P}^3$ *with* $c_1 = -1$, $c_2 = 2.$, Math. Research Report 3, Canberra (1982).

[217] ———, *Linear equivalence of Tango bundles on* $\mathbf{P}^4$, J. Reine Angew. Math. **351** (1984), 12–19.

[218] ———, *Heisenberg-invariant quintic 3-folds and sections of the Horrocks-Mumford bundle*, Math. Research Report 33, Canberra (1985).

[219] R. Moore and R. Wardelmann, **PGL**(4) *acts transitively on* $M(-1, 2)$, J. Reine Angew. Math. **346** (1984), 48–53.

[220] H. Morikawa, *A note on holomorphic vector bundles over complex tori*, Nagoya Math. J. **41** (1971), 101–106.

[221] A. Morimoto, *Sur la classification des espaces fibrés vectoriels holomorphes sur un tore complexe admettant des connexions holomorphes*, Nagoya Math. J **15** (1959), 83–154.

[222] S. Mukai, *Semi-homogeneous vector bundles on an Abelian variety*, J. Math. Kyoto Univ. **18** (1978), no. 2, 239–272.

[223] ———, *Symplectic structure of the moduli space of sheaves on an abelian or* $K3$ *surface*, Invent. Math. **77** (1984), no. 1, 101–116.

[224] ———, *On the moduli space of bundles on* $K3$ *surfaces. I*, Vector bundles on algebraic varieties (Bombay, 1984), Tata Inst. Fund. Res. Stud. Math., vol. 11, Tata Inst. Fund. Res., Bombay, 1987, pp. 341–413.

[225] M. Mulase, *Poles of instantons and jumping lines of algebraic vector bundles on* $\mathbf{P}^3$, Proc. Japan Acad. Ser. A Math. Sci. **55** (1979), no. 5, 185–189.

[226] R. Mulczinski, *Eine neue algebraische Methode zur Untersuchung von Vektorbündeln auf der projektiven Ebene*, Bonner Mathematische Schriften, 134, Mathematisches Institut der Universität Bonn, Bonn, 1981, Dissertation 1978.

[227] D. Mumford and J. Fogarty, *Geometric invariant theory*, second ed., Ergebnisse der Mathematik und ihrer Grenzgebiete, vol. 34, Springer-Verlag, Berlin, 1982.

[228] M.S. Narasimhan and G. Trautmann, *Compactification of* $M(0,2)$, Vector bundles on algebraic varieties (Bombay, 1984), Tata Inst. Fund. Res. Stud. Math., vol. 11, Tata Inst. Fund. Res., Bombay, 1987, pp. 429–443.

[229] P.E. Newstead, *Introduction to moduli problems and orbit spaces*, Tata Institute of Fundamental Research Lectures on Mathematics and Physics, vol. 51, Tata Institute of Fundamental Research, Bombay, 1978.

[230] ———, *The fundamental group of a moduli space of bundles on* $\mathbf{P}^3$, Topology **19** (1980), no. 4, 419–426.

[231] ———, *On the cohomology and the Picard group of a moduli space of bundles on* $\mathbf{P}^3$, Quart. J. Math. Oxford Ser. (2) **33** (1982), no. 131, 349–355.

[232] C. Okonek, *Homotopiegruppen des Modulraumes* $M_{\mathbf{P}_2}(-1, c_2)$, Math. Ann. **258** (1981/82), no. 3, 253–266.

[233] ———, *3-Mannigfaltigkeiten im* $\mathbf{P}^5$ *und ihre zugehörigen stabilen Garben*, Manuscripta Math. **38** (1982), no. 2, 175–199.

[234] ———, *Reflexive Garben auf* $\mathbf{P}^4$, Math. Ann. **260** (1982), no. 2, 211–237.

[235] ———, *Moduli extremer reflexiver Garben auf* $\mathbf{P}^n$, J. Reine Angew. Math. **338** (1983), 183–194.

[236] ———, *Moduli reflexiver Garben und Flächen von kleinem Grad in* $\mathbf{P}^4$, Math. Z. **184** (1983), no. 4, 549–572.

[237] ———, *Über 2-codimensionale Untermannigfaltigkeiten vom Grad 7 in* $\mathbf{P}^4$ *und* $\mathbf{P}^5$, Math. Z. **187** (1984), no. 2, 209–219.

[238] ———, *Flächen vom Grad 8 im* $\mathbf{P}^4$, Math. Z. **191** (1986), no. 2, 207–223.

[239] ———, *On dimension-2 submanifolds in* $\mathbf{P}^4$ *and* $\mathbf{P}^5$, Mathematica Gottingensis 50 (1986).

[240] ———, *Fake Enriques surfaces*, Topology **27** (1988), no. 4, 415–427.

[241] C. Okonek and H. Spindler, *Reflexive Garben vom Rang* $r > 2$ *auf* $\mathbf{P}^n$, J. Reine Angew. Math. **344** (1983), 38–64.

[242] ———, *Stabile reflexive Garben vom Rang 3 auf* $\mathbf{P}^3$ *mit kleinen Chernklassen*, Math. Ann. **264** (1983), no. 1, 91–118.

[243] ———, *Das Spektrum torsionsfreier Garben. I*, Manuscripta Math. **47** (1984), no. 1-3, 187–228.

[244] ———, *Die Modulräume* ${}^3M^{\mathrm{st}}_{\mathbf{P}^3}(-2, 3, c_3)$, Math. Ann. **267** (1984), no. 3, 365–375.

[245] ———, *Das Spektrum torsionsfreier Garben. II*, Seminar on deformations (Łódź/Warsaw, 1982/84), Lecture Notes in Math., vol. 1165, Springer, Berlin, 1985, pp. 211–234.

[246] ———, *Mathematical instanton bundles on* $\mathbf{P}^{2n+1}$, J. Reine Angew. Math. **364** (1986), 35–50.

[247] C. Okonek and A. Van de Ven, *Stable bundles and differentiable structures on certain elliptic surfaces*, Invent. Math. **86** (1986), no. 2, 357–370.

[248] T. Ono, *Simple vector bundles of rank 2 on a del Pezzo surface*, TRU Math. **19** (1983), no. 2, 125–131.

[249] G. Ottaviani, *A class of n-bundles on* Gr(k, n), J. Reine Angew. Math. **379** (1987), 182–208.

[250] ______, *Some extensions of Horrocks criterion to vector bundles on Grassmannians and quadrics*, Ann. Mat. Pura Appl. (4) **155** (1989), 317–341.

[251] D.A. Ponomarev, *Germs on* $\mathbf{C}\mathrm{P}^1 \times \mathbf{C}\mathrm{P}^1$ *of holomorphic vector bundles on* $\mathbf{C}\mathrm{P}^3 \times \mathbf{C}\mathrm{P}^3$ *trivial on* $\mathbf{C}\mathrm{P}^1 \times \mathbf{C}\mathrm{P}^1$, Dokl. Akad. Nauk SSSR **276** (1984), no. 2, 292–295.

[252] S. Ramanan, *Holomorphic vector bundles on homogeneous spaces*, Topology **5** (1966), 159–177.

[253] Z. Ran, *On projective varieties of codimension* 2, Invent. Math. **73** (1983), no. 2, 333–336.

[254] P. Rao, *Liaison among curves in* $\mathbf{P}^3$, Invent. Math. **50** (1978/79), no. 3, 205–217.

[255] ______, *Liaison equivalence classes*, Math. Ann. **258** (1981/82), no. 2, 169–173.

[256] ______, *A note on cohomology modules of rank-two bundles*, J. Algebra **86** (1984), no. 1, 23–34.

[257] S. Roan, *Homogeneous vector bundles over projective varieties*, Chinese J. Math. **12** (1984), no. 3, 137–169.

[258] E. Sato, *On the decomposability of infinitely extendable vector bundles on projective spaces and Grassmann varieties*, J. Math. Kyoto Univ. **17** (1977), no. 1, 127–150.

[259] ______, *On infinitely extendable vector bundles on* G/P, J. Math. Kyoto Univ. **19** (1979), no. 1, 171–189.

[260] T. Sauer, *Nonstable reflexive sheaves on* $\mathbf{P}^3$, Trans. Amer. Math. Soc. **281** (1984), no. 2, 633–655.

[261] U. Schafft, *Nichtsepariertheit instabiler Rang-*2*-Vektorbündel auf* $\mathbf{P}_2$, Bayreuth. Math. Schr. (1983), no. 13, 155–212.

[262] ______, *Nichtsepariertheit instabiler Rang-*2*-Vektorbündel auf* $\mathbf{P}_2$, J. Reine Angew. Math. **338** (1983), 136–143.

[263] ______, *Chernklassen semi-stabiler Rang-*3*-Vektorbündel auf kubischen Dreimannigfaltigkeiten*, Manuscripta Math. **57** (1986), no. 1, 33–48.

[264] M. Schneider, *Chernklassen semi-stabiler Vektorraumbündel vom Rang* 3 *auf dem komplex-projektiven Raum*, J. Reine Angew. Math. **315** (1980), 211–220.

[265] ______, *Einschränkung stabiler Vektorraumbündel vom Rang* 3 *auf Hyperebenen des projektiven Raumes*, J. Reine Angew. Math. **323** (1981), 177–192.

[266] ______, *Stable vector bundles of rank* 3 *on* $\mathbf{P}_n$, Conference on Complex Analysis, Nancy 80 (Nancy, 1980), Inst. Élie Cartan, vol. 3, Univ. Nancy, Nancy, 1981, pp. 1–25.

[267] ______, *Submanifolds of projective space with semistable normal bundle*, Several complex variables (Hangzhou, 1981), Birkhäuser Boston, Boston, MA, 1984, pp. 151–160.

[268] ______, *Vector bundles and submanifolds of projective space: nine open problems*, Algebraic geometry, Bowdoin, 1985 (Brunswick, Maine, 1985), Proc. Sympos. Pure Math., vol. 46, Amer. Math. Soc., Providence, RI, 1987, pp. 101–107.

[269] C. Schoen, *On the geometry of a special determinantal hypersurface associated to the Mumford-Horrocks vector bundle*, J. Reine Angew. Math. **364** (1986), 85–111.

[270] H.W. Schuster, *Locally free resolutions of coherent sheaves on surfaces*, J. Reine Angew. Math. **337** (1982), 159–165.

[271] R.L.E. Schwarzenberger, *Reducible vector bundles on a quadric surface*, Proc. Cambridge Philos. Soc. **58** (1962), 209–216.

[272] C.S. Seshadri, *Fibrés vectoriels sur les courbes algébriques*, Astérisque, vol. 96, Société Mathématique de France, Paris, 1982, Notes written by J.-M. Drezet from a course at the École Normale Supérieure, June 1980.

[273] S.S. Shatz, *Coverings of ruled surfaces and applications to vector bundles. I*, Proc. London Math. Soc. (3) **35** (1977), no. 1, 89–112.

[274] ———, *The decomposition and specialization of algebraic families of vector bundles*, Compositio Math. **35** (1977), no. 2, 163–187.

[275] ———, *Fibre equivalence of vector bundles on ruled surfaces*, J. Pure Appl. Algebra **12** (1978), no. 3, 201–205.

[276] S. Soberon, *Vector bundles and the Yang-Mills equations in algebraic geometry*, Proceedings of the Special Seminar on Topology, Vol. I (Mexico City, 1980/1981) (Mexico City), Univ. Nac. Autónoma México, 1981, pp. 231–270.

[277] S. Soberon-Chavez, *Rank* 2 *vector bundles over a complex quadric surface*, Quart. J. Math. Oxford Ser. (2) **36** (1985), no. 142, 159–172.

[278] I. Sols, *On spinor bundles*, J. Pure Appl. Algebra **35** (1985), no. 1, 85–94.

[279] H. Spindler, *Die Modulräume stabiler* 3*-Bündel auf* $\mathbf{P}_3$ *mit den Chernklassen* $c_1 = 0$, $c_3 = c_2^2 - c_2$, Math. Ann. **256** (1981), no. 1, 133–143.

[280] ———, *Ein Satz über die Einschränkung holomorpher Vektorbündel auf* $\mathbf{P}_n$ *mit* $c_1 = 0$ *auf Hyperebenen*, J. Reine Angew. Math. **327** (1981), 93–118.

[281] ———, *Holomorphe Vektorbündel auf* $\mathbf{P}_n$ *mit* $c_1 = 0$ *und* $c_2 = 1$, Manuscripta Math. **42** (1983), no. 2-3, 171–198.

[282] ———, *Uniforme Vektorbündel vom Typ* $(0, -1, \ldots, -r+1)$ *auf* $\mathbf{P}_n$, Mathematica Gottingensis **76** (1986).

[283] H. Spindler and G. Trautmann, *Rational normal curves and the geometry of special instanton bundles on* P^{2n+1}, Mathematica Gottingensis **18** (1987).

[284] P.F. Stiller, *Certain reflexive sheaves on* $\mathbf{P}^n_{\mathbf{C}}$ *and a problem in approximation theory*, Trans. Amer. Math. Soc. **279** (1983), no. 1, 125–142.

[285] ———, *Vector bundles on complex projective spaces and systems of partial differential equations. I*, Trans. Amer. Math. Soc. **298** (1986), no. 2, 537–548.

[286] S.A. Strømme, *Deforming vector bundles on the projective plane*, Math. Ann. **263** (1983), no. 3, 385–397.

[287] ———, *Ample divisors on fine moduli spaces on the projective plane*, Math. Z. **187** (1984), no. 3, 405–423.

[288] A.S. Tihomirov, *A four-dimensional family of quadrics and a monad*, Izv. Akad. Nauk SSSR Ser. Mat. **44** (1980), no. 1, 219–232, 240.

[289] A.N. Tjurin, *On the superpositions of mathematical instantons*, Arithmetic and geometry, Vol. II, Progr. Math., vol. 36, Birkhäuser Boston, Boston, MA, 1983, pp. 433–450.

[290] G. Trautmann, *Poncelet curves and associated theta characteristics*, Exposition. Math. **6** (1988), no. 1, 29–64.

[291] K. Uhlenbeck and S.-T. Yau, *On the existence of Hermitian-Yang-Mills connections in stable vector bundles*, Comm. Pure Appl. Math. **39** (1986), no. S,

suppl., S257–S293, Frontiers of the mathematical sciences: 1985 (New York, 1985).

[292] H. Umemura, *A theorem of Matsushima*, Nagoya Math. J. **54** (1974), 123–134.

[293] ———, *On a certain type of vector bundles over an Abelian variety*, Nagoya Math. J. **64** (1976), 31–45.

[294] ———, *On a theorem of Ramanan*, Nagoya Math. J. **69** (1978), 131–138.

[295] ———, *Moduli spaces of the stable vector bundles over abelian surfaces*, Nagoya Math. J. **77** (1980), 47–60.

[296] T. Urabe, *Some properties of vector bundles on the flag variety* $\mathrm{Fl}(r, s; n)$, Tokyo J. Math. **6** (1983), no. 2, 335–362.

[297] A. Van de Ven, *Twenty years of classifying algebraic vector bundles*, Journées de Géometrie Algébrique d'Angers, Juillet 1979/Algebraic Geometry, Angers, 1979, Sijthoff & Noordhoff, Alphen aan den Rijn, 1980, pp. 3–20.

[298] ———, *On the differentiable structure of certain algebraic surfaces*, Astérisque (1987), no. 145-146, 5, 299–312, Séminaire Bourbaki, Vol. 1985/86.

[299] V.K. Vedernikov, *Moduli of stable vector bundles of rank* 2 *on* $\mathbf{P}_3$ *with a fixed spectrum*, Izv. Akad. Nauk SSSR Ser. Mat. **48** (1984), no. 5, 986–998.

[300] ———, *The moduli of super-null-correlation bundles on* $\mathbf{P}_3$, Math. Ann. **276** (1987), no. 3, 365–383.

[301] J. Wehler, *Moduli space and versal deformation of stable vector bundles*, Rev. Roumaine Math. Pures Appl. **30** (1985), no. 1, 69–78.

[302] R.O. Wells, Jr., *Cohomology and the Penrose transform*, Complex manifold techniques in theoretical physics (Proc. Workshop, Lawrence, Kan., 1978), Res. Notes in Math., vol. 32, Pitman, Boston, Mass., 1979, pp. 92–114.

[303] G.P. Wever, *The moduli of a class of rank* 2 *vector bundles on* P^3, Nagoya Math. J. **84** (1981), 9–30.

Index

APPENDIX A

Sheaves on $\mathbb{P}_n$ and problems in linear algebra

by S. I. Gelfand (translated by R. Zeinstra)

The purpose of this supplement is to present with proofs the results of the note [1*] on the reduction of the classification of algebraic vector bundles over $\mathbb{P}_n$ to a certain problem in linear algebra, namely the classification of finite-dimensional graded modules over the Grassmann algebra Λ with $n+1$ generators. The main result—theorem 2.4—admits a rather elementary formulation. But actually the possibility of such a reduction is based on a deep and unexpected relation between the category of coherent sheaves on $\mathbb{P}_n$ and the derived category of finite-dimensional Λ-modules. The reader may find the needed information on derived categories in the works of Hartshorne [6*] and Verdier [2*]. Unfortunately, lack of space did not allow to mention here all the necessary results.

A different approach to the classification of bundles over $\mathbb{P}_n$ was proposed by A. A. Beilinson [18]. The translation of his results from the language of derived categories into the language of spectral sequences is contained in the main text (theorems 3.1.3 and 3.1.4). Below we shall formulate the result from [18] and indicate the relation of the classification from [1*] and [18].

We remark that recently the approach using derived categories has led to a number of new results. In particular, M. M. Kapranov [3*] has described the category of coherent sheaves on Grassmann manifolds.

As already said, the present supplement is an extended presentation of the results of [1*]. I would like to express my deep acknowledgement to my co-authors, in the first place to I. N. Bernstein, who proposed many technical improvements compared to [1*]. Very useful were also conversations with A. A. Beilinson and Y. I. Manin.

§1. The exterior algebra and modules over the exterior algebra

1.1. Let Ξ be the $(n+1)$-dimensional vector space over the field of complex numbers (or over an arbitrary algebraically closed field of characteristic 0). We denote by $\Lambda = \Lambda(\Xi)$ the exterior (Grassmann) algebra of the space Ξ. We introduce a grading on Λ by putting $\deg \xi = 1$ if $0 \neq \xi \in \Xi$, so that

$$\Lambda(\Xi) = \bigoplus_{i=0}^{n+1} \Lambda^i(\Xi), \qquad \dim \Lambda^i(\Xi) = \binom{n+1}{i}.$$

By a left Λ-module we shall understand (unless otherwise mentioned) a left unitary graded module $V = \bigoplus V_j$ over the graded algebra Λ. We denote by $\mathcal{M}(\Lambda)$ the category consisting of all Λ-modules and degree preserving homomorphisms. By $\mathcal{M}^b(\Lambda)$ we denote the full subcategory of $\mathcal{M}(\Lambda)$ which consists of the Λ-modules with a finite number of generators. It is clear that $V \in \mathcal{M}^b(\Lambda)$ if and only if $\dim V < \infty$.

1.2. Operations on Λ-modules.

a) Let $V = \bigoplus V_j \in \mathcal{M}(\Lambda)$ and r an integer. We define $V[r] \in \mathcal{M}(\Lambda)$ by putting

$$(V[r])_j = V_{j-r}$$

with the previous action of Λ. It is clear that

$$V[r][s] = V[r+s].$$

b) Let $\mathbb{1}$ be the trivial one-dimensional module (i.e., $\dim(\mathbb{1})_0 = 1$, $(\mathbb{1})_j = \{0\}$ for $j \neq 0$). Then $\mathbb{1}[k]$ is a one-dimensional module with non-zero component of degree k.

Further, we consider Λ as a free left Λ-module, generated by one generator of degree 0. Then $\Lambda[k]$ is a free Λ-module generated by one generator of degree k.

c) Let V, $V' \in \mathcal{M}(\Lambda)$. We put $(V \oplus V')_j = V_j \oplus V'_j$ with the natural action of Λ on $V \oplus V' = \bigoplus_j (V \oplus V')_j$.

Further we put $(V \otimes V')_j = \bigoplus_\alpha (V_\alpha \otimes V'_{j-\alpha})$ and define the action of Λ on $V \otimes V' = \bigoplus_j (V \otimes V')_j$ by putting

$$\xi(v \otimes v') = \xi v \otimes v' + (-1)^\alpha v \otimes \xi v'$$

for $v \in V_\alpha$, $v' \in V'_\beta$, $\xi \in \Xi \subset \Lambda$.

There exist natural isomorphisms

$$V \otimes V'[k] \cong V[k] \otimes V' \cong (V \otimes V')[k].$$

For arbitrary $V \in \mathcal{M}(\Lambda)$ the module $V \otimes \Lambda$ is free,

$$V \otimes \Lambda = \bigoplus_\ell \Lambda[i_\ell]$$

and the number of modules $\Lambda[j]$ in this decomposition is equal to $\dim V_j$.

d) We define the symmetric and exterior powers of the Λ-module V. On $V^{\otimes k} = V \otimes \cdots \otimes V$ (k times) we put the action of the symmetric group Σ_k of degree k in the following way: if $\sigma_i \in \Sigma_k$ is the transposition of the elements i and $i+1$ $(1 \le i \le k-1)$ then

$$\sigma_i(v_1 \otimes \cdots \otimes v_i \otimes v_{i+1} \otimes \cdots \otimes v_k) = (-1)^{\alpha_i \alpha_{i+1}}(v_1 \otimes \cdots \otimes v_{i+1} \otimes v_i \otimes \cdots \otimes v_k)$$

for $v_i \in V_{\alpha_i}$, $v_{i+1} \in V_{\alpha_{i+1}}$. It is easy to check that this action of the generators σ_i of the group Σ_k extends to an action of all of Σ_k.

We call the submodule of $V^{\otimes k}$ consisting of the Σ_k-invariant elements of the k th symmetric power $S^k V$ of the module V, and we call the submodule of $V^{\otimes k}$ which consists of the (completely) skew symmetric elements with respect to Σ_k the kth exterior power $\Lambda^k V$ of the module V.

We remark that in general $\Lambda^k V \neq 0$ for all k, even for finite dimensional V. For example, if $V = \mathbb{1}[1]$, then $S^k V = 0$ for $k > 0$, $\Lambda^k V = \mathbb{1}[k]$.

e) Let $V \in \mathcal{M}(\Lambda)$ be a left Λ-module. We construct with respect to V a right Λ-module $V^{(r)}$ by putting

$$v\lambda = (-1)^{j\ell}\lambda v \quad \text{for} \quad \lambda \in \Lambda_\ell, \quad v \in V_j.$$

It is clear that the mapping $V \mapsto V^{(r)}$ extends to a functor establishing an equivalence between the categories of left and right Λ-modules.

Further, for $V \in \mathcal{M}(\Lambda)$ we put $V^* = \operatorname{Hom}_{\mathbb{C}}(V^{(r)}, \mathbb{C})$; the space V^* is provided with the structure of a left Λ-module according to the formula

$$(\lambda\phi)(v) = \phi(v\lambda).$$

1.3. Let $V = \bigoplus V_j \in \mathcal{M}^b(\Lambda)$ and $\xi \in \Xi \subset \Lambda$. Since $\deg \xi = 1$ we have for each j a linear map $V_j \to V_{j+1}$; $v \mapsto \xi v$. Since $\xi^2 = 0$, we obtain a complex of linear spaces

$$\cdots \to V_{j-1} \xrightarrow{\xi} V_j \xrightarrow{\xi} V_{j+1} \to \cdots,$$

which will be denoted by $L_\xi(V)$.

1.4. Definition. The module $V \in \mathcal{M}^b(\Lambda)$ is called suitable if for arbitrary $\xi \neq 0$ we have $H^j(L_\xi(V)) = 0$ when $j \neq 0$.

It follows at once from Euler's formula that for V suitable the dimension $\dim H^0(L_\xi(V))$ is independent of $\xi \in \Xi$, $\xi \neq 0$.

1.5. Examples of suitable Λ-modules.

(i) The algebra Λ itself, considered as a left Λ-module, is suitable. And what is more, $H^j(L_\xi(\Lambda)) = 0$ for all j and for all $\xi \neq 0$. Therefore all modules $\Lambda[r]$ are also suitable.

(ii) The one-dimensional trivial Λ-module $\mathbb{1}$ is suitable.

(iii) Let $\xi_0, \dots, \xi_n$ be a basis in Ξ and let $\tilde{V}$ be the submodule of Λ generated by the elements $\xi_0, \dots, \xi_n$. Then the module $\tilde{V}[-1]$ is suitable.

(iv) The element $\omega = \xi_0 \dots \xi_n \in \Lambda$ generates a one-dimensional submodule (ω) in Λ. The module $\Lambda/(\omega))[-n]$ is suitable.

(v) If V, V' are suitable Λ-modules, then $V \oplus V'$, $V \otimes V'$, V^*, $S^k V$, $\Lambda^k V$ are suitable Λ-modules.

(vi) If in the exact sequence $0 \to V' \to V \to V'' \to 0$ the modules V', V'' are suitable, then the module V is suitable.

1.6. In the remaining part of this section we mention without proof a number of properties of the category $\mathcal{M}^b(\Lambda)$.

Proposition. Let $V \in \mathcal{M}^b(\Lambda)$. The following properties are equivalent:

(i) V is a free Λ-module.
(ii) V is a projective Λ-module.
(iii) V is an injective Λ-module.

1.7. We fix a non-zero element ω from the one dimensional space Λ_{n+1}. We shall call $V \in \mathcal{M}^b(\Lambda)$ reduced if $\omega V = 0$.

Proposition. Every module $V \in \mathcal{M}^b(\Lambda)$ can be represented in the form $V = F \oplus V_{\min}$, where F is free and $V_{\min}$ is reduced. F and $V_{\min}$ are uniquely determined by V up to isomorphism (although the decomposition itself into a direct sum is not unique).

1.8. We define the quotient category $\mathcal{M}^0(\Lambda) = \mathcal{M}^b(\Lambda)/\mathcal{P}$ of the category $\mathcal{M}^b(\Lambda)$ with respect to the class of free objects $\mathcal{P}$ in the following way. The objects in $\mathcal{M}^0(\Lambda)$ are the same as in $\mathcal{M}^b(\Lambda)$. Furthermore we shall say that a morphism of Λ-modules, $\alpha\colon V \to V'$, is $\mathcal{P}$-equivalent to 0 if there exists a free module $F \in \mathcal{P}$ and morphisms $\gamma\colon V \to F$, $\beta\colon F \to V'$ such that $\alpha = \beta \circ \gamma$.

Let $L(V, V') \subset \operatorname{Hom}_{\mathcal{M}^b(\Lambda)}(V, V')$ be the set of all morphisms $\mathcal{P}$-equivalent to 0. It is clear that $L(V, V')$ is a subspace in

$$\operatorname{Hom}_{\mathcal{M}^b(\Lambda)}(V, V').$$

It is also clear that if $\alpha \in L(V, V')$, $\beta\colon V' \to V''$, $\gamma\colon V''' \to V$, then

$$\beta \circ \alpha \circ \gamma \in L(V''', V''). \tag{1.1}$$

We now define the space of morphisms in $\mathcal{M}^0(\Lambda)$ by the formula

$$\mathrm{Hom}_{\mathcal{M}^0(\Lambda)}(V, V') = \mathrm{Hom}_{\mathcal{M}^b(\Lambda)}(V, V')/L(V, V').$$

From (1.1) it is clear that the composition of morphisms in $\mathcal{M}^b(\Lambda)$ gives the composition of morphisms in $\mathcal{M}^0(\Lambda)$. Thus the category $\mathcal{M}^0(\Lambda)$ is completely described.

§2. Sheaves on $\mathbb{P}$ and Λ-modules. Formulation of the main theorem

2.1. Let $\mathbb{P}(\Xi)$ (or simply $\mathbb{P}$) be the projectivization of the linear space Ξ. By $\bar{\xi}$ we denote the point in $\mathbb{P}$ corresponding to the non-zero element $\xi \in \Xi$. The main theorem of this appendix describes the algebraic vector bundles over $\mathbb{P}$, that is the locally free coherent sheaves of $\mathcal{O}_{\mathbb{P}}$-modules. Roughly speaking, it asserts that an arbitrary such bundle is a family of vector spaces $E_{\bar{\xi}} = H^0(L_{\bar{\xi}}(V))$, where V is some suitable $\Lambda(\Xi)$-module (see §1.4).

2.2. We turn to precise definitions. Let $V = \bigoplus V_j \in \mathcal{M}^b(\Lambda)$. For every j we put

$$L_j = L_j(V) \underset{\text{def}}{=} \mathcal{V}_j \otimes \mathcal{O}(j),$$

where $\mathcal{V}_j$ is the constant sheaf with fibre V_j. We fix a basis $\xi_0, \dots, \xi_n$ in Ξ. Every element ξ_i determines a map $\xi_i\colon \mathcal{V}_j \to \mathcal{V}_{j+1}$. Furthermore, the space of sections $H^0(\mathbb{P}, \mathcal{O}(1))$ is canonically isomorphic to the dual space $X = \Xi^*$.

Let $x_0, \dots, x_n$ be the basis in $H^0(\mathbb{P}, \mathcal{O}(1)) = \Xi^*$ dual to the basis $\xi_0, \dots, \xi_n$. Since $\mathcal{O}(j+1) = \mathcal{O}(j) \otimes \mathcal{O}(1)$, every x_i determines a morphism $x_i\colon \mathcal{O}(j) \to \mathcal{O}(j+1)$. We put $\partial_j = \sum \xi_i \otimes x_i\colon L_j(V) \to L_{j+1}(V)$. Clearly ∂_j does not depend on the choice of the basis $\xi_0, \dots, \xi_n$. From the commutation relations

$$\xi_i \xi_k = -\xi_k \xi_i, \qquad x_i x_k = x_k x_i$$

it follows immediately that $\partial_{j+1}\partial_j = 0$. In this way to every finite dimensional graded Λ-module V there is associated a complex of locally free sheaves

$$\mathcal{L}(V) = \{\cdots \to L_j(V) \xrightarrow{\partial_j} L_{j+1}(V) \to \cdots\}.$$

Let us consider two subsheaves in $L_0(V)$: $B = \operatorname{im} \partial_{-1}$, $Z = \ker \partial_0$. Since $\partial_0 \circ \partial_{-1} = 0$ we have $B \subset Z$.

Definition: We put $\Phi(V) = Z/B$.

2.3. Lemma. Let V be a suitable Λ-module. Then Z, B and $\Phi(V)$ are locally free sheaves.

The proof follows immediately from the assertions in §1.4, for the geometric fibre of the complex $\mathcal{L}(V)$ at the point $\bar{\xi} \in \mathbb{P}$ coincides with the complex $L_\xi(V)$.

Let $V = F$ be a free Λ-module. Then it is clear that $\Phi(F) = 0$. Therefore for arbitrary $V \in \mathcal{M}^b(\Lambda)$ we have $\Phi(V \oplus F) = \Phi(V)$. The main theorem of the article can be formulated as follows.

2.4. Theorem. (i) An arbitrary bundle $\mathcal{L}$ over $\mathbb{P}(\Xi)$ has the form $\mathcal{L} = \Phi(V)$ for some suitable $V \in \mathcal{M}^b(\Lambda)$.

(ii) $\Phi(V) \cong \Phi(V')$ if and only if there exist free modules F, F' such that $V \oplus F \cong V' \oplus F'$.

2.5. We formulate a number of properties of the correspondence $V \leftrightarrow \Phi(V)$.

a) In the course of the proof it will actually be shown that Φ is a functor from the full subcategory $\mathrm{Pr}/\mathcal{P}$ of suitable modules in $\mathcal{M}^b(\Lambda)/\mathcal{P}$ into the category $\mathrm{Vect}(\mathbb{P})$ of algebraic vector bundles over $\mathbb{P}$ that establishes an equivalence of these categories.

b) The functor Φ commutes with the operations of direct sum and tensor product in $\mathcal{M}^b(\Lambda)$ and $\mathrm{Vect}(\mathbb{P})$, with the passage to dual object, with taking symmetric and exterior powers.

c) We shall describe the bundles corresponding to the suitable modules in the examples of §1.5.

(i) $V = \Lambda \Rightarrow \Phi(V) = 0$.

(ii) $V =$ one-dimensional trivial module $\mathbb{1} \Rightarrow \Phi(V) = \mathcal{O}_{\mathbb{P}}$, the trivial one dimensional vector bundle over $\mathbb{P}$.

(iii) $V = (\xi_0, \dots, \xi_n)\Lambda[-1] =$ the submodule in $\Lambda[-1]$ generated by all ξ_i. Then it is easy to see that $\Phi(V) = \mathcal{O}_{\mathbb{P}}(-1)$.

(iv) $V = (\Lambda/(\omega))[-n]$, where $\omega = \xi_0 \dots \xi_n \in \Lambda$. Then $\Phi(V) = \mathcal{O}_{\mathbb{P}}(1)$.

(v) Let $0 \leq k \leq n$. We consider in $\Lambda[-n-1+k]$ the submodule L_k generated by all the elements of degree ≥ 0. Then L_k is a suitable module and $\Phi(L_k) = \Omega^k_{\mathbb{P}}(k)$.

(vi) Similarly, let T_k be the submodule in $\Lambda[-k]$ generated by all elements of degree > 0, and let $Q_k = \Lambda[-k]/T_k$. Then Q_k is a suitable module and $\Phi(Q_k) = \Omega^{n-k}_{\mathbb{P}}(n-k+1)$.

(vii) (generalization of (iv)). Let V be a suitable module, F a free module and

$$0 \to V \to F \to \tilde{V} \to 0$$

an exact sequence of Λ-modules. Then $\tilde{V}[-1]$ is also a suitable module and $\Phi(\tilde{V}[-1]) = \Phi(V) \otimes \mathcal{O}_{\mathbb{P}}(1)$.

2.6. The proof of the main theorem occupies §3–5. In the remaining part of this section we shall briefly describe its structure and point out the single steps.

First it is necessary to consider the category Sh of coherent sheaves of $\mathcal{O}_{\mathbb{P}}$-modules and the corresponding bounded derived category $\mathsf{D}^b(\mathsf{Sh})$ (see [2*,6*]). We recall that complexes of coherent sheaves

$$\mathcal{L}_\bullet = \{\cdots \to \mathcal{L}_{-1} \to \mathcal{L}_0 \to \mathcal{L}_1 \to \cdots\}$$

are the objects of the category $\mathsf{D}^b(\mathsf{Sh})$, where in each complex there is only a finite number of non-zero terms. A morphism of complexes $F\colon \mathcal{L}_\bullet \to \tilde{\mathcal{L}}_\bullet$ is called a quasi-isomorphism if the corresponding mapping in cohomology $H^*(F)\colon H^*(\mathcal{L}_\bullet) \to H^*(\tilde{\mathcal{L}}_\bullet)$ is an isomorphism. Roughly speaking, the derived category $\mathsf{D}^b(\mathsf{Sh})$ results from the category of complexes by addition of morphisms which are inverse to all quasi-isomorphisms.

2.7. We recall that in §1.8 we have defined the quotient category $\mathcal{M}^0(\Lambda) = \mathcal{M}^b(\Lambda)/\mathcal{P}$ of the category $\mathcal{M}^b(\Lambda)$ with respect to the class $\mathcal{P}$ of free Λ-modules.

We shall construct a functor $\Phi_D\colon \mathcal{M}^0(\Lambda) \to \mathsf{D}^b(\mathsf{Sh})$, generalizing the functor Φ from §2.2. Let $V \in \mathcal{M}^b(\Lambda)$, $\mathcal{L}(V)$ the complex constructed in §2.2. Since V is finite-dimensional, only a finite number of the $L_j(V)$ is different from 0. It is easy to check that if $V \in \mathcal{P}$ then $\mathcal{L}(V)$ is an acyclic complex (that is $H^*(\mathcal{L}(V)) = 0$). We denote by $L_D(V)$ the object of $\mathsf{D}^b(\mathsf{Sh})$ which corresponds to $\mathcal{L}(V)$. From the previous statement it follows that the mapping $V \mapsto L_D(V)$ extends to a functor $\Phi_D\colon \mathcal{M}^0(\Lambda) \to \mathsf{D}^b(\mathsf{Sh})$.

2.8. Theorem. The functor $\Phi_D\colon \mathcal{M}^0(\Lambda) \to \mathsf{D}^b(\mathsf{Sh})$ is an equivalence of categories.

2.9. This theorem is the main result on the relation of sheaves on $\mathbb{P}(\Xi)$ with $\Lambda(\Xi)$-modules. Its proof consists of several practically mutually unrelated steps each of which establishes an equivalence of certain categories. We first introduce all categories needed.

a) $\mathsf{D}^b(\mathsf{Sh})$, the derived category of the category Sh of coherent sheaves of $\mathcal{O}_{\mathbb{P}}$-modules.

b) $\mathcal{M}^b(\Lambda)$, the category of finite-dimensional graded Λ-modules and $\mathsf{D}^b(\Lambda)$ the corresponding derived category.

c) Let $X = \Xi^*$ be the space dual to Ξ, and $S = S(X)$ the symmetric algebra on the space X. We denote by $\mathcal{M}^b(S)$ the category of finitely generated graded S-modules and by $\mathsf{D}^b(S)$ the corresponding derived category.

d) Let $\mathcal{I}$ be the full subcategory in $\mathsf{D}^b(\Lambda)$, consisting of complexes, isomorphic (in $\mathsf{D}^b(\Lambda)$) to complexes

$$\cdots \to F_{-1} \to F_0 \to F_1 \to \cdots ,$$

for which all $F_i \in \mathcal{P}$ (that is, the F_i are free Λ-modules with a finite number of generators).

e) Let $\mathcal{F}$ be the full subcategory in $\mathsf{D}^b(S)$, consisting of complexes, isomorphic (in $\mathsf{D}^b(S)$) to complexes

$$\cdots \to M_{-1} \to M_0 \to M_1 \to \cdots ,$$

for which all M_i are finite-dimensional graded S-modules.

2.10. The plan of the proof consists in proving the following four assertions:

a) The triangulated categories $\mathsf{D}^b(\Lambda)$ and $\mathsf{D}^b(S)$ are equivalent;

b) $\mathcal{I}$ and $\mathcal{F}$ are thick subcategories in $\mathsf{D}^b(\Lambda)$ and $\mathsf{D}^b(S)$ respectively (that is, it is possible to take the corresponding quotient categories, see [6*, §I.3, I.4]), and the quotient categories $\mathsf{D}^b(\Lambda)/\mathcal{I}$ and $\mathsf{D}^b(S)/\mathcal{F}$ are equivalent;

c) The categories $\mathsf{D}^b(\Lambda)/\mathcal{I}$ and $\mathcal{M}^0(\Lambda)$ are equivalent;

d) The categories $\mathsf{D}^b(S)/\mathcal{F}$ and $D^b(\mathsf{Sh})$ are equivalent.

Among these four statements a) and c) are the more complicated ones. §4 and §5, respectively, are dedicated to their proof (theorem 4.5 and theorem 5.1). The statement b) easily follows from a) (Corollary 4.12). Statement d) happens to be an easy consequence of a theorem of Serre on the description of coherent sheaves on $\mathbb{P}$.

§3. Supplement to the main theorem

3.1. First we derive theorem 2.4 from theorem 2.8. We shall associate to each sheaf $\mathcal{E}$ on $\mathbb{P}$ the complex of sheaves $d(\mathcal{E})$ whose zeroth term is equal to $\mathcal{E}$ and all of whose remaining ones are 0. Let $\mathcal{I}(\mathcal{E}) \in \mathsf{D}^b(\mathsf{Sh})$ be the element of the derived category, corresponding to this complex. The functor $\mathcal{E} \mapsto \mathcal{I}(\mathcal{E})$ identifies the category Sh with the full subcategory of $\mathsf{D}^b(\mathsf{Sh})$ consisting of the objects X for which $H^i(X) = 0$ for $i \neq 0$. Therefore in view of theorem 2.8, the category Sh of all coherent sheaves on $\mathbb{P}$ is equivalent to the subcategory of $\mathcal{M}^0(\Lambda)$ consisting of those modules V such that $H^i(\mathcal{L}(V)) = 0$ for

$i \neq 0$. Theorem 2.4 follows now from the following simple lemma whose proof we omit here.

Lemma. Let $V \in \mathcal{M}^b(\Lambda)$ be a module such that $H^i(\mathcal{L}(V)) = 0$ for $i \neq 0$ and such that $H^0(\mathcal{L}(V))$ is a locally free sheaf. Then V is a suitable module, that is $H^i(L_\xi(V)) = 0$ for all $i \neq 0$, $0 \neq \xi \in \Xi$.

3.2. The equivalence of categories $\Phi_D\colon \mathcal{M}^0(\Lambda) \to \mathsf{D}^b(\mathsf{Sh})$ defines on $\mathcal{M}^0(\Lambda)$ the structure of a triangulated category. This structure may be described in the following way.

a) The translation functor T.

Let $V \in \mathcal{M}^b(\Lambda)$. We put $V' = V \otimes \Lambda$ with the natural embedding $V \to V'$ $(v \mapsto v \otimes 1)$ and $T(V) = V'/V$. It is clear that T defines a functor on $\mathcal{M}^0(\Lambda)$. It is easy to see that if

$$0 \to V \to F \to \tilde{V} \to 0$$

is an exact sequence in $\mathcal{M}^b(\Lambda)$ and $F \in \mathcal{P}$, then in $\mathcal{M}^0(\Lambda)$ there is a canonical isomorphism $\tilde{V} \xrightarrow{\sim} T(V)$.

b) Distinguished triangles.

Let $0 \to V' \to V \to V'' \to 0$ be an exact sequence in $\mathcal{M}^b(\Lambda)$; then the morphisms $V' \to V \to V''$ (more precisely, their images in $\mathcal{M}^0(\Lambda)$) are enclosed in a distinguished triangle in $\mathcal{M}^0(\Lambda)$ and in this way one obtains (up to isomorphism) all pairs of morphisms which are enclosed in a distinguished triangle.

The proof of the formulated statements can be carried out simultaneously with the proof of theorem 2.8. It will be useful for the reader to prove directly that all axioms of a triangulated category are satisfied.

3.3. Let V, W be two suitable Λ-modules, $\Phi(V)$, $\Phi(W)$ the corresponding locally free sheaves. We shall compute in terms of the modules V, W the groups $\mathrm{Ext}^i(\Phi(V), \Phi(W))$, where Ext is taken in the category Sh of coherent sheaves on $\mathbb{P}$.

Proposition. (i) For $i \geq 1$ we have

$$\mathrm{Ext}^i(\Phi(V), \Phi(W)) = \mathrm{Ext}^i_{\mathcal{M}^b(\Lambda)}(V, W)$$

(ii) $\mathrm{Hom}(\Phi(V), \Phi(W)) = \mathrm{Hom}_{\mathcal{M}^b(\Lambda)}(\mathbb{1}, (V^* \otimes W)_{\min})$, where $(V^* \otimes W)_{\min}$ is the unique reduced Λ-module $\mathcal{P}$-equivalent to $V^* \otimes W$.

3.4. Corollary. For an arbitrary suitable Λ-module V we have

a) $H^0(\mathbb{P}, \Phi(V)) = \mathrm{Hom}_{\mathcal{M}^b(\Lambda)}(\mathbb{1}, V_{\min})$, where $V_{\min}$ is the unique reduced Λ-module $\mathcal{P}$-equivalent to V.

b) $H^i(\mathbb{P}, \Phi(V)) = \mathrm{Hom}_{\mathcal{M}^0(\Lambda)}(\mathbb{1}, T^i(V)) = \mathrm{Ext}^i_{\mathcal{M}^b(\Lambda)}(\mathbb{1}, V)$
$= H^i(\Lambda, V)$ for $i \geq 1$.

3.5. Proposition. Let $V = \bigoplus V_j$ be a suitable module. The total Chern class $c(\Phi(V))$ of the bundle $\Phi(V)$ (see §1.2 Ch. 1) may be computed by the formula

$$c(\Phi(V)) = \prod_j (1 + jh)^{(-1)^j \dim V_j},$$

where $h \in H^2(\mathbb{P}, \mathbb{Z})$ is the class of a hyperplane section.

3.6. Every locally free sheaf $\mathcal{E}$ on $\mathbb{P}$ has the form $\mathcal{E} = \Phi(V)$ for some (unique up to isomorphism) suitable reduced Λ-module V. We shall try to construct the module V starting from $\mathcal{E}$. Let first $\mathcal{E}$ be globally generated, that is the fibre of $\mathcal{E}$ in every point is generated by global sections of $\mathcal{E}$ over $\mathbb{P}$. In this case the corresponding Λ-module V can be constructed explicitly. We consider the linear spaces $L_0 = H^0(\mathbb{P}, \mathcal{E})$, $L_1 = H^0(\mathbb{P}, \mathcal{E}(1))$. Let

$$V^{(0)} = \mathrm{Hom}_{\mathbb{C}}(\Lambda, L_0), \qquad V^{(1)} = \mathrm{Hom}_{\mathbb{C}}(\Lambda, L_1).$$

The spaces $V^{(0)}$ and $V^{(1)}$ are provided with the structure of Λ-modules by the formula

$$\xi\phi(\lambda) = \phi(\lambda\xi), \quad \xi \in \Xi, \quad \lambda \in \Lambda, \quad \phi \in V^{(0)} \text{ or } V^{(1)}.$$

We introduce a grading on $V^{(i)}$ by putting

$$V_j^{(i)} = \mathrm{Hom}_{\mathbb{C}}(\Lambda_{-i-j}, L_i), \qquad i = 0, 1.$$

In this way the $V^{(i)}$ become graded Λ-modules. We fix a basis $\xi_0, \dots, \xi_n$ in Ξ; let $x_0, \dots, x_n$ be the dual basis in $X = \Xi^* = H^0(\mathbb{P}, \mathcal{O}_{\mathbb{P}}(1))$. Every element x_ℓ defines a linear mapping $x_\ell \colon L_0 \to L_1$. We define a morphism of Λ-modules $\Delta \colon V^{(0)} \to V^{(1)}$ by the formula

$$(\Delta\phi)(\lambda) = \sum x_\ell \phi(\xi_\ell \lambda), \quad \phi \in V^{(0)}, \quad \lambda \in \Lambda.$$

We put $V = \ker \Delta$.

Proposition. Let $\mathcal{E}$ be a globally generated locally free sheaf. Then V is a suitable reduced Λ-module and $\mathcal{E} \cong \Phi(V)$.

3.7. In the general case it is apparently rather complicated to construct explicitly a module V with $E \cong \Phi(V)$. It is possible, however, to find the dimensions $\dim V_j$ of the homogeneous components of V.

Let $\Omega_{\mathbb{P}}^k$ be the bundle of k-forms over $\mathbb{P}$ and

$$0 \to \Omega_{\mathbb{P}}^k(k) \to \Lambda^k(X) \otimes \mathcal{O}_{\mathbb{P}} \to \Omega_{\mathbb{P}}^{k-1}(k) \to 0$$

the exact Euler sequence (see formula (3) in §1.1, Ch. I). Tensoring by $\mathcal{E}$ and $\mathcal{O}_{\mathbb{P}}(-j-k)$ we obtain an exact sequence (since $\mathcal{E}$ is locally free)

$$0 \to \mathcal{E} \otimes \Omega_{\mathbb{P}}^k(-j) \to \Lambda^k(X) \otimes \mathcal{E}(-j-k) \to \mathcal{E} \otimes \Omega_{\mathbb{P}}^{k-1}(-j) \to 0.$$

Let

$$\delta\colon H^{k+j-1}(\mathbb{P}, \mathcal{E} \otimes \Omega_{\mathbb{P}}^{k-1}(-j)) \to H^{k+j}(\mathbb{P}, \mathcal{E} \otimes \Omega_{\mathbb{P}}^k(-j))$$

be the connecting homomorphism in the corresponding exact cohomology sequence.

Proposition. Let $\mathcal{E} = \Phi(V)$ for a suitable reduced module $V = \bigoplus V_j$. Then

$$\dim V_j = \sum_{k=0}^{n} \dim(H^{k+j}(\mathbb{P}, E \otimes \Omega_{\mathbb{P}}^k(-j))/\operatorname{Im}\delta).$$

One can formulate an analogous statement (with the replacement of cohomology by hypercohomology) for the derived object $Z = \Phi_D(V)$ in $\mathsf{D}^b(\mathsf{Sh})$, where $V \in \mathcal{M}^b(\Lambda)$ is a reduced module (see §2.7).

§4. Equivalence of derived categories

4.1. Let $X = \Xi^*$ be the dual space of Ξ, $S = S(X)$ its symmetric algebra, graded so that $\deg x = 1$ for $0 \neq x \in X$. By $\mathcal{M}(S)$ we denote the category of graded S-modules where the morphisms are the morphisms of graded S-modules of degree 0. Let $\mathcal{M}^b(S)$ be the full subcategory of $\mathcal{M}(S)$ consisting of finitely generated S-modules. The aim of this section is to prove an algebraic theorem which establishes a connection between the category of graded Λ-modules and the graded S-modules (theorem 4.5). It is the algebraic foundation of the geometric theorem 2.4.

4.2. We shall introduce the categories of complexes of Λ- and S-modules which we shall need later.

A complex of Λ-modules is defined to be a sequence

$$\cdots \to V^{(i-1)} \xrightarrow{\partial_{i-1}} V^{(i)} \xrightarrow{\partial_i} V^{(i+1)} \to \cdots$$

of modules $V^{(i)}$ from $\mathcal{M}(\Lambda)$, where the $\partial_i\colon V^{(i)} \to V^{(i+1)}$ are linear maps of graded vector spaces of degree 0, anticommuting with the action of $\xi \in \Xi$ (that is, $\partial_i \xi = -\xi \partial_i$) and satisfying the condition

$\partial_{i+1} \circ \partial_i = 0$ (thus ∂_i is not a morphism in $\mathcal{M}(\Lambda)$). A morphism $\{V^{(i)}, \partial_i\} \to \{\tilde{V}^{(i)}, \tilde{\partial}_i\}$ is defined to be a collection of morphisms $f_i \colon V^{(i)} \to \tilde{V}^{(i)}$ in $\mathcal{M}(V)$, such that $f_{i+1}\partial_i = \tilde{\partial}_i f_i$. Let $\mathcal{C}(\Lambda)$ be the category of all complexes of Λ-modules with the indicated morphisms.

Let $\mathcal{C}^b(\Lambda)$ be the full subcategory consisting of bounded complexes of finite-dimensional modules.

Analogously a complex of S-modules is defined to be a sequence

$$\cdots \to W^{(i-1)} \xrightarrow{d_{i-1}} W^{(i)} \xrightarrow{d_i} W^{(i+1)} \to \cdots$$

of modules in $\mathcal{M}(S)$, where $d_i \colon W^{(i)} \to W^{(i+1)}$ are morphisms in $\mathcal{M}(S)$ and $d_{i+1} \circ d_i = 0$. (Thus the d_i commute with the action of S).

A morphism from a complex $\{W^{(i)}, d_i\}$ to a complex $\{\tilde{W}^{(i)}, \tilde{d}_i\}$ is defined to be a set of morphisms $f_i \colon W^{(i)} \to \tilde{W}^{(i)}$ in $\mathcal{M}(S)$, such that $f_{i+1}d_i = \tilde{d}_i f_i$ for all i. Let $\mathcal{C}(S)$ be the category of all complexes of S-modules with the indicated morphisms. We shall denote by $\mathcal{C}^b(S)$ the subcategory of $\mathcal{C}(S)$ consisting of bounded complexes of finitely generated modules.

4.3. It is convenient to describe complexes $\{V^{(i)}, \partial_i\} \in \mathcal{C}(\Lambda)$ in the following way. Let $V_j^{(i)}$ be the space of homogeneous elements of degree j in $V^{(i)}$. On the bigraded space $V = \bigoplus_{i,j} V_j^{(i)}$ the algebra Λ and the linear operator $\partial = \bigoplus \partial_i$ act, where

(i) the multiplication operator $\xi \in \Xi$ has bidegree $(0,1)$;

(ii) the operator ∂ has bidegree $(1,0)$;

(iii) $\xi\partial = -\partial\xi$ for $\xi \in \Xi$;

(iv) $\partial^2 = 0$.

A morphism $\{f_i\}$ from one complex of Λ-modules $\{V^{(i)}, \partial_i\}$ to another complex $\{\tilde{V}^{(i)}, \tilde{\partial}_i\}$ defines a morphism of bigraded Λ-modules $f \colon V \to \tilde{V}$ of bidegree $(0,0)$ (that is, $f(V_j^{(i)}) \subset \tilde{V}_j^{(i)}$) such that $f\partial = \tilde{\partial} f$.

It is easy to see that the category $\mathcal{C}(\Lambda)$ is equivalent to the category of bigraded Λ-modules V, equipped with an operator $\partial \colon V \to V$ satisfying the conditions (i)–(iv). Furthermore V corresponds to a complex in $\mathcal{C}^b(\Lambda)$ if and only if $\dim V < \infty$.

Similarly the category $\mathcal{C}(S)$ is equivalent to the category of bigraded S-modules $W = \bigoplus_{i,j} W_j^{(i)}$, equipped with a linear operator $d \colon W \to W$ where

(i′) the elements $x \in X \subset S$ have bidegree $(0,1)$;

(ii′) d has bidegree $(1,0)$;

(iii′) $sd = ds$ for $s \in S$;

(iv′) $d^2 = 0$.

A morphism $\{f_i\}$ from a complex $\{W^{(i)}, d_i\}$ to a complex $\{\tilde{W}^{(i)}, \tilde{d}_i\}$ defines a morphism $f\colon W \to \tilde{W}$ of bigraded S-modules of bidegree $(0,0)$ such that $fd = \tilde{d}f$.

Furthermore, the complexes in $\mathcal{C}^b(S)$ correspond to the finitely generated S-modules W.

4.4. The functor $F\colon \mathcal{C}(\Lambda) \to \mathcal{C}(S)$**.** Let $V = \bigoplus_{i,j} V_j^{(i)} \in \mathcal{C}(\Lambda)$. We put $W = S \otimes_{\mathbb{C}} V$. We define an action of S on W by putting $s(s_1 \otimes v) = ss_1 \otimes v$. We define the differential $d\colon W \to W$ by the formula

$$d(s \otimes v) = \sum_{\ell} x_\ell s \otimes \xi_\ell v \;+\; s \otimes \partial v,$$

where $\{\xi_\ell\}$, $\{x_\ell\}$ are dual bases in Ξ and X. We define a bigrading on W in the following way: if $s \in S_k$, $v \in V_j^{(i)}$ then we put $s \otimes v \in W_{-j+k}^{(i+j)}$. It is easy to check that the action of S on W and the operator $d\colon W \to W$ satisfy the conditions (i′)–(iv′) from §4.3.

Therefore W defines a complex in $\mathcal{C}(S)$. The mapping $V \mapsto W$ determines a functor $F\colon \mathcal{C}(\Lambda) \to \mathcal{C}(S)$. Clearly F commutes with the translation T in the categories of complexes $\mathcal{C}(\Lambda)$ and $\mathcal{C}(S)$. It is also easy to verify that if $V \in \mathcal{C}^b(\Lambda)$ then $F(V) \in \mathcal{C}^b(S)$.

Let $\mathsf{D}(\Lambda)$, $\mathsf{D}^b(\Lambda)$, $\mathsf{D}(S)$, $\mathsf{D}^b(S)$ be the derived categories corresponding to $\mathcal{C}(\Lambda)$, $\mathcal{C}^b(\Lambda)$, $\mathcal{C}(S)$, $\mathcal{C}^b(S)$. The categories $\mathsf{D}(\Lambda)$, $\mathsf{D}(S)$ are triangulated categories, and $\mathsf{D}^b(\Lambda)$, $\mathsf{D}^b(S)$ are full triangulated subcategories (see [2*] §II.1). The main theorem of this section consists of the following

4.5. Theorem. (i) The functor $F\colon \mathcal{C}^b(\Lambda) \to \mathcal{C}^b(S)$ extends to a functor $F_{\mathsf{D}}\colon \mathsf{D}^b(\Lambda) \to \mathsf{D}^b(S)$.

(ii) F_{D} defines an equivalence of the triangulated categories $\mathsf{D}^b(\Lambda)$ and $\mathsf{D}^b(S)$.

4.6. Lemma. (i) The functor F is exact, commutes with the functor T and transforms the cone of an arbitrary morphism $f\colon V \to \tilde{V}$ into the cone of the morphism $F(f)\colon F(V) \to F(\tilde{V})$.

(ii) The restriction of F to the category $\mathcal{C}^b(\Lambda)$ transforms quasi-isomorphisms into quasi-isomorphisms.

Proof. All assertions of (i) can easily be checked immediately. To prove (ii), it suffices to prove that F transforms an arbitrary acyclic object Z into an acyclic object. We consider $S \otimes Z$ as a bicomplex of S-modules, by putting $(S \otimes Z)^{p,q} = S \otimes Z_p^{(q)}$ with the two differentials $d' = \sum_\ell x_\ell \otimes \xi_\ell$ and $d'' = 1 \otimes \partial$ of bidegree $(1,0)$ and $(0,1)$, respectively. Then the differential d in $F(Z)$ is $d' + d''$. Since Z is acyclic, $H_{\mathrm{II}}(S \otimes Z) = 0$. Furthermore, since $Z \in \mathcal{C}^b(\Lambda)$, $(S \otimes Z)^{p,q} = 0$ for all pairs (p,q) except for a finite number. Hence it is easy to prove that $F(Z)$ is acyclic (either directly, or by using the spectral sequence of the bicomplex [4*, theorem XI.6.1]).

From the lemma it follows that the functor $F\colon \mathcal{C}^b(\Lambda) \to \mathcal{C}^b(S)$ extends to a functor $F_{\mathsf{D}}\colon \mathsf{D}^b(\Lambda) \to \mathsf{D}^b(S)$, that is, the first part of theorem 4.5 is proved.

4.7. Lemma. (i) There exists a functor $G\colon \mathcal{C}(S) \to \mathcal{C}(\Lambda)$ right adjoint to the functor F. In particular, for every $V \in \mathcal{C}(\Lambda)$, $W \in \mathcal{C}(S)$ there exists an isomorphism

$$\alpha_{V,W}\colon \mathrm{Hom}(F(V), W) \xrightarrow{\sim} \mathrm{Hom}(V, G(W)).$$

(ii) The functor G commutes with translations.

(iii) Under the isomorphism $\alpha_{V,W}$ homotopic mappings transform into homotopic mappings.

Proof. (i) We will construct the functor G explicitly. We shall identify, as before, $\{W^{(i)}, d_i\} \in \mathcal{C}(S)$ with the bigraded S-module $W = \bigoplus_{i,j} W_j^{(i)}$. We put $V = G(W) = \mathrm{Hom}_{\mathbb{C}}(\Lambda, W)$. We define a bigrading on V in the following way:

If $\phi(\Lambda_k) \subseteq W_{-j-k}^{(i+k+j)}$ for all k, then $\phi \in V_j^{(i)}$.

We define an action of Λ on V by the formula $\xi\phi(\lambda) = \phi(\lambda\xi)$ for $\phi \in V$, $\xi \in \Xi \subset \Lambda$, $\lambda \in \Lambda$. We define a differential $\partial\colon V \to V$ by means of the formula

$$\partial\phi(\lambda) = -\sum_\ell x_\ell \phi(\lambda\xi_\ell) + d(\phi(\lambda)), \quad \phi \in V.$$

For the construction of the isomorphism

$$\alpha_{V,W}\colon \mathrm{Hom}_{\mathcal{C}(S)}(F(V), W) \xrightarrow{\sim} \mathrm{Hom}_{\mathcal{C}(\Lambda)}(V, G(W))$$

it is easiest to verify directly that both parts may naturally be identified with the space of $\mathbb{C}$-linear maps $\psi\colon V \to W$, satisfying the conditions

a) $\psi(V^{(i)}) \subset W_{-j}^{(i-j)}$

b) $\psi(\partial v) + \sum_\ell x_\ell \psi(\xi_\ell v) = d\psi(v)$, $v \in V$.

We leave this simple verification to the reader.

(ii) Clear from the construction.

(iii) Let $V \in \mathcal{C}(\Lambda)$, let C_V be the cone over V (i.e., the mapping cone of the identity morphism id_V). Then for arbitrary $\tilde{V} \in \mathcal{C}(\Lambda)$ the image of the homomorphism $\mathrm{Hom}(C_V, \tilde{V}) \to \mathrm{Hom}(V, \tilde{V})$ induced by the natural embeddding $V \to C_V$, coincides with the subgroup $\mathrm{Ht}(V, \tilde{V}) \subset \mathrm{Hom}(V, \tilde{V})$, consisting of the morphisms which are homotopic to 0. In view of lemma 4.6 (i) and part (i) of lemma 4.7, for $V \in \mathcal{C}(\Lambda)$, $W \in \mathcal{C}(S)$ we have:

$$\begin{aligned}\mathrm{Ht}(F(V), W) &= \mathrm{Im}\,(\mathrm{Hom}(C_{F(V)}, W) \to \mathrm{Hom}(F(V), W))\\ &= \mathrm{Im}\,(\mathrm{Hom}(F(C_V), W) \to \mathrm{Hom}(F(V), W))\\ &= \mathrm{Im}\,(\mathrm{Hom}(C_V, G(W)) \to \mathrm{Hom}(V, G(W))\\ &= \mathrm{Ht}(V, G(W)).\end{aligned}$$

□

4.8. Lemma. If $V \in \mathcal{C}^b(\Lambda)$, $W \in \mathcal{C}^b(S)$, then

$$\mathrm{Hom}_{\mathsf{D}^b(S)}(F(V), W) = \mathrm{Hom}_{\mathsf{D}(\Lambda)}(V, G(W)).$$

Proof. Since $V \in \mathcal{C}^b(\Lambda)$, the complex $F(V)$ consists of projective S-modules and is bounded. Therefore (see [6*] proposition I.4.7)

$$\mathrm{Hom}_{\mathsf{D}^b(S)}(F(V), W) = \mathrm{Hom}_{\mathcal{C}(S)}(F(V), W)/\mathrm{Ht}(F(V), W).$$

Analogously, the complex $G(W)$ for $W \in \mathcal{C}^b(S)$ consists of injective Λ-modules and is bounded from the right, so that

$$\mathrm{Hom}_{\mathsf{D}(\Lambda)}(V, G(W)) = \mathrm{Hom}_{\mathcal{C}(\Lambda)}(V, G(W))/\mathrm{Ht}(V, G(W)).$$

Lemma 4.8 follows now from lemma 4.7 (i), (iii). □

4.9. Let $V \in \mathcal{C}(\Lambda)$. Let $\mathrm{id}_{F(V)}$ be the identity morphism $F(V) \to F(V)$. We define $i_V \in \mathrm{Hom}_{\mathcal{C}(\Lambda)}(V, G(F(V)))$ by

$$i_V = \alpha_{V,F(V)}(\mathrm{id}_{F(V)}).$$

Clearly the family of morphisms i_V, $V \in \mathcal{C}(\Lambda)$ defines a morphism from the identity functor $\mathcal{C}(\Lambda) \to \mathcal{C}(\Lambda)$ to the functor $G \circ F\colon \mathcal{C}(\Lambda) \to \mathcal{C}(\Lambda)$.

Lemma. If $V \in \mathcal{C}^b(V)$, then $i_V\colon V \to G \circ F(V)$ is a quasi-isomorphism.

Proof. a) First let V be the trivial one-dimensional complex, i.e., $V^{(0)} = \mathbb{1}$, $V^{(i)} = (0)$ for the remaining values of i. We put $U = G \circ F(V) = \bigoplus U_j^{(i)}$. From the explicit definition of the functors F and G we have $U_j^{(i)} = \mathrm{Hom}_{\mathbb{C}}(\Lambda_{j-i}, S_i)$, and the differential $\partial\colon U_j^{(i)} \to U_j^{(i+1)}$ acts according to the formula $\partial\phi(\lambda) = -\Sigma_\ell x_\ell \phi(\lambda\xi_\ell)$.

The morphism $i_V \colon V \to U$ transforms the element $e \in V_0^{(0)}$ into the identity map in

$$U_0^{(0} = \mathrm{Hom}_{\mathbb{C}}(\Lambda_0, S_0) = \mathrm{Hom}_{\mathbb{C}}(\mathbb{C}, \mathbb{C}).$$

It is easy to see that we have obtained the standard Koszul complex for $S(X)$, corresponding to the choice of the regular sequence $(x_0, \dots, x_n)$ (see [4*] §III.3). Therefore $i_V \colon V \to U = G \circ F(V)$ is a quasi-isomorphism for the trivial one-dimensional complex.

b) From a) it follows that i_V is a quasi-isomorphism for an arbitrary one-dimensional complex. Since every complex $V \in C^b(\Lambda)$ is finite-dimensional, it is sufficient for the completion of the proof to carry out a standard induction with respect to $\dim V$, using the five lemma. □

4.10. We will now finish the proof of theorem 4.5.

a) $F_{\mathsf{D}} \colon \mathsf{D}^b(\Lambda) \to \mathsf{D}^b(S)$ is a full and faithful functor. Let $V, \tilde{V} \in \mathsf{D}^b(\Lambda)$. We consider the sequence of mappings

$$\mathrm{Hom}_{\mathsf{D}^b(\Lambda)}(V, \tilde{V}) \xrightarrow{F_{\mathsf{D}}} \mathrm{Hom}_{\mathsf{D}^b(S)}(F(V), F(\tilde{V})) \overset{\alpha}{\xrightarrow{\sim}} \mathrm{Hom}_{\mathsf{D}^b(\Lambda)}(V, G \circ F(\tilde{V})).$$

The composite evidently has the form $\phi \mapsto i_{\tilde{V}} \circ \phi$, (where

$$i_{\tilde{V}} \in \mathrm{Hom}_{\mathsf{D}^b(\Lambda)}(\tilde{V}, G \circ F(\tilde{V}))$$

is an isomorphism by virtue of lemma 4.9) and is therefore an isomorphism. On the other hand, α is an isomorphism according to lemma 4.8. Hence

$$F_{\mathsf{D}} \colon \mathrm{Hom}_{\mathsf{D}^b(\Lambda)}(V, \tilde{V}) \to \mathrm{Hom}_{\mathsf{D}^b(S)}(F(V), F(\tilde{V}))$$

is also an isomorphism.

b) In view of a), the subcategory $F_{\mathsf{D}}(\mathsf{D}^b(\Lambda))$ is a full triangulated subcategory in $\mathsf{D}^b(S)$. Moreover $F_{\mathsf{D}}(\mathsf{D}^b(\Lambda))$ contains all the objects which are represented by free complexes W with one generator (they have the form $F_{\mathsf{D}}(V)$, where $\dim V = 1$).

Such objects are generators in $\mathsf{D}^b(S)$ (Syzygy-theorem), so that $F_{\mathsf{D}}(\mathsf{D}^b(\Lambda)) = \mathsf{D}^b(S)$ and the theorem is proved.

4.11. We finish this section with one additional comment. Let $\mathcal{F}$ be the full subcategory of $\mathsf{D}^b(S)$ consisting of the objects which are isomorphic (in $\mathsf{D}^b(S)$) to finite dimensional complexes of S-modules. On the one hand, let $\mathcal{I}$ be the full subcategory of $\mathsf{D}^b(\Lambda)$ consisting of objects which are isomorphic (in $\mathsf{D}^b(\Lambda)$) to complexes of free Λ-modules.

Proposition. The functors F and G establish an equivalence between the categories $\mathcal{F}$ and $\mathcal{I}$.

From Serre's theorem it follows easily that $\mathcal{F}$ is a thick subcategory of $\mathsf{D}^b(S)$.

4.12. Corollary. $\mathcal{I}$ is a thick subcategory of $\mathsf{D}^b(\Lambda)$, and the categories $\mathsf{D}^b(S)/\mathcal{F}$ and $\mathsf{D}^b(\Lambda)/\mathcal{I}$ are equivalent.

§5. The category $\mathcal{A}$

5.1. The main result of this section is the following

Theorem. The categories $\mathsf{D}^b(\Lambda)/\mathcal{I}$ and $\mathcal{M}^0(\Lambda) = \mathcal{M}^b(\Lambda)/\mathcal{P}$ are equivalent.

Following a proposal of A. A. Beilinson, for the proof of this theorem we shall introduce one more triangulated category $\mathcal{A}$ and we shall construct functors

$$\mathsf{D}^b(\Lambda) \xrightarrow{\alpha} \mathcal{A} \xrightarrow{\delta} \mathcal{M}^0(\Lambda).$$

Furthermore it turns out that the functor δ defines an equivalence of categories. The functor α has a kernel, exactly coinciding with $\mathcal{I}$, and α defines an equivalence between $\mathsf{D}^b(\Lambda)/\mathcal{I}$ and $\mathcal{A}$.

With the aid of the category $\mathcal{A}$, one can also very naturally establish the connection between our description of the sheaves on $\mathbb{P}$ and the description of [18].

5.2. The category $\mathcal{A}$.

The objects of the category $\mathcal{A}$ are complexes $A = \{A_i\}_{i=-\infty}^{\infty}$, satisfying the following conditions:

a) Every A_i is a free Λ-module with a finite number of generators, $A_i = \bigoplus_{\ell=1}^{n_i} \Lambda[r_{i,\ell}]$;

b) the complex A is acyclic, i.e., $H^i(A) = 0$ for all i.

As in §4, it is convenient to assume that the differentials $\partial\colon A_i \to A_{i+1}$ anti-commute with the action of Λ, i.e., $\partial_i \xi = -\xi \partial_i$.

The morphisms of $\mathcal{A}$ are the morphisms of complexes up to homotopy equivalence.

The distinguished triangles in $\mathcal{A}$ are, as usual, triangles which are isomorphic to triangles

$$A \xrightarrow{u} B \to C_u \to T(A),$$

where $u\colon A \to B$ is an arbitrary morphism, C_u is the mapping cone of u, and T is the translation functor.

5.3. The category $\mathcal{A}$ owes its non-triviality to the fact that we consider complexes A which are infinite on both sides. More precisely, we have the simple

Lemma. Let the complex $A \in \mathcal{A}$ be bounded on one side. Then A is isomorphic to 0 in $\mathcal{A}$.

Proof. For example, let $A_i = 0$ for $i \leq 0$. Then

$$0 \to A_1 \to A_2 \to \cdots$$

is an injective resolution of the zero Λ-module 0. Since two arbitrary injective resolutions are homotopic, A is isomorphic to 0 in $\mathcal{A}$.

5.4. In the following we require one fixed complex $\Delta \in \mathcal{A}$ which we now construct.

The right part Δ^+ of this complex is a shifted Koszul complex (see [4*] §III, 3). Namely, for $\ell \geq 0$ we put

$$\Delta_\ell = \Lambda[-n-1] \otimes S^\ell(X).$$

Δ_ℓ is a free Λ-module with $\binom{n+\ell}{\ell}$ generators of degree $\ell - n - 1$. We define the operator $\partial_\ell \colon \Delta_\ell \to \Delta_{\ell+1}$ by putting

$$\partial_\ell(\lambda \otimes s) = \sum \xi_i \lambda \otimes x_i s$$

where $\{\xi_i\}$, $\{x_i\}$ are dual bases of Ξ, X.

The left part Δ^- of the complex is dual to the right part. Namely, we put

$$\Delta_{-\ell} = (\Delta_{\ell-1})^*, \qquad \ell > 0$$
$$\partial_{-\ell} = \partial^*_{\ell-2}, \qquad \ell > 1.$$

Thus all modules Δ_ℓ and all operators ∂_ℓ are defined except ∂_{-1}. It remains to define ∂_{-1}. To this end we remark that

$$\Delta_0 = \Lambda[-n-1], \qquad \Delta_{-1} = (\Lambda[-n-1])^* = \Lambda.$$

We define the operator $\partial_{-1} \colon \Lambda \to \Lambda[-n-1]$ by putting

$$\partial_{-1}(1) = \xi_0 \dots \xi_n e$$

($1 \in \Lambda$, e is a generator of $\Lambda[-n-1]$).

The following statements may be verified immediately.

Lemma. (i) $\Delta \in \mathcal{A}$.

(ii) In Δ we have $\ker \partial_0 = \operatorname{im} \partial_{-1} = \mathbb{1}$. □

5.5. Construction of the functor $\alpha\colon \mathsf{D}^b(\Lambda) \to \mathcal{A}$**.** For every complex $X \in \mathcal{C}^b(\Lambda)$ we put $\alpha(X) = X \otimes \Delta$. It is clear that $\alpha(X)$ is an object of $\mathcal{A}$.

Let $\phi\colon X \to Y$ be a morphism of complexes. We put $\alpha(\phi) = \phi \otimes 1\colon \alpha(X) \to \alpha(Y)$ and prove that α gives a functor from $\mathsf{D}^b(\Lambda)$ to $\mathcal{A}$.

It is clear that if ϕ is homotopic to ψ, then $\alpha(\phi)$ is homotopic to $\alpha(\psi)$. Hence it is sufficient to prove that if $\phi\colon X \to Y$ is a quasi-isomorphism, then $\alpha(\phi)\colon \alpha(X) \to \alpha(Y)$ is a homotopy equivalence. A more general result is contained in the lemma below, which we will need subsequently.

We shall say that a morphism of complexes $i\colon X \to Y^+$ is an injective resolution of $X \in \mathcal{C}^b(\Lambda)$ if

1) Y^+ is a left-bounded complex, consisting of free finite-dimensional Λ-modules.

2) i is a quasi-isomorphism.

Similarly we define a projective resolution $j\colon Y^- \to X$.

Lemma. Let $X \to Y^+$, $Y^- \to X$ be a choice of injective and projective resolutions of a complex $X \in \mathcal{C}^b(\Lambda)$, C_u the mapping cone of the composition $u\colon Y^- \to Y^+$. Then

(i) $C_u \in \mathcal{A}$

(ii) if X' is quasiisomorphic to X and $C_{u'}$ is defined with respect to X' in the same way as C_u with respect to X, then C_u and $C_{u'}$ are isomorphic in $\mathcal{A}$; in particular $C_u \cong \alpha(X)$ in $\mathcal{A}$.

Proof. For the proof of (i) we need to verify conditions a)–b) of §5.2. Because injective and projective objects coincide in $\mathcal{M}^b(\Lambda)$, a) is obvious. Furthermore, b) follows from the fact that the mapping cone of the quasi-isomorphism u is acyclic. The proof of (ii) follows from the fact that two arbitrary projective (injective) resolutions of quasiisomorphic objects are homotopic (see [6*] §1.4). Since

$$X \otimes \Delta^- \to X, \qquad X \to X \otimes \Delta^+ \tag{5.1}$$

(where Δ^- and Δ^+ are the negative and positive parts of Δ, see §5.4) are projective and injective resolutions of X and since $\alpha(X)$ is the mapping cone of the composite map

$$X \otimes \Delta^- \to X \otimes \Delta^+,$$

we obtain a functor $\alpha\colon \mathsf{D}^b(\Lambda) \to \mathcal{A}$.

5.6. We shall now prove that the kernel of the functor α coincides with the subcategory $\mathcal{I} \subset \mathsf{D}^b(\Lambda)$.

Proposition. The object $\alpha(X) \in \mathcal{A}$ is isomorphic to 0 in $\mathcal{A}$ if and only if $X \in \mathcal{I}$.

Proof. a) Let $X \in \mathcal{I}$. We shall prove that $\alpha(X)$ is isomorphic to 0 in $\mathcal{A}$. We may assume that the projective and injective resolutions of X coincide with X. In view of lemma 5.5, $\alpha(X)$ is isomorphic in $\mathcal{A}$ to a bounded complex (the mapping cone of the identity map from X to X), so that in view of lemma 5.3, $\alpha(X) \cong 0$.

b) Let $\alpha(X) \cong 0$. We prove that $X \in \mathcal{I}$. Let $Y^- \to X$, $X \to Y^+$ be projective and injective resolutions of X (for example the resolutions (5.1)). By the lemma of §5.5, we can replace $\alpha(X)$ by the mapping cone C of the composite map $Y^- \to Y^+$ and assume that there exists a homotopy $\{k_i \colon C_i \to C_{i-1}\}$ between the identity and zero homomorphisms from C to C, i.e.,

$$\partial_{i-1} k_i + k_{i+1} \partial_i = \mathrm{id}. \tag{5.2}$$

Let i be such that for $j \geq i-1$ we have $X_j = 0$ and $Y^-_{j+1} = 0$ so that $C_j = Y^+_j$. We put $U = \operatorname{im} \partial_i$. Then in the first place the complex

$$\tau_i Y^+ \underset{\text{def}}{=} \{\cdots \to Y^+_{i-1} \to Y^+_i \to U \to 0 \to 0\}$$

is quasiisomorphic to Y^+ and hence, quasiisomorphic to X. On the other hand, since $C_j = Y^+_j$ for $j \geq i-1$, we obtain from (5.2) that the morphisms

$$k_i \colon Y^+_i \to Y^+_{i-1}, \qquad k_{i+1} \colon Y^+_{i+1} \to Y^+_i$$

satisfy the condition

$$\partial_{i-1} k_i + k_{i+1} \partial_i = \mathrm{id} \quad \text{on } Y^+_i.$$

From this it follows that U is a direct summand of Y^+_i (with complementary module $\operatorname{im} \partial_{i-1} k_i$). Since $Y^+_i \in \mathcal{P}$ we have $U \in \mathcal{P}$, so that $\tau_i Y^+ \in \mathcal{I}$ and $X \in \mathcal{I}$. □

5.7. Lemma. Let $\zeta \colon \mathcal{C}_1 \to \mathcal{C}_2$ be an exact functor of triangulated categories and $\mathcal{D} \subset \mathcal{C}_1$ the full subcategory consisting of all X for which $\zeta(X)$ is isomorphic to 0. Then $\mathcal{D}$ is a thick subcategory in $\mathcal{C}$, and ζ defines a functor $\mathcal{C}_1/\mathcal{D} \to \mathcal{C}_2$.

The proof follows easily from the definition of a thick subcategory. The details are left to the reader. □

From proposition 5.6 and lemma 5.7 it follows that α defines a functor $\alpha' \colon \mathsf{D}^b(\Lambda)/\mathcal{I} \to \mathcal{A}$.

5.8. We shall prove that α' is an equivalence of categories.

This can be done in several ways. The method presented here consists of an explicit construction of the functor which is inverse to α'.

For every pair of integers $i \leq j$ we construct a functor $\beta_{i,j}\colon \mathcal{A} \to \mathsf{D}^b(\Lambda)/\mathcal{I}$ in the following way. Let

$$A = \{\cdots \to A_{-1} \to A_0 \to A_1 \to \cdots\} \in \mathcal{A}.$$

We consider the complex

$$\{0 \to A_i \to \cdots \to A_{j-1} \to B \to 0\} \tag{5.3}$$

where $B = \ker(\partial_j\colon A_j \to A_{j+1}) = \operatorname{im}(\partial_{j-1}\colon A_{j-1} \to A_j)$. We shall denote by $\beta_{ij}(A)$ the image of this complex in $\mathsf{D}^b(\Lambda)/\mathcal{I}$.

Lemma. (i) β_{ij} defines a functor from $\mathcal{A}$ to $\mathsf{D}^b(\Lambda)/\mathcal{I}$.

(ii) For different i, j the functors β_{ij} are mutually isomorphic.

(iii) β_{ij} is an exact functor of triangulated categories.

Proof. For the proof of (i) it suffices to show that if $A \in \mathcal{A}$ is homotopic to 0, then the complex (5.3) lies in $\mathcal{I}$. Let $\{k_\ell\colon A_\ell \to A_{\ell-1}\}$ be a homotopy between the identity and the zero mapping. As for the proof of proposition 5.6, it is easy to show that the module B in the complex (5.3) is a direct summand of A_j (with complementary summand $\operatorname{im} k_{j+1}\partial_j$). Therefore $B \in \mathcal{P}$ and (5.3)$\in \mathcal{I}$.

We prove (ii). It suffices to show that $\beta_{ij} \sim \beta_{i+1,j}$ for $i < j$ and $\beta_{ij} \sim \beta_{i,j+1}$ for $i \leq j$. It is easy to verify that for the isomorphisms

$$\theta_A \in \operatorname{Hom}_{\mathsf{D}^b(\Lambda)/\mathcal{I}}(\beta_{ij}(A), \beta_{i+1,j}(A)),$$

giving an isomorphism $\theta\colon \beta_{ij} \to \beta_{i+1,j}$, one can take the natural morphisms

$$\theta_A\colon \{A_i \to A_{i+1} \to \cdots \to A_{j-1} \to B\} \to \{A_{i+1} \to \cdots \to A_{j-1} \to B\};$$

also θ_A is an isomorphism in $\mathsf{D}^b(\Lambda)/\mathcal{I}$, for A_i is free. An isomorphism $\beta_{ij} \sim \beta_{i,j+1}$ is constructed analogously.

Now it is sufficient to prove the statements in (iii) for any of the morphisms β_{ij}, e.g., for $\beta_{0,0}$. We prove, for example, that if $\phi\colon A \to A'$ is homotopic to 0, then $\beta_{0,0}(\phi) = 0$ in $\mathsf{D}^b(\Lambda)/\mathcal{I}$. Let $k_i\colon A_i \to A'_{i-1}$ be a homotopy, i.e.,

$$k_{i+1}\partial_i + \partial'_{i-1}k_i = \phi_i\colon A_i \to A'_i.$$

On $B = \ker(\partial_0\colon A_0 \to A_1)$ we have $\phi_0 = \partial'_{-1}k_0$. Therefore the morphism

$$\beta_{0,0}(\phi)\colon \{\cdots \to 0 \to B \to 0 \to \cdots\} \to \{\cdots \to 0 \to B' \to 0 \to \cdots\}$$

factorizes through the complex

$$\{\cdots \to 0 \to A'_{-1} \to A'_{-1} \to 0 \to \cdots\} \in \mathcal{I},$$

i.e., $\beta_{0,0}(\phi) = 0$ in $\mathsf{D}^b(\Lambda)/\mathcal{I}$.

The other assertions of (iii) are proved similarly.

5.9. Proposition. The functors $\beta_{ij} \circ \alpha$ and $\alpha \circ \beta_{ij}$ are isomorphic to the identity functors in $\mathsf{D}^b(\Lambda)/\mathcal{I}$ and $\mathcal{A}$, respectively.

Proof. Let $A \in \mathcal{A}$ and $B = \ker(\partial_0 \colon A_0 \to A_1) = \operatorname{im}(\partial_{-1} \colon A_1 \to A_0)$. Then

$$\cdots \to A_{-2} \to A_{-1} = Y^-, \qquad A_0 \to A_1 \to A_2 \to \cdots = Y^+$$

are respectively projective and injective resolutions of the complex

$$\beta_{00}(A) = \{\cdots \to 0 \to B \to 0 \to \cdots\},$$

and A is the cone of the composition $Y^- \to \beta_{00}(A) \to Y^+$. Therefore, according to the lemma of §5.5, $\alpha \circ \beta_{00}(A)$ is isomorphic to A in $\mathcal{A}$. It is easy to show that this isomorphism is functorial in A, i.e., $\alpha \circ \beta \sim \mathrm{id}_{\mathcal{A}}$. According to the lemma in 5.8, $\alpha \circ \beta_{ij} \sim \mathrm{id}_{\mathcal{A}}$ for arbitrary $i \le j$.

Conversely, let $X = \{X_k \to \cdots \to X_\ell\} \in \mathsf{D}^b(\Lambda)/\mathcal{I}$. Let $Y^- = X \otimes \Delta^-$, $Y^+ = X \otimes \Delta^+$ be two resolutions of X such that $\alpha(X)$ is the mapping cone of the composition $Y^- \to X \to Y^+$. We choose $i < k$, $j > \ell$ and consider the complexes

$$\{Y_i^- \to \cdots \to Y_\ell^-\} = \tilde{Y}^-, \qquad \{Y_k^+ \to \cdots \to Y_{j-1}^+ \to B\} = \tilde{Y}^+$$

where $B = \operatorname{im}(Y_{j-1}^+ \to Y_j^+)$. Then as before we have mappings $\tilde{Y}^- \to X$, $X \to \tilde{Y}^+$. Furthermore, $\beta_{ij} \circ \alpha(X)$ is clearly the mapping cone of the composite $\tilde{Y}^- \to \tilde{Y}^+$, so that in $\mathsf{D}^b(\Lambda)/\mathcal{I}$ there is a distinguished triangle

$$\tilde{Y}^- \to \tilde{Y}^+ \to \beta_{ij} \circ \alpha(X) \to T(\tilde{Y}^-).$$

Since $\tilde{Y}^- \in \mathcal{I}$, $\tilde{Y}^+$ and $\beta_{ij} \circ \alpha(X)$ are isomorphic in $\mathsf{D}^b(\Lambda)/\mathcal{I}$. Furthermore (since $j > \ell$), it is clear that $X \to \tilde{Y}^+$ is a quasi-isomorphism. Therefore $\tilde{Y}^+$ is isomorphic to X in $\mathsf{D}^b(\Lambda)$ and in $\mathsf{D}^b(\Lambda)/\mathcal{I}$. Consequently X is isomorphic to $\beta_{ij} \circ \alpha(X)$ in $\mathsf{D}^b(\Lambda)/\mathcal{I}$. Again it is easy to show that the constructed isomorphism $X \to \beta_{ij} \circ \alpha(X)$ is functorial in X. Thus proposition 5.9 has been proved. □

5.10. Proposition 5.9 establishes an equivalence of the categories $\mathsf{D}^b(\Lambda)/\mathcal{I}$ and $\mathcal{A}$. The equivalence of the categories $\mathcal{A}$ and $\mathcal{M}^0(\Lambda)$ is proved similarly (but essentially more easily). We restrict ourselves to indicating the corresponding functors on the objects, leaving all details to the reader.

The functor $\gamma\colon \mathcal{A} \to \mathcal{M}^0(\Lambda) = \mathcal{M}^b(\Lambda)/\mathcal{P}$ associates to the complex $A = \{A_i\}$ the module $M = \ker(\partial_0\colon A_0 \to A_1)$.

The functor $\delta\colon \mathcal{M}^0(\Lambda) \to \mathcal{A}$ is given by the formula $\delta(V) = V \otimes \Delta$, Δ is the complex defined in §5.4.

5.11. At the end of this paragraph we shall explain the connection of our results with those of A. A. Beilinson [18] who has given a somewhat different description of the derived category $\mathsf{D}^b(\mathrm{Sh})$.

We recall that $\Lambda[i]$ is the free graded module with one generator of degree i. We denote by $\mathcal{M}_{[0,n]}(\Lambda)$ the full subcategory of the category $\mathcal{M}^b(\Lambda)$, consisting of finite direct sums of modules $\Lambda[i]$, $0 \le i \le n$. Let $\mathcal{C}_{[0,n]}(\Lambda)$ be the full subcategory of $\mathcal{C}^b(\Lambda)$ consisting of finite complexes of modules from $\mathcal{M}_{[0,n]}(\Lambda)$ and $\mathcal{K}_{[0,n]}(\Lambda)$ the corresponding homotopy category.

Similarly, let $S[i]$ be the graded S-module with one generator of degree i, $\mathcal{M}_{[0,n]}(S)$ the full subcategory of $\mathcal{M}^b(S)$ consisting of finite direct sums of modules $S[i]$, $0 \le i \le n$. We define the homotopy category $\mathcal{K}_{[0,n]}(S)$, consisting of finite complexes of S-modules, similarly to the definition of the category $\mathcal{K}_{[0,n]}(\Lambda)$.

The categories $\mathcal{K}_{[0,n]}(\Lambda)$ and $\mathcal{K}_{[0,n]}(S)$ are full triangulated subcategories of the categories $\mathsf{D}^b(\Lambda)$ and $\mathsf{D}^b(S)$, respectively. The main theorem from [18] asserts that each of the categories $\mathcal{K}_{[0,n]}(\Lambda)$, $\mathcal{K}_{[0,n]}(S)$ is equivalent to the category $\mathsf{D}^b(\mathrm{Sh})$ as a triangulated category.

According to the results of sections 5.2–5.10 the category $\mathcal{A}$ is equivalent to the category $\mathsf{D}^b(\mathrm{Sh})$. We will replace the category $\mathcal{A}$ by its full subcategory $\mathcal{A}'$ which consists of the complexes $A = \{A_i\}$ satisfying conditions a), b) of §5.2 and the condition

c) for every r the equality $r_{i,\ell} = r$ (see 5.2a)) holds only for a finite number of pairs (i, ℓ).

It is easy to check that in all the arguments of §5.2–5.10 one can replace $\mathcal{A}$ by $\mathcal{A}'$ (since the complex Δ of §5.4 lies in $\mathcal{A}'$). Actually the functor $\mathcal{A}' \to \mathcal{A}$ realizes an equivalence between $\mathcal{A}$ and $\mathcal{A}'$.

We shall construct functors $\gamma_\Lambda\colon \mathcal{A}' \to \mathcal{K}_{[0,n]}(\Lambda)$, $\gamma_S\colon \mathcal{A}' \to \mathcal{K}_{[0,n]}(S)$, which are equivalences of the corresponding categories. We shall not give the proof of the equivalence; it is similar to the arguments carried out earlier in this paragraph.

5.12. For each Λ-module $V \in \mathcal{M}^b(\Lambda)$ we introduce two filtrations by Λ-submodules $V\{\ell\}$ and $V(\ell)$ as follows. Let $V = \bigoplus V_j$ be the decomposition into homogeneous components. We put

$$V(\ell) = \bigoplus_{j \geq \ell} V_j, \qquad V\{\ell\} = \Lambda\Big(\bigoplus_{j \leq \ell} V_j\Big).$$

The filtration $V\{\ell\}$ is increasing, whereas the filtration $V(\ell)$ is decreasing. For every morphism $\phi\colon V \to W$ in $\mathcal{M}^b(\Lambda)$ we have $\phi(V\{\ell\}) \subset W\{\ell\}$, $\phi(V(\ell)) \subset W(\ell)$. Moreover,

$$V\{\ell_1\} \cap V(\ell_2) = \{0\} \qquad \text{for} \qquad \ell_2 - \ell_1 > n + 1.$$

For every $V \in \mathcal{M}(\Lambda)$ we put

$$\eta(V) = V\{n\}/V\{-1\}, \qquad \eta'(V) = V\{-1\} \cap V(0).$$

It is clear that η, η' define two functors from $\mathcal{M}^b(\Lambda)$ into itself.

5.13. Construction of the functor γ_Λ. Let

$$A = \{\cdots \to A_{-1} \to A_0 \to A_1 \to \cdots\} \in \mathcal{A}'.$$

We denote by $\gamma_\Lambda(A)$ the complex

$$\gamma_\Lambda(A) = \{\cdots \to \eta(A_{-1}) \to \eta(A_0) \to \cdots\}.$$

From property c) of the category $\mathcal{A}'$ (see §5.11) and property a) above it easily follows that $\gamma_\Lambda(A) \in \mathcal{C}_{[0,n]}(\Lambda)$. The action of γ_Λ on the morphisms of complexes is defined in the natural way, and it is clear that γ_Λ preserves the relation of homotopy equivalence of morphisms. Therefore γ_Λ defines a functor (denoted by the same letter)

$$\gamma_\Lambda\colon \mathcal{A}' \to \mathcal{K}_{[0,n]}(\Lambda).$$

Proposition. γ_Λ is an equivalence of triangulated categories.

5.14. Construction of the functor γ_S. First of all we denote by $\tilde{\mathsf{D}}^b(\Lambda)$ the full subcategory of $\mathsf{D}^b(\Lambda)$ formed by complexes X such that $H^*(X)$ consists of elements whose homogeneous components have degree j, $0 \leq j \leq n$.

Proposition. The functor $F_{\mathsf{D}}\colon \mathsf{D}^b(\Lambda) \to \mathsf{D}^b(S)$ identifies $\tilde{\mathsf{D}}^b(\Lambda)$ with the full subcategory $\mathcal{K}_{[0,n]}(S)$ of $\mathsf{D}^b(S)$.

By analogy with the definition of γ_Λ, but using η' instead of η, we construct a functor $\mathcal{A} \to \mathcal{K}^b(\Lambda)$. Let $\gamma'\colon \mathcal{A} \to \mathsf{D}^b(\Lambda)$ be the composite of this functor with the natural functor $\mathcal{K}^b(\Lambda) \to \mathsf{D}^b(\Lambda)$. It is easy to check that $\gamma'(A) \in \tilde{\mathsf{D}}^b(\Lambda)$ for arbitrary $A \in \mathcal{A}$.

Proposition. γ' realizes an equivalence between $\mathcal{A}$ and $\tilde{\mathsf{D}}^b(\Lambda)$.

We now put $\gamma_S = F_{\mathsf{D}} \circ \gamma'$.

Corollary. γ_S realizes an equivalence between $\mathcal{A}$ and $\mathcal{K}_{[0,n]}(S)$.

Bibliography for Appendix A

[1*] Bernstein, I. N., Gelfand, I. M., Gelfand, S. I.: Algebraic Bundles over $\mathbb{P}^n$ and Problems of Linear Algebra. Funct. Anal. and Appl. 1978, Vol. 12, No. 3, pp. 66–67.

[2*] Verdier, J. L.: Catégories dérivées. In: SGA $4\frac{1}{2}$, Lecture Notes in Mathematics 569, Springer 1977, pp. 262–311.

[3*] Kapranov, M. M.: Derived category of coherent sheaves on Grassmann manifolds. Funct. Anal. and Appl. 1983, Vol. 17, No. 2, pp. 78–79.

[4*] Cartan, H., Eilenberg, S.: Homological algebra. Princeton Univ. Press, 1956.

[5*] Hartshorne, R.: Algebraic geometry. Graduate Texts in Mathematics 52, Springer 1977.

[6*] Hartshorne, R.: Residues and duality. Lecture Notes in Mathematics 20, Springer 1966.